Honoring Prof. Dr. Valentin A. Stonik for His Outstanding Contribution to Marine Natural Product Chemistry on the Occasion of His 80th Birthday

Honoring Prof. Dr. Valentin A. Stonik for His Outstanding Contribution to Marine Natural Product Chemistry on the Occasion of His 80th Birthday

Editors

Vladimir I. Kalinin
Pavel S. Dmitrenok
Natalia V. Ivanchina

MDPI

Basel • Beijing • Wuhan • Barcelona • Belgrade • Novi Sad • Cluj • Manchester

Editors

Vladimir I. Kalinin
G.B. Elyakov Pacific Institute
of Bioorganic Chemistry of
the Far East Branch of the
Russian Academy of Sciences
Vladivostok
Russia

Pavel S. Dmitrenok
G.B. Elyakov Pacific Institute
of Bioorganic Chemistry of
the Far East Branch of the
Russian Academy of Sciences
Vladivostok
Russia

Natalia V. Ivanchina
G.B. Elyakov Pacific Institute
of Bioorganic Chemistry of
the Far East Branch of the
Russian Academy of Sciences
Vladivostok
Russia

Editorial Office
MDPI
St. Alban-Anlage 66
4052 Basel, Switzerland

This is a reprint of articles from the Special Issue published online in the open access journal *Marine Drugs* (ISSN 1660-3397) (available at: https://www.mdpi.com/journal/marinedrugs/special_issues/Valentin_A_Stonik_80th_Birthday).

For citation purposes, cite each article independently as indicated on the article page online and as indicated below:

Lastname, A.A.; Lastname, B.B. Article Title. *Journal Name* **Year**, *Volume Number*, Page Range.

ISBN 978-3-7258-0347-7 (Hbk)
ISBN 978-3-7258-0348-4 (PDF)
doi.org/10.3390/books978-3-7258-0348-4

Cover image courtesy of Valentin A. Stonik and Tatiana N. Makarieva

Contents

About the Editors

Vladimir I. Kalinin

Vladimir I. Kalinin was born in Vladivostok, Soviet Union, now the Russian Federation, in 1957. He graduated Far Eastern State University, Chemical Department (Vladivostok), in 1979. From 1979, he was in the G.B. Elyakov Pacific Institute of Bioorganic Chemistry of the Far East Branch of the Russian Academy of Sciences. From 1979 to 1981 he was a Research Intern; from 1981 to 1987 a Junior Scientist; from 1987 to 1989 a Scientist; from 1989 to 1998 a Senior Scientist; and from 1998 a Leading Scientist. He has a PhD in chemistry (1989) and Dr.Sc. in biology (biochemistry) (1998). His scientific interests include structure, biological activities, taxonomical distribution, chemotaxonomy significance, and evolution of sea cucumber triterpene glycosides. He is the author and co-author of more than 110 chapters and scientific articles, etc., including in international scientific journals indexed by SCOPUS and WOS such as Marine Drugs, Molecules, Tetrahedron, Journal of Natural Products (Lloydia), Natural Product Research, Natural Product Communications, Carbohydrate Research, Journal of Theoretical Biology, Toxicon, Biological Systematics and Ecology, etc. He is Associate Editor of the Natural Product Communications, an Editorial Board Member of Marine Drugs, Guest Editor of the SIs "Marine Glycoconjugates, Trends and Perspectives", "Echinoderms Metabolites: Structure, Function, Biomedical Perspectives", "Echinoderms Metabolites: Structure, Function, Biomedical Perspectives II", and Guest Editor of the Natural Product Communications Special Collection "Marine Natural Products: Structure, Biosynthesis, Biological Activities, Synthetic Derivatives and Analogs".

Pavel S. Dmitrenok

Pavel S. Dmitrenok was born in Vladivostok, Soviet Union, now the Russian Federation, in 1957. He graduated from the Far East State University, Faculty of Physics (Vladivostok) in 1979. From 1980, he was part of G.B. Elyakov Pacific Institute of Bioorganic Chemistry of the Far East Branch of the Russian Academy of Sciences. From 1980 to 1982, he worked as an engineer of the Laboratory of Instrumental and Radioisotope Methods of Analysis; from 1982 to 1984 he served in the Soviet Army; from 1984 to 1986 he was an engineer for the Laboratory of Instrumental and Radioisotope Methods of Analysis; from 1986 to 1991 he was a Junior Scientist; from 1991 to 2003 he was a Staff Scientist; and from 2003 to 2018 he was Head of the Laboratory of Instrumental and Radioisotope Methods. Furthermore, from 2018 to 2019 he was an Interim Director, and from 2019 until present he has been the Director of G.B. Elyakov Pacific Institute of Bioorganic Chemistry. He has a PhD (2008) and Dr.Sc. (2023) in chemistry. His scientific interests include marine and other natural products, mass-spectrometry, chemical structures, metabolomics, and proteomics. He is the author or co-author of more than 340 articles in international scientific journals indexed by SCOPUS and WOS such as the Journal of Natural Products (Lloydia), Marine Drugs, Natural Product Communications, Steroids, Phytochemistry, Organic Letters, Metabolomics, Molecules, Carbohydrates Research, Carbohydrate Polymers, Tetrahedron, Journal of the American Society for Mass Spectrometry etc. He is an Editorial Board Member of Marine Drugs, and a Guest Editor of the SI for the Marine Drugs Special Issue "Dedicated to the 55th Anniversary of G.B. Elyakov Pacific Institute of Bioorganic Chemistry of the Far Eastern Branch of the Russian Academy of Sciences".

Natalia V. Ivanchina

Natalia V. Ivanchina, PhD. Born in Vladivostok, Russia on August 31, 1971. Graduated from the Department of Chemistry of the Far Eastern State University (Vladivostok) in 1993. Earned PhD degree (Candidat of Science) in bioorganic chemistry in 2000; title of dissertation "Isolation and structure elucidation of starfish polyhydroxysteroids and polyhydroxysteroidal glycosides". From 1996 worked for G.B. Elyakov Pacific Institute of Bioorganic Chemistry, Far Eastern Branch of the Russian Academy of Sciences. Junior Scientist from 1996; Research Scientist from 2000; Senior Researcher from 2005; and Head of the Laboratory of the Chemistry of Marine Natural Products from 2018. Winner of G.B. Elyakov Far Eastern Branch of the Russian Academy of Sciences Prize for work in the field of organic and bioorganic chemistry (2018). Scientific interests: starfish polyhydroxysteroids and polyhydroxysteroidal glycosides, structures, biosynthesis, biological activities, metabolomics. Author and co-author of more than 100 scientific articles in Russian and international journals. Member of the Editorial Board of Marine Drugs, Guest Editor of the SIs "Marine Glycoconjugates, Trends and Perspectives", "Carbohydrate-containing marine compounds of mixed biogenesis", and "Carbohydrate-containing marine compounds of mixed biogenesis II".

Preface

Professor, Dr.Sc. Valentin A. Stonik is one of the pioneers in the study of marine natural products. The scope of his research includes the metabolites of sponges, echinoderms and other marine invertebrates. He is the author or co-author of approximately 500 scientific articles across journals, four monographs and two-dozen patents. He is the founder of scientific schools and has supervised and consulted for dozens of PhDs and Dr.Scs.

To mark the occasion of Professor Valentin Stonik's 80th birthday in December 2023, Dr. Pavel S. Dmitrenok, Dr. Natalia V. Ivanchina, and Dr. Vladimir I. Kalinin edited a Themed Issue entitled "Honoring Prof. Dr. Valentin A. Stonik for His Outstanding Contribution to Marine Natural Product Chemistry on the Occasion of His 80th Birthday", in order to congratulate their scientific mentor. The Special Issue is a gift to Professor Valentin A. Stonik from his colleagues and friends and world experts in marine natural products chemistry and other related disciplines that are thriving in the G.B. Elyakov Pacific Institute of Bioorganic Chemistry under his supervision.

The Editors are very appreciative to all contributors who submitted a manuscript for publication.

Vladimir I. Kalinin, Pavel S. Dmitrenok, and Natalia V. Ivanchina
Editors

marine drugs

Editorial

Honoring Prof. Dr. Valentin A. Stonik for His Outstanding Contribution to Marine Natural Product Chemistry on the Occasion of His 80th Birthday

Pavel S. Dmitrenok, Natalia V. Ivanchina and Vladimir I. Kalinin *

G.B. Elyakov Pacific Institute of Bioorganic Chemistry, Far Eastern Branch of the Russian Academy of Sciences, Pr. 100-Letya Vladivostoka 159, 690022 Vladivostok, Russia; paveldmt@piboc.dvo.ru (P.S.D.); ivanchina@piboc.dvo.ru (N.V.I.)
* Correspondence: kalininv@piboc.dvo.ru; Tel.: +7-914-705-08-45

Citation: Dmitrenok, P.S.; Ivanchina, N.V.; Kalinin, V.I. Honoring Prof. Dr. Valentin A. Stonik for His Outstanding Contribution to Marine Natural Product Chemistry on the Occasion of His 80th Birthday. *Mar. Drugs* **2024**, *22*, 56. https://doi.org/10.3390/md22020056

Received: 19 January 2024
Accepted: 23 January 2024
Published: 24 January 2024

Marine natural products are a very structurally diverse group of preferably low-weight organic molecules. They began to be intensively studied in the 1960s. Except for their wide structural diversity, these metabolites are interesting as modulators of chemoecological interactions, possible active substances in medicinal and other useful preparations, and as taxonomic and food chain markers. Their biosynthesis and evolution of biosynthesis are also very interesting.

Prof. Dr. Sc. Valentin A. Stonik is one of the pioneers in this field; he began his impressive research career in marine natural products in the early 1970s and has continued his research activities in this field ever since. He was born in Vladivostok, Russia, on 4 December 1942 and graduated from the Department of Chemistry of the Far Eastern State University (Vladivostok) in 1965. Valentin A. Stonik received his PhD degree in Organic Chemistry in 1969 and Dr. Sc. degree in Bioorganic Chemistry: Chemistry of Natural and Physiologically Active Compounds in 1988. He became a Corresponding Member of the Russian Academy of Sciences in 1997 and a Full Member of the Russian Academy of Sciences (Academician) in 2003. He began his scientific activities in the synthesis of hydroacrydines and relative compounds [1]. Since 1970, he has worked in the G.B. Elyakov Pacific Institute of Bioorganic Chemistry, Far Eastern Branch of the Russian Academy of Sciences (PIBOC). Professor Stonik became Head of the Laboratory of Biosynthesis in 1976, Head of the Laboratory of the Chemistry of Marine Natural Products in 1985, Deputy Director of the Institute from 1990 to 2002, and Director from 2002 to 2017. Since 2018, he has been the Scientific Advisor of the Institute.

His general research scopes include the structure and properties of biphylic physiologically active natural products from marine invertebrates, especially echinoderms and sponges, and his specific interests include alkaloids [2], unusual lipids [3], isoprenoids [4], polyhydroxysteroids [5], glycosides of polyhydroxysteroids [6], steroidal [7] and triterpenoidal oligoglycosides [8], investigation of biological activities [9], biosynthesis [10], chemotaxonomy [11], and the chemical evolution of secondary metabolites [12]. In 1970–1990s, he led numerous scientific expeditions on the research vessels "Kallisto", "Professor Bogorov", and "Akademic Oparin" in different regions of the World Ocean. He is an author and co-author of more than 400 scientific articles in Russian and international journals, with more than 7500 citations, 4 monographs, and over 20 patents. He is a member of the Editorial Boards of the following Journals: *Marine Drugs*, *Natural Product Communications*, *Natural Product Letters*, *Russian Journal of Bioorganic Chemistry*, and others.

This Special Issue includes 12 contributions:

1. Silchenko, A.S.; Avilov, S.A.; Popov, R.S.; Dmitrenok, P.S.; Chingizova, E.A.; Grebnev, B.B.; Rasin, A.B.; Kalinin, V.I. Chilensosides E, F, and G—new tetrasulfated triterpene glycosides from the sea cucumber *Paracaudina chilensis* (Caudinidae, Molpadida): structures, activity, and biogenesis. *Mar. Drugs* **2023**, *21*, 114.

2. Menchinskaya, E.S.; Dyshlovoy, S.A.; Venz, S.; Jacobsen, C.; Hauschild, J.; Rohlfing, T.; Silchenko, A.S.; Avilov, S.A.; Balabanov, S.; Bokemeyer, C.; Aminin, D.L.; von Amsberg, G.; Honecker, F. Anticancer activity of the marine triterpene glycoside cucumarioside A_2-2 in human prostate cancer cells. *Mar. Drugs* **2024**, *22*, 20.

3. Malyarenko, T.V.; Zakharenko, V.M.; Kicha, A.A.; Kuzmich, A.S.; Malyarenko, O.S.; Kalinovsky, A.I.; Popov, R.S.; Svetashev, V.I.; Ivanchina, N.V. New ceramides and cerebrosides from the deep-sea Far Eastern starfish *Ceramaster patagonicus*. *Mar. Drugs* **2022**, *20*, 641.

4. Popov, A.M.; Kozlovskaya, E.P.; Klimovich, A.A.; Rutckova, T.A.; Vakhrushev, A.I.; Hushpulian, D.M.; Gazaryan, I.G.; Makhankov, V.V.; Son, O.M.; Tekutyeva, L.A. Carotenoids from starfish *Patiria pectinifera*: therapeutic activity in models of inflammatory diseases. *Mar. Drugs* **2023**, *21*, 470.

5. Kikionis, S.; Papakyriakopoulou, P.; Mavrogiorgis, P.; Vasileva, E.A.; Mishchenko, N.P.; Fedoreyev, S.A.; Valsami, G.; Ioannou, E.; Roussis, V. Development of novel pharmaceutical forms of the marine bioactive pigment echinochrome A enabling alternative routes of administration. *Mar. Drugs* **2023**, 21, 250.

6. Dyshlovoy, S.A.; Fedorov, S.N.; Svetashev, V.I.; Makarieva, T.N.; Kalinovsky, A.I.; Moiseenko, O.P.; Krasokhin, V.B.; Shubina, L.K.; Guzii, A.G.; von Amsberg, G.; Stonik, V.A. 1-O-Alkylglycerol ethers from the marine sponge *Guitarra abbotti* and their cytotoxic activity. *Mar. Drugs* **2022**, *20*, 409.

7. Gartshore, C.J.; Wang, X.; Su, Y.; Molinski, T.F. Petrosamine revisited. Experimental and computational investigation of solvatochromism, tautomerism and free energy landscapes of a pyridoacridinium quaternary salt. *Mar. Drugs* **2023**, *21*, 446.

8. Dyshlovoy, S.A.; Shubina, L.K.; Makarieva, T.N.; Guzii, A.G.; Hauschild, J.; Strewinsky, N.; Berdyshev, D.V.; Kudryashova, E.K.; Menshov, A.S.; Popov, R.S.; Dmitrenok, P.S.; Graefen, M.; Bokemeyer, C.; von Amsberg, G. New guanidine alkaloids batzelladines O and P from the marine sponge *Monanchora pulchra* induce apoptosis and autophagy in prostate cancer cells. *Mar. Drugs* **2022**, *20*, 738.

9. Ivanov, I.A.; Siniavin, A.E.; Palikov, V.A.; Senko, D.A.; Shelukhina, I.V.; Epifanova, L.A.; Ojomoko, L.O.; Belukhina, S.Y.; Prokopev, N.A.; Landau, M.A.; Palikova, Y.A.; Kazakov, V.A.; Borozdina, N.A. Bervinova, A.V.; Dyachenko, I.A.; Kasheverov I.E.; Tsetlin V.I.; Kudryavtsev D.S. Analogs of 6-bromohypaphorine with increased agonist potency for α7 nicotinic receptor as anti-Inflammatory analgesic agents. *Mar. Drugs* **2023**, *21*, 368.

10. Usov, A.I.; Bilan, M.I.; Ustyuzhanina, N.E.; Nifantiev, N.E. Fucoidans of brown algae: comparison of sulfated polysaccharides from *Fucus vesiculosus* and *Ascophyllum nodosum*. *Mar. Drugs* **2022**, *20*, 638.

11. Davydova, V.N.; Krylova, N.V.; Iunikhina, O.V.; Volod'ko, A.V.; Pimenova, E.A.; Shchelkanov, M.Y.; Yermak, I.M. Physicochemical properties and antiherpetic activity of α-carrageenan complex with chitosan. *Mar. Drugs* **2023**, *21*, 238.

12. Solov'eva, T.F.; Bakholdina, S.I.; Naberezhnykh, G.A. Host defense proteins and peptides with lipopolysaccharide-binding activity from marine invertebrates and their therapeutic potential in Gram-negative sepsis. *Mar. Drugs* **2023**, *21*, 581.

Contributions 1–5 concern the metabolites from echnoderms, the representatives of the phylum Echinodermata, one of priority interests of Professor Valentin A. Stonik. The first contribution concerns the isolation of three new tetrasulfated triterpene glycosides, chilensosides E–G, isolated from the Far-Eastern sea cucumber *Paracaudina chilensis*. Their structures were elucidated using 2D NMR and ESIMS procedures. The compounds differ in carbohydrate chains; while chilensosides E and F are tetrasulfated pentaosides with the position of one of the sulfate groups at C-3 of the glucose residues which occupied third position in carbohydrate chains, chilensoside G is a tetrasulfated hexaoside. The isolation of tetrasulfated glycosides is rare.

The second contribution is dedicated to the study of the anticancer activity of triterpene glycoside cucumarioside A_2-2 from the Far Eastern sea cucumber *Cucumaria japonica* in human castrate-resistant cancer. The authors found that the glycoside induced a G2/M-phase cell cycle arrest and caspase-dependent apoptosis via an intrinsic pathway. The glycoside also inhibited the formation of cancer cell colonies and growth in non-cytotoxic concentrations. A proteome analysis, using the 2D-PAGE technique and MALDI-MS followed by bioinformatical evaluation, revealed alterations in proteins involved in metastatic potential, invasion, and apoptosis. The regulation of keratin 81, CrkII, IL-1β, and cathepsin B was found. The results demonstrated the promising anticancer activity of cucumarioside A_2-2 in a prostate cancer model.

The third contribution concerns the study of ceramides and cerebrosides from the deep-sea Far Eastern starfish *Ceramaster patagonicus*. The authors isolated three new ceramides and three new cerebrosides along with three known cerebrosides. The structures of the new compounds were elucidated using NMR and ESIMS procedures and the necessary chemical transformations. All the new cerebrosides have β-D-glucopyranose as a monosaccharide moiety. Most of the new compounds exhibited slight to moderate cytotoxic activity against human cancer cells (HT-29, SK-MEL-28, and MDA-MB-231) and normal embryonic kidney cells HEK293. They also inhibited the colony formation of MDA-MB-231.

The fourth contribution is dedicated to the study of the therapeutic activity of carotenoids from the Far Eastern starfish *Patiria pectinifera* in models of inflammatory diseases. The authors found that the carotenoids mixture isolated from the starfish *P. pectinifera* contains about 50% astaxanthin, 4–6% zeaxanthine and lutein, as well as free fatty acids and their glycerides. The complex exhibited anti-inflammatory, anti-allergic, and cancer-preventive activity without any toxicity at a dose of 500 mg/kg. The carotenoid mixture effectively improves the clinical picture of disease progression, as well as normalizing the cytokine profile and the antioxidant defense system in the in vivo animal models of inflammatory diseases, including skin carcinogenesis, allergic contact dermatitis, and systemic inflammation. The results show that the carotenoid complex from the starfish *P. pectinifera* may be effective for the treatment or prevention of different inflammations.

The fifth contribution concerns the development of novel pharmaceutical forms of the naphtaquinoid pigment echinochrome A, a characteristic metabolite of different sea urchins that possesses useful biological activities as an antioxidant. The authors incorporated echinochrome A isolated from *Diadema* sea urchins harvested near that island of Kastellorizo (Mediterranean Sea) in electrospun micro-/nanofibrous matrices composed of polycaprolactone and polyvinylpyrrolidone in different combinations. Ex vivo permeability studies using echinochrome A-loaded micro-/nanofibrous matrices showed an increased permeation of the pigment across the duodenum barrier. The results reveal that electrospun polymeric micro-/nanofibers may be promising carriers for new pharmaceutical formulations with controlled release, as well as increased echinochrome A stability and solubility useful for oral administration, and have the potential for the targeted delivery of echinochrome A.

The next three contributions cover articles concerning the metabolites of sponges that also are in the sphere of the primary scientific scopes of Professor Valentin A. Stonik. Contribution 6 concerns the investigation of 1-*O*-alkylglycerol ethers of marine sponge *Guitarra abbotti* and their cytotoxic activity. The authors determined the composition of the mixture of 1-*O*-alkylglycerol ethers using ^{1}H and ^{13}C NMR, GLC/MS, and chemical derivatization and found 6 new and 22 previously known 1-*O*-alkylglycerol ethers. The mixture reveals weak cytotoxic activity on HL-60, THP-1, DLD-1, HeLa, SKMEL-28, MDA-MB-231, and SNU C4 human cancer cells. Further cytotoxicity studies in JB6 P+ Cl41 cells bearing mutated MAP kinase genes revealed that JNK1 and ERK2 play a cytoprotective role in the cellular response to the 1-*O*-alkylglycerol ethers-induced cytotoxic effects.

Contribution 7 concerns the reinvestigation of the chemical and physical properties of a colored pyridoacridine alkaloid petrosamine isolated from the Belizean sponge, *Petrosia* sp., a potent inhibitor of acetylcholine esterase. The properties were investigated using spectroscopic and computational methods. The authors found that by analyzing petrosamine-free energy landscapes, pKa, and tautomerism, an accurate electronic depiction of the molecular structure is the di-keto form, with a net charge of $q = +1$, but this is not a dication (q = +2). Such molecular structure complements have been published in computational docking studies to define the contact points in the enzyme active site. This finding may improve the design of new acetylcholine esterase inhibitors based on such a molecular skeleton.

Contribution 8 is dedicated to the isolation and structural elucidation of two new guanidine alkaloids, batzelladines O and P, from the deep-water marine sponge *Monanchora pulchra* and studies of the induction of apoptosis and autophagy in prostate cancer cells by these substances. The structures of these alkaloids were elucidated using NMR spectroscopy, mass spectrometry, and ECD. The isolated substances revealed cytotoxic activity on human prostate cancer cells 22Rv1, PC3, as well as PC3-DR and inhibited colony formation and cancer cells survival. The alkaloids induced apoptosis detected by Western blotting as caspase-3 and PARP cleavage. The pro-survival autophagy, indicated as the upregulation of LC3B-II and downregulation of mTOR, was found to be treated by alkaloids cells. The use of the autophagy inhibitor 3-methyladenine in combination with alkaloids synergistically increased cytotoxic activity. In a combination of alkaloids with docetaxel, an additive effect was determined. The isolated new guanidine alkaloids may be promising drug candidates for the treatment of taxane-resistant prostate cancer.

The alkaloid 6-bromohypaphorine was isolated from the marine sponges *Pachymatisma johnstoni*, *Aplysina* sp., *Aplidium conicum*, the nudibranch *Hermissenda crassicornis* [13], and sea cucumber *Apostichopus japonicus* [14]. Such a wide distribution of this substance in marine invertebrates from so far taxa allows for its microbial origination to be suggested. L-6-bromohypaphorine acts as an agonist of the $\alpha7$ nicotinic acetylcholine receptor (nAChR) involved in anti-inflammatory regulation [13]. The ninth contribution of this Special Issue concerns the synthesis of analogs of natural 6-bromohypaphorine with increased agonist potency for the $\alpha7$ nicotinic receptor as anti-inflammatory analgesic agents. The authors used virtual screening to synthesize the analogs binding to the $\alpha7$ nAChR molecular model. They synthesized fourteen analogs and tested them in vitro using a calcium fluorescence assay on the $\alpha7$ nAChR expressed in neuro 2a cells. The synthesized methoxy ester of D-6-iodohypaphorine (6ID) revealed the highest potency being almost inactive toward $\alpha9\alpha10$ nAChR. The macrophages cytometry showed anti-inflammatory activity because of the decrease in the expression of TLR4 and increasing CD86, similar to the action of PNU282987, known as the selective $\alpha7$ nAChR agonist. The methoxy ester of D-6-nitrohypaphorine revealed anti-oedemic and analgesic effects in an arthritis rat model. The tested compounds showed excellent tolerability with no acute in vivo toxicity. Thus, the combination of molecular modeling with a natural product-inspired drug design allowed for the preparation to be obtained with the desired activity of the chosen nAChR ligand. Hence, the isolation of the biological active natural product by the members of Professor Valentin A. Stonik's group allows for the creation of new perspective candidates for the development of anti-inflammation preparations.

Contribution 10 is a review that concerns fucoidans of brown algae, namely the comparison of sulfated polysaccharides from *Fucus vesiculosus* and *Ascophyllum nodosum*. The authors noted that these biopolymers have many biological activities that may be used in practical applications. These two species from the family Phaeophyceae are known sources of commercial fucoidans. The authors conclude that fucoidans from these species are very complicated mixtures, and only fractions with carefully characterized structures prepared from both fucoidans may be used for drug development.

Contribution 11 concerns the obtaining of the κ-carrageenan polyelectrolyte complex with chitosan and studying its physicochemical properties and antiherpetic activity. The authors showed a two-fold increase in the antiherpetic activity (selective index) of the obtained polyelectrolyte complex compared to κ-carrageenan that may be caused by a change in the physicochemical characteristics of κ-carrageenan in the complex.

Contribution 12 is a review summarizing the data on lipopolysaccharide-binding proteins from marine invertebrates that may inhibit the toxic effects of bacterial lipopolysaccharides and are possible potential drugs for the treatment of lipopolysaccharide-induced sepsis. The structure of the proteins and peptides and their synthetic analogs, physicochemical properties, antimicrobial, and LPS-binding/neutralizing activity are described in detail. The problems which arise during the clinical trials of these substances are discussed.

Hence, all the contributions of this Special Issue revealed the wide scientific scopes and interests of Professor Valentin A. Stonik as a researcher and as the Director and Scientific Advisor of the G.B. Elyakov Pacific Institute of Bioorganic Chemistry.

Conflicts of Interest: The authors declare no conflict interests.

References

1. Stonik, V.A.; Vysotski, V.I.; Tilichenko, M.N. Hydroacridines and related compounds. IX. Stepwise alkylation of 1,2,3,4-tetrahydroacridine and sym-octahydroacridine. *Chem. Heterocycl. Compd.* **1972**, *8*, 611–613. [CrossRef]
2. Guzii, A.G.; Makarieva, T.N.; Denisenko, V.A.; Dmitrenok, P.S.; Kuzmich, A.S.; Dyshlovoy, S.A.; Krasokhin, V.B.; Stonik, V.A. Monanchocidin: A new apoptosis-inducing polycyclic guanidine alkaloid from the marine sponge *Monanchora pulchra*. *Org. Lett.* **2010**, *12*, 4292–4295. [CrossRef] [PubMed]
3. Makarieva, T.N.; Denisenko, V.A.; Stonik, V.A.; Milgrom, Y.M.; Rashkes, Y.V. Rhizochalin, a novel secondary metabolite of mixed biosynthesis from the sponge *Rhizochalina incrustata*. *Tetrahedron Lett.* **1989**, *30*, 6581–6584. [CrossRef]
4. Fedorov, S.N.; Radchenko, O.S.; Shubina, L.K.; Kalinovsky, A.I.; Gerasimenko, A.V.; Popov, D.Y.; Stonik, V.A. Aplydactone, a new sesquiterpenoid with an unprecedented carbon skeleton from the sea hare *Aplysia dactylomela*, and its Cargill-like rearrangement. *J. Am. Chem. Soc.* **2001**, *123*, 504–505. [CrossRef] [PubMed]
5. Makarieva, T.N.; Shubina, L.K.; Kalinovsky, A.I.; Stonik, V.A.; Elyakov, G.B. Steroids in porifera. II. Steroid derivatives from two sponges of the family Halichondriidae. Sokotrasterol sulfate, a marine steroid with a new pattern of side chain alkylation. *Steroids* **1983**, *42*, 267–281. [CrossRef] [PubMed]
6. Kicha, A.A.; Kalinovsky, A.I.; Levina, E.V.; Stonik, V.A.; Elyakov, G.B. Asterosaponin P_1 from *Patiria pectinifera*. *Tetrahedron Lett.* **1983**, *24*, 3893–3896. [CrossRef]
7. Ivanchina, N.V.; Kicha, A.A.; Stonik, V.A. Steroid glycosides from marine organisms. *Steroids* **2011**, *76*, 425–454. [CrossRef] [PubMed]
8. Silchenko, A.S.; Kalinovsky, A.I.; Avilov, S.A.; Dmitrenok, P.S.; Kalinin, V.I.; Berdyshev, D.V.; Chingizova, E.A.; Andryjaschenko, P.V.; Minin, K.V.; Stonik, V.A. Fallaxosides B_1 and D_3, triterpene glycosides with novel skeleton types of aglycones from the sea cucumber *Cucumaria fallax*. *Tetrahedron* **2017**, *73*, 2335–2341. [CrossRef]
9. Aminin, D.; Pislyagin, E.; Astashev, M.; Es'kov, A.; Kozhemyako, V.; Avilov, S.; Zelepuga, E.; Yurchenko, E.; Kaluzhskiy, L.; Kozlovskaya, E.; et al. Glycosides from edible sea cucumbers stimulate macrophages via purinergic receptors. *Sci. Rep.* **2016**, *6*, 39683. [CrossRef]
10. Ivanchina, N.V.; Kicha, A.A.; Malyarenko, T.V.; Kalinovsky, A.I.; Dmitrenok, P.S.; Stonik, V.A. Biosynthesis of polar steroids from the Far Eastern starfish *Patiria (=Asterina) pectinifera*. Cholesterol and cholesterol sulfate are converted into polyhydroxylated sterols and monoglycoside asterosaponin P_1 in feeding experiments. *Steroids* **2013**, *78*, 1183–1191. [CrossRef] [PubMed]
11. Kalinin, V.I.; Silchenko, A.S.; Avilov, S.A.; Stonik, V.A.; Smirnov, A.V. Sea cucumbers triterpene glycosides, the recent progress in structural elucidation and chemotaxonomy. *Phytochem. Rev.* **2005**, *4*, 221–236. [CrossRef]
12. Stonik, V.A.; Kalinin, V.I.; Avilov, S.A. Toxins from sea cucumbers (holothuroids): Chemical structures, properties, taxonomic distribution, biosynthesis and evolution. *J. Nat. Toxins* **1999**, *8*, 235–248. [PubMed]
13. Kasheverov, I.; Shelukhina, I.; Kudryavtsev, D.; Makarieva, T.; Spirova, E.; Guzii, A.; Stonik, V.; Tsetlin, V. 6-Bromohypaphorine from marine nudibranch mollusk is an agonist of human α7 nicotinic acetylcholine receptor. *Mar. Drugs* **2015**, *13*, 1255–1266. [CrossRef] [PubMed]
14. Dai, Y.-L.; Zhou, D.-Y.; Jiang, Y.-F.; Zheng, F.; Yue, H.; Jeon, Y.-J. 6-Bromohypaphorine isolated from red sea cucumbers *Apostichopus japonicus* exhibits potent anticancer activity in A549 cancer cell line. *Chin. J. Analyt. Chem.* **2021**, *49*, 37–42. [CrossRef]

marine drugs

Article

Chilensosides E, F, and G—New Tetrasulfated Triterpene Glycosides from the Sea Cucumber *Paracaudina chilensis* (Caudinidae, Molpadida): Structures, Activity, and Biogenesis

Alexandra S. Silchenko, Sergey A. Avilov, Roman S. Popov, Pavel S. Dmitrenok, Ekaterina A. Chingizova, Boris B. Grebnev, Anton B. Rasin and Vladimir I. Kalinin *

G.B. Elyakov Pacific Institute of Bioorganic Chemistry, Far Eastern Branch of the Russian Academy of Sciences, Pr. 100-letya Vladivostoka 159, 690022 Vladivostok, Russia
* Correspondence: kalininv@piboc.dvo.ru; Tel./Fax: +7-(423)2-31-40-50

Abstract: Three new tetrasulfated triterpene glycosides, chilensosides E (**1**), F (**2**), and G (**3**), have been isolated from the Far-Eastern sea cucumber *Paracaudina chilensis* (Caudinidae, Molpadida). The structures were established based on extensive analysis of 1D and 2D NMR spectra and confirmed by HR-ESI-MS data. The compounds differ in their carbohydrate chains, namely in the number of monosaccharide residues (five or six) and in the positions of sulfate groups. Chilensosides E (**1**) and F (**2**) are tetrasulfated pentaosides with the position of one of the sulfate groups at C-3 Glc3, and chilensoside G (**3**) is a tetrasulfated hexaoside. The biogenetic analysis of the glycosides of *P. chilensis* has revealed that the structures form a network due to the attachment of sulfate groups to almost all possible positions. The upper semi-chain is sulfated earlier in the biosynthetic process than the lower one. Noticeably, the presence of a sulfate group at C-3 Glc3—a terminal monosaccharide residue in the bottom semi-chain of compounds **1** and **2**—excludes the possibility of this sugar chain's further elongation. Presumably, the processes of glycosylation and sulfation are concurrent biosynthetic stages. They can be shifted in time in relation to each other, which is a characteristic feature of the mosaic type of biosynthesis. The hemolytic action of compounds **1–3** against human erythrocytes and cytotoxic activities against five human cancer cell lines were tested. The compounds showed moderate hemolytic activity but were inactive against cancer cells, probably because of their structural peculiarities, such as the combination of positions of four sulfate groups.

Keywords: *Paracaudina chilensis*; Caudinidae; Molpadida; triterpene glycosides; chilensosides; sea cucumber; hemolytic and cytotoxic activity

Citation: Silchenko, A.S.; Avilov, S.A.; Popov, R.S.; Dmitrenok, P.S.; Chingizova, E.A.; Grebnev, B.B.; Rasin, A.B.; Kalinin, V.I. Chilensosides E, F, and G—New Tetrasulfated Triterpene Glycosides from the Sea Cucumber *Paracaudina chilensis* (Caudinidae, Molpadida): Structures, Activity, and Biogenesis. *Mar. Drugs* **2023**, *21*, 114. https://doi.org/10.3390/md21020114

Academic Editor: Bill J. Baker

Received: 11 January 2023
Revised: 1 February 2023
Accepted: 3 February 2023
Published: 5 February 2023

1. Introduction

Triterpene glycosides from sea cucumbers are long-studied metabolites that still draw interest from researchers from different scientific fields. The structural studies of the compounds from holothuroids—representatives of different taxonomic groups—are predominating [1–4], but they are in close connection with biological activity research [4–8] and, therefore, structure-activity relationships analyses [3,9]. An additional direction of the glycosides' applied investigations is their use as chemotaxonomic markers to clarify the systematic position of the producing species of sea cucumbers [10–12]. This issue is also relevant for the species under investigation—*Paracaudina chilensis*, belonging to the family Molpadida. The systematics and phylogeny of this taxonomic group have raised questions until now [13,14]. Our recent study concerning the isolation and structural elucidation of chilensosides A–D—the glycosides from *P. chilensis*—demonstrated their structural similarity to the glycosides isolated from different representatives of the order Dendrochirotida, indicating the closeness of Molpadida to Dendrochirotida [15]. Chilensosides A–D contains two different aglycones and four types of carbohydrate chains, differing in the positions and quantity of sulfate groups from two (chilensosides A, A_1, B) to three in chilensoside

C and four in chilensoside D. Three out of five glycosides demonstrated relatively high hemolytic and cytotoxic activity.

In continuation of the structural research on the glycosides from *P. chilensis*, the isolation, structure elucidation, biologic activity testing, and carbohydrate chain biogenesis of new sulfated highly polar glycosides chilensosides E (**1**), F (**2**), and G (**3**) are reported. Each of the three new compounds contains four sulfate groups. The first finding of two tetrasulfated glycosides in sea cucumbers has occurred fairly recently in *Psolus fabricii* [16]. Three years later, three other glycosides containing four sulfate groups were found in the sea cucumber *Psolus chitonoides* [17] and one in *P. chilensis* [15]. The finding of tetrasulfated glycosides in recent years could be explained by the increased ability of HPLC separation techniques, including the use of different sorbents (immobile phases) to separate earlier inseparable polar substances [18]. The expansion of knowledge concerning the structural variability of the glycosides enables the clarification of the biosynthetic pathways of these metabolites.

The chemical structures of **1**–**3** were elucidated by the analyses of the ^{1}H, ^{13}C NMR, 1D TOCSY, and 2D NMR (^{1}H,^{1}H-COSY, HMBC, HSQC, ROESY) spectra, as well as HR-ESI mass spectra. All the original spectra are displayed in Figures S1–S24 in the Supplementary data. The hemolytic activity on human erythrocytes and cytotoxic activities on leukemia promyeloblast HL-60, adenocarcinoma HeLa, colorectal adenocarcinoma DLD-1, human neuroblastoma SH-SY5Y, and monocytic THP-1 cells were tested.

2. Results and Discussion

2.1. Structure Elucidation of the Glycosides

The crude glycosidic mixture of *Paracaudina chilensis* was isolated by hydrophobic chromatography of the concentrated ethanolic extract on a Polychrom-1 column (powdered Teflon, Biolar, Latvia). Its further separation using chromatography on Si gel columns with a stepped gradient of the system of eluents CHCl3/EtOH/H2O in the ratios of 100:100:17, 100:125:25, and 100:150:50 was followed by the additional purification of the obtained fractions, yielding the subfractions I.0, I.1, II, III.1, and III.2. The individual glycosides **1**–**3** (Figure 1) have been isolated using HPLC of subfractions II and III.2 on silica-based columns Supelcosil LC-Si (4.6 × 150 mm), and reversed-phase semipreparative columns Supelco Ascentis RP-Amide (10 × 250 mm) and Diasfer 110 C-8 (4.6 × 250 mm).

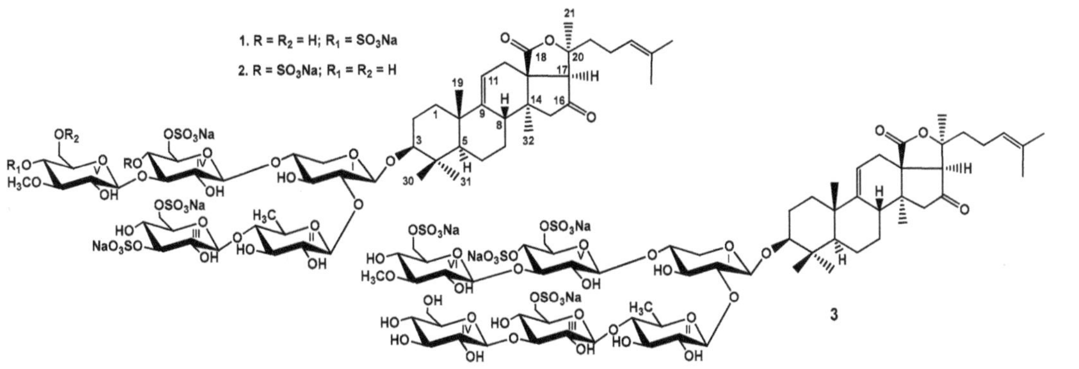

Figure 1. Chemical structures of glycosides isolated from *Paracaudina chilensis*: **1**—chilensoside E; **2**—chilensoside F; **3**—chilensoside G.

The sugar configurations in the glycosides **1**–**3** were assigned as *D*, along with the biogenetic analogies with all other known triterpene glycosides from the sea cucumber.

The holostane aglycones of chilensosides E (**1**), F (**2**), and G (**3**) (Table 1, Tables S1 and S2, Figures S1–S7, S9–S15 and S17–S23) have 9(11)- and 24(25)-double bonds as well as a 16-oxo-group, and are identical to each other and those of chilensosides A_1, B, C, and D, isolated earlier [15]. The identity of the aglycones of compounds **1–3** was evidenced by the coincidence of their NMR spectroscopic data. The structure of the aglycone moiety of **1** was established based on 2D NMR spectra analyses, where the characteristic features were found: 18(20)-lactone signals at δ_C 176.9 (C-18) and δ_C 83.1 (C-20), 9(11)-double bond signals at δ_C 151.1 (C-9), δ_C 111.3 (C-11) and δ_H 5.37 (brs, H-11), and the downfield signal of quaternary carbon at δ_C 214.6 that corresponded to a carbonyl group at C-16. This position was corroborated by the cross-peaks H_2-15/C-16 and H-17/C-16 in the HMBC spectrum of **1**. The signals characteristic of 24(25)-double bonds was observed in the downfield region of the ^{13}C NMR spectrum at δ_C 124.1 (C-24) and 132.1 (C-25). The correlations between H-24/H-22, H-26/H-24, and H-27/H-23 in the ROESY spectrum and H-26/C: 24, 25, and 27 cross-peaks in the HMBC spectrum confirmed the structure of the side chain (Figure 2).

Table 1. ^{13}C and ^{1}H NMR chemical shifts, HMBC and ROESY correlations of aglycone moiety of chilensoside E (**1**).

Position	δ_C Mult. [a]	δ_H Mult. (*J* in Hz) [b]	HMBC	ROESY
1	36.0 CH_2	1.73 m		H-11
		1.30 m		H-3
2	26.8 CH_2	2.06 m		
		1.84 m		H-19, H-30
3	88.4 CH	3.11 dd (4.7; 11.8)		H-1, H-5, H-31, H1-Xyl1
4	39.5 C			
5	52.7 CH	0.79 brd (11.8)	C: 4, 19, 30	H-1, H-3, H-7
6	20.9 CH_2	1.59 m		
7	28.3 CH_2	1.60 m		H-15
		1.12 m		
8	38.6 CH	3.14 m		H-6
9	151.1 C			
10	39.5 C			
11	111.3 CH	5.27 brs	C: 10, 13	H-1
12	31.9 CH_2	2.64 brd (16.5)	C: 11, 18	H-17
		2.48 dd (5.9; 16.5)	C: 11, 14	
13	56.0 C			
14	42.0 C			
15	51.8 CH_2	2.42 d (16.0)	C: 13, 16, 17, 32	
		2.11 d (16.0)	C: 14, 16, 32	H-8
16	214.6 C			
17	61.8 CH	2.89 s	C: 12, 13, 16, 18, 20, 21	H-12, H-23, H-32
18	176.9 C			
19	21.9 CH_3	1.27 s	C: 1, 5, 9, 10	H-1, H-2, H-8, H-30
20	83.1 C			
21	26.6 CH_3	1.48 s	C: 17, 20, 22	H-12, H-17, H-23
22	38.6 CH_2	1.80 m		
		1.59 m		
23	23.0 CH_2	2.27 m		
		2.08 m		
24	124.1 CH	5.03 m		H-22
25	132.1 C			
26	25.5 CH_3	1.56 s	C: 24, 25, 27	H-24
27	17.4 CH_3	1.54 s	C: 24, 25, 26	H-23
30	16.4 CH_3	0.89 s	C: 3, 4, 5, 31	H-2, H-6, H-19, H-31
31	27.8 CH_3	1.10 s	C: 3, 4, 5, 30	H-3, H-5, H-6, H-30
32	20.5 CH_3	0.89 s	C: 8, 13, 14, 15	H-7, H-12, H-15, H-17

[a] Recorded at 176.04 MHz in C_5D_5N/D_2O (4/1). [b] Recorded at 700.13 MHz in C_5D_5N/D_2O (4/1). The original spectra of **1** are provided in Figures S1–S7 (Supplementary Material).

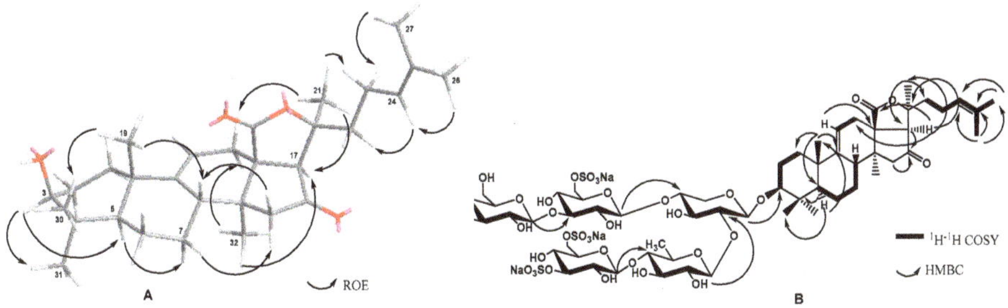

Figure 2. Key ROE- (**A**), HMBC and COSY (**B**) correlations of chilensoside E. The oxygen atoms have marked with red color.

Extensive analysis of the ^{1}H,^{1}H-COSY, 1D TOCSY, HSQC, and ROESY spectra of carbohydrate parts of compounds **1–2** (Tables 2 and 3) indicated the same monosaccharide composition: one xylose (Xyl1), one quinovose (Qui2), two glucose (Glc3 and Glc4), and 3-*O*-methylglucose (MeGlc5) residues. The positions of glycosidic linkages established by the ROESY and HMBC (Figure 2) correlations were typical for the sea cucumber glycosides (*β*-(1→4) and *β*-(1→3) bonds) and corresponded to the oligosaccharide chains branched by C-4 Xyl1 having a bottom semi-chain composed of Qui2 and Glc3 and an upper semi-chain composed of Xyl1, Glc4, and MeGlc5.

Table 2. ^{13}C and ^{1}H NMR chemical shifts, HMBC and ROESY correlations of the carbohydrate moiety of chilensoside E (**1**).

Atom	δ_C Mult. [a,b,c]	δ_H Mult. (*J* in Hz) [d]	HMBC	ROESY
Xyl1 (1→C-3)				
1	104.6 CH	4.61 d (7.4)	C: 3	H-3; H-5 Xyl1
2	**83.4** CH	3.70 m	C: 3 Xyl1	H-1 Qui2
3	75.0 CH	3.93 m		
4	**80.8** CH	3.94 m	C: 3 Xyl1; 1 Glc4	H-1 Glc4; H-2 Xyl1
5	63.4 CH_2	4.33 dd (6.0; 12.3)		
		3.58 m		H-1 Xyl1
Qui2 (1→2Xyl1)				
1	104.6 CH	4.74 d (8.3)	C: 2 Xyl1	H-2 Xyl1; H-3, 5 Qui2
2	75.5 CH	3.92 t (9.4)	C: 3 Qui2	H-4 Qui2
3	74.4 CH	4.06 t (9.4)		H-1, 5 Qui2
4	**86.0** CH	3.30 t (9.4)	C: 1 Glc3	H-1 Glc3
5	71.7 CH	3.60 dd (6.0; 9.4)		H-1, 3 Qui2
6	17.4 CH_3	1.54 d (6.0)	C: 4, 5 Qui2	H-4 Qui2
Glc3 (1→4Qui2)				
1	104.1 CH	4.67 d (7.9)	C: 4 Qui2	H-4 Qui2; H-3, 5 Glc3
2	72.8 CH	3.81 t (7.9)	C: 1, 3 Glc3	
3	*83.8* CH	4.98 t (7.9)		H-1, 5 Glc3
4	69.5 CH	3.80 t (7.9)	C: 3, 5 Glc3	H-6 Glc3
5	74.2 CH	4.07 t (7.9)		
6	*67.4* CH_2	4.91 brd (9.8)		
		4.52 dd (7.3; 11.0)		
Glc4 (1→4Xyl1)				
1	103.4 CH	4.85 d (7.9)	C: 4 Xyl1	H-4 Xyl1; H-3, 5 Glc4
2	74.0 CH	3.83 t (7.9)		
3	**85.8** CH	4.17 t (9.2)	C: 1 MeGlc5	H-1 MeGlc5; H-1 Glc4
4	69.5 CH	3.68 t (9.2)	C: 5 Glc4	
5	74.4 CH	4.16 m		
6	*67.7* CH_2	4.95 brd (10.5)		
		4.44 brdd (7.2; 11.2)		

Table 2. *Cont.*

Atom	δ_C Mult. [a,b,c]	δ_H Mult. (*J* in Hz) [d]	HMBC	ROESY
MeGlc5 (1→3Glc4)				
1	104.3 CH	5.16 d (8.3)	C: 3 Glc4	H-3 Glc4; H-5 MeGlc5
2	74.0 CH	3.87 t (8.3)	C: 1 MeGlc5	H-4 MeGlc5
3	85.3 CH	3.71 t (8.3)	C: 4 MeGlc5; OMe	H-1 Me Glc5
4	*76.1* CH	4.88 t (8.8)	C: 5 MeGlc5	
5	76.4 CH	3.86 m		H-1 MeGlc5
6	61.7 CH_2	4.49 d (11.6)		
		4.33 m		
OMe	60.8 CH_3	3.93 s	C: 3 MeGlc5	H-3 MeGlc5

[a] Recorded at 125.67 MHz in C_5D_5N/D_2O (4/1). [b] Bold = interglycosidic positions. [c] Italic = sulfate position. [d] Recorded at 500.12 MHz in C_5D_5N/D_2O (4/1). Multiplicity by 1D TOCSY. The original spectra of **1** are provided in Figures S1–S7.

Table 3. ^{13}C and ^{1}H NMR chemical shifts, HMBC and ROESY correlations of the carbohydrate moiety of chilensoside F (**2**).

Atom	δ_C Mult. [a,b,c]	δ_H Mult. (*J* in Hz) [d]	HMBC	ROESY
Xyl1 (1→C-3)				
1	104.6 CH	4.62 d (7.7)	C: 3	H-3; H-3, 5 Xyl1
2	**82.9** CH	3.71 t (7.7)	C: 1, 3 Xyl1	H-1 Qui2
3	75.1 CH	3.93 t (7.7)	C: 2, 4 Qui2	H-1 Xyl1
4	**81.2** CH	3.92 m	C: 1 Glc4	H-1 Glc4
5	63.4 CH_2	4.28 dd (5.1; 11.5)	C: 3 Xyl1	
		3.57 m		H-1, 3 Xyl1
Qui2 (1→2Xyl1)				
1	104.7 CH	4.75 d (7.3)	C: 2 Xyl1	H-2 Xyl1; H-3, 5 Qui2
2	75.3 CH	3.94 t (9.3)		H-4 Qui2
3	74.4 CH	4.06 t (9.3)	C: 2, 4 Qui2	H-1, 5 Qui2
4	**85.9** CH	3.31 t (9.3)	C: 1 Glc3; 5 Qui2	H-1 Glc3; H-2 Qui2
5	71.7 CH	3.60 m		H-1, 3 Qui2
6	17.4 CH_3	1.53 d (6.4)	C: 4, 5 Qui2	H-4 Qui2
Glc3 (1→4Qui2)				
1	104.1 CH	4.68 d (7.5)	C: 4 Qui2	H-4 Qui2; H-3, 5 Glc3
2	72.8 CH	3.80 t (9.0)	C: 1, 3 Glc3	
3	*83.9* CH	4.98 t (9.0)	C: 2, 4 Glc3	H-1, 5 Glc3
4	69.5 CH	3.82 t (9.0)	C: 3, 5, 6 Glc3	
5	74.3 CH	4.08 t (9.0)		H-1, 3 Glc3
6	*67.4* CH_2	4.92 brd (11.7)		
		4.53 dd (7.6; 11.7)	C: 5 Glc3	H-4 Glc3
Glc4 (1→4Xyl1)				
1	103.5 CH	4.81 d (7.8)	C: 4 Xyl1	H-4 Xyl1; H-3, 5 Glc4
2	74.0 CH	3.92 t (8.4)	C: 1, 3 Glc4	
3	**82.8** CH	4.37 t (8.4)	C: 1 MeGlc5; 2, 4 Glc4	H-1 MeGlc5
4	*75.5* CH	4.74 t (8.4)	C: 5, 6 Glc4	H-2 Glc4
5	73.8 CH	4.33 m		H-1 Glc4
6	*68.5* CH_2	5.50 m		
		4.62 t (10.9)		H-4 Glc4
MeGlc5 (1→3Glc4)				
1	104.5 CH	5.20 d (8.2)	C: 3 Glc4	H-3 Glc4; H-3, 5 MeGlc5
2	74.5 CH	4.00 t (9.4)	C: 1, 3 MeGlc5	
3	86.9 CH	3.65 t (9.4)	C: 2, 4 MeGlc5; OMe	H-1, 5 Me Glc5; OMe
4	70.0 CH	3.91 t (9.4)	C: 3, 5, 6 MeGlc5	H-6 MeGlc5
5	77.5 CH	3.87 m		H-1 MeGlc5
6	62.0 CH_2	4.34 brd (11.7)		
		4.09 dd (7.0; 11.7)	C: 5 MeGlc5	

Table 3. *Cont.*

Atom	δ_C Mult. [a,b,c]	δ_H Mult. (*J* in Hz) [d]	HMBC	ROESY
OMe	60.3 CH_3	3.76 s	C: 3 MeGlc5	

[a] Recorded at 125.67 MHz in C_5D_5N/D_2O (4/1). [b] Bold = interglycosidic positions. [c] Italic = sulfate position. [d] Recorded at 500.12 MHz in C_5D_5N/D_2O (4/1). Multiplicity by 1D TOCSY. The original spectra of **2** are provided in Figures S9–S15.

The molecular formula of chilensoside E (**1**) was determined to be $C_{60}H_{90}O_{39}S_4Na_4$ from the $[M_{4Na}–2Na]^{2-}$ ion peak at *m/z* 804.1857 (calc. 804.1874), $[M_{4Na}–3Na]^{3-}$ ion peak at *m/z* 528.4622 (calc. 528.4619), and $[M_{4Na}–4Na]^{4-}$ ion peak at *m/z* 390.5999 (calc. 390.5991) in the (−)HR-ESI-MS (Figure S8). The ^{1}H and ^{13}C NMR spectra of the carbohydrate chain of chilensoside E (**1**) (Table 2, Figures S1–S7) demonstrated five characteristic doublets of anomeric protons at δ_H 4.61–5.16 (*J* = 7.4–8.3 Hz) and the signals of anomeric carbons at δ_C 103.4–104.6, indicating the presence of a pentasaccharide chain and *β*-configurations of glycosidic bonds. The MS data indicated the presence of four sulfate groups in the sugar chain of **1** due to the registration of a four-charged ion. The analysis of the signals of the isolated spin system corresponding to the Glc3 residue deduced by 1D TOCSY and ^{1}H,^{1}H-COSY spectra showed three strongly deshielded proton signals (δ_H 4.98 (H-3 Glc3); 4.91 and 4.52 (2H-6 Glc3)). The HSQC spectrum correlated them with two carbon signals at δ_C 83.8 (C-3 Glc3) and 67.4 (C-6 Glc3). These data indicated that two sulfate groups were linked to the Glc3 residue at C-3 and C-6. The ROESY and HMBC correlations of H-3 Glc3 with any protons or carbons of neighboring monosaccharide residues were absent. These data indicated that Glc3 is a terminal residue of the bottom semi-chain. The attachment of the third sulfate group to C-6 Glc4 was deduced from the characteristic signal at δ_C 67.7 assigned in the same manner as for Glc3. The position of the last sulfate group was established to be at C-4 MeGlc5 due to the deshielding of its signal to δ_C 76.1 when compared with corresponding signals in the spectra of chilensosides A and C observed at δ_C 70.0 [15]. Moreover, all the signals of this monosaccharide unit in the ^{13}C NMR spectrum of **1** coincided with the signals of 3-*O*-methylglucose residue, sulfated by C-4, in the spectrum of chilensoside B [15]. Hence, chilensoside E (**1**) is characterized by a tetrasulfated carbohydrate chain with a new combination of sulfate group positions.

The (−)ESI-MS/MS of **1** (Figure S8) demonstrated the fragmentation of $[M_{4Na}–2Na]^{2-}$ ion with *m/z* 804.2 giving the fragment ion-peak at *m/z* 665.7 $[M_{4Na}–2Na–MeGlcSO_3Na+H]^{2-}$, as well as the fragmentation of $[M_{4Na}–3Na]^{3-}$ ion with *m/z* 528.5 resulting in the ion peaks $[M_{4Na}–3Na–HSO_4Na–SO_3Na+2H]^{3-}$ and $[M_{4Na}–3Na–MeGlcSO_3Na+H]^{3-}$ observed at *m/z* 456.8 and 430.1, correspondingly.

These data indicate that chilensoside E (**1**) is 3*β*-*O*-{3,6-*O*-sodium disulfate-*β*-D-glucopyranosyl-(1→4)-*β*-D-quinovopyranosyl-(1→2)-[4-*O*-sodium sulfate-3-*O*-methyl-*β*-D-glucopyranosyl-(1→3)-6-*O*-sodium sulfate-*β*-D-glucopyranosyl-(1→4)]-*β*-D-xylopyranosyl}-16-oxoholosta-9(11),24(25)-diene.

The molecular formula of chilensoside F (**2**) was determined to be $C_{60}H_{90}O_{39}S_4Na_4$ from the $[M_{4Na}–2Na]^{2-}$ ion peak at *m/z* 804.1875 (calc. 804.1874), $[M_{4Na}–3Na]^{3-}$ ion peak at *m/z* 528.4628 (calc. 528.4619) and $[M_{4Na}–4Na]^{4-}$ ion peak at *m/z* 390.5998 (calc. 390.5991) in the (−)HR-ESI-MS (Figure S16), indicating compound **2** is isomeric to **1** presumably by the sulfate group positions. The carbohydrate chain of **2** consisted of five sugar residues (from five signals of anomeric protons at δ_H 4.62–5.20 (*J* = 7.3–8.2 Hz)) (Table 3, Figures S9–S15). Two glucose residues (Glc3 and Glc4) were sulfated by C-6, which was deduced from the presence of characteristic signals of the sulfated hydroxy methylene groups at δ_C 67.4 and 68.5, while the 3-*O*-methylglucose residue did not bear any sulfates (δc 70.0 (C-4 MeGlc5) and δc 62.0 (C-6 MeGlc5)). An additional two sulfates were attached to C-3 Glc3 and C-4 Glc4, deduced from the downfield shifting of their signals to 83.9 and 75.5, respectively. The comparison of the signals corresponding to the Glc3 residue in the ^{13}C NMR spectra of chilensosides E (**1**) and F (**2**) showed their coincidence corroborating the presence of sulfate groups at C-3 Glc3 and C-6 Glc3 in **2**. The same procedure was conducted for the signals of the Glc4 residue in the ^{13}C NMR spectra of chilensosides A, A_1, C [15], and F (**2**) and confirmed the sulfate groups attachment to C-4 Glc4 and C-6 Glc4 in **2**. The deshielding of the signal of C-4 Glc4 to δ_C 75.5 due to the α-shifting effect of the sulfate group in the spectrum of **2** compared with the same signal in the spectrum of **1** (δ_C 69.5) and corroborated the position of the sulfate group. Therefore, chilensoside F (**2**) has a tetrasulfated sugar chain with two disulfated glucose residues.

The (−)ESI-MS/MS of **2** (Figure S16) demonstrated the fragmentation of $[M_{4Na}–2Na]^{2-}$ ion at *m/z* 804.2 that led to the ion peak at *m/z* 533.7 $[M_{4Na}–2Na–MeGlc–GlcSO_3Na+H]^{2-}$ and fragmentation of $[M_{4Na}–3Na]^{3-}$ ion at *m/z* 528.5 resulted in the ion peaks at *m/z* 488.8 $[M_{4Na}–3Na–HSO_4Na]^{3-}$ and 430.1 $[M_{4Na}–3Na–MeGlc–SO_4Na+H]^{3-}$.

All these data indicate that chilensoside F (**2**) is 3*β*-*O*-{3,6-*O*-sodium disulfate-*β*-D-glucopyranosyl-(1→4)-*β*-D-quinovopyranosyl-(1→2)-[3-*O*-methyl-*β*-D-glucopyranosyl-(1→3)-4,6-*O*-sodium disulfate-*β*-D-glucopyranosyl-(1→4)]-*β*-D-xylopyranosyl}-16-oxoholosta-9(11),24(25)-diene.

The molecular formula of chilensoside G (**3**) was determined to be $C_{66}H_{100}O_{44}S_4Na_4$ from the $[M_{4Na}–2Na]^{2-}$ ion peak at *m/z* 885.2161 (calc. 885.2138), $[M_{4Na}–3Na]^{3-}$ ion peak at *m/z* 582.4816 (calc. 582.4795), and $[M_{4Na}–4Na]^{4-}$ ion peak at *m/z* 431.1139 (calc. 431.1123) in the (−)HR-ESI-MS (Figure S24).

The ^{1}H and ^{13}C NMR spectra of the carbohydrate chain of chilensoside G (**3**) (Table 4, Figures S17–S22) demonstrated six characteristic doublets of anomeric protons at δ_H 4.65–5.38 (J = 6.8–8.7 Hz) and six signals of anomeric carbons at δ_C 102.3–104.7 that corresponded to hexasaccharide chain having *β*-configurations of glycosidic bonds. The extensive analysis of the ^{1}H,^{1}H-COSY, 1D TOCSY, HSQC, ROESY, and HMBC spectra of **3** indicated the presence of one xylose (Xyl1), one quinovose (Qui2), three glucose (Glc3, Glc4, Glc5), and one 3-*O*-methylglucose (MeGlc6) residue. The monosaccharides were connected to each other and to the aglycone by the glycosidic linkages located at typical sea cucumber glycoside positions, which was confirmed by the correlations in the ROESY and HMBC spectra of **3**: H-1 Xyl1/H-3 (C-3) of the aglycone, H-1 Qui2/H-2 (C-2) Xyl1, H-1 Glc3/H-4 (C-4) Qui2, H-1 Glc4/H-3 (C-3) Glc3, H-1 Glc5/H-4 (C-4) Xyl1, and H-1 MeGlc6/H-3 (C 3) Glc5 (Table 4, Figures S17–S23).

Table 4. ^{13}C and ^{1}H NMR chemical shifts, HMBC and ROESY correlations of the carbohydrate moiety of chilensoside G (**3**).

Atom	δ_C Mult. ^{a,b,c}	δ_H Mult. (J in Hz) ^{d}	HMBC	ROESY
Xyl1 (1→C-3)				
1	104.7 CH	4.65 d (6.8)	C: 3	H-3; H-3, 5 Xyl1
2	**82.2** CH	3.85 t (8.7)	C: 1 Xyl1	H-1 Qui2
3	75.0 CH	4.10 t (8.7)	C: 2, 4 Qui2	H-5 Xyl1
4	**79.5** CH	4.05 m		H-1 Glc5
5	63.4 CH_2	4.34 dd (4.9; 11.7)	C: 3 Xyl1	
		3.63 brdd (8.7; 11.7)		H-1 Xyl1
Qui2 (1→2Xyl1)				
1	104.5 CH	4.89 d (8.0)	C: 2 Xyl1	H-2 Xyl1; H-3, 5 Qui2
2	75.4 CH	3.83 t (8.9)	C: 1, 3 Qui2	H-4 Qui2
3	74.7 CH	3.92 t (8.3)	C: 2 Qui2	H-1 Qui2
4	**86.9** CH	3.34 t (8.3)	C: 1 Glc3; 3, 5 Qui2	H-1 Glc3; H-2 Qui2
5	71.4 CH	3.61 dd (6.2; 8.3)		H-1, 3 Qui2
6	17.6 CH_3	1.57 d (6.5)	C: 4, 5 Qui2	H-4 Qui2

Table 4. *Cont.*

Atom	δ_C Mult. [a,b,c]	δ_H Mult. (*J* in Hz) [d]	HMBC	ROESY
Glc3 (1→4Qui2)				
1	104.2 CH	4.71 d (7.8)	C: 4 Qui2	H-4 Qui2; H-3, 5 Glc3
2	73.4 CH	3.83 t (8.6)	C: 1, 3 Glc3	
3	**86.1** CH	4.16 t (8.6)	C: 1 Glc4; 2, 4 Glc3	H-1 Glc4; H-1 Glc3
4	69.3 CH	3.75 t (8.6)	C: 3, 5, 6 Glc3	H-6 Glc3
5	74.7 CH	4.08 t (8.6)		H-1 Glc3
6	*67.4* CH_2	4.95 brd (10.2)	C: 4 Glc3	
		4.54 dd (7.1; 11.0)		H-4 Glc3
Glc4 (1→3Glc3)				
1	104.6 CH	5.19 d (7.8)	C: 3 Glc3	H-3 Glc3; H-3, 5 Glc4
2	74.9 CH	3.91 t (8.6)	C: 3 Glc4	
3	77.2 CH	4.09 t (8.6)	C: 2, 4 Glc4	H-1 Glc4
4	71.0 CH	3.88 t (8.6)		
5	77.7 CH	3.90 t (8.6)		H-1 Glc4
6	61.9 CH_2	4.35 d (11.8)		
		4.05 dd (5.5; 11.8)	C: 5 Glc4	
Glc5 (1→4Xyl1)				
1	102.7 CH	4.81 d (8.7)	C: 4 Xyl1	H-4 Xyl1; H-3, 5 Glc5
2	73.5 CH	3.96 t (8.7)	C: 1, 3 Glc5	
3	**80.1** CH	4.52 t (8.7)	C: 1 MeGlc6; 2, 4 Glc5	H-1 MeGlc6; H-1, 5 Glc5
4	75.3 CH	4.76 t (8.7)	C: 3, 5, 6 Glc5	H-6 Glc5
5	73.8 CH	4.25 t (8.7)	C: 4, 6 Glc5	H-1, 3 Glc5
6	*68.2* CH_2	5.39 d (10.9)		
		4.66 dd (7.6; 10.9)	C: 5 MeGlc5	
MeGlc6 (1→3Glc5)				
1	102.3 CH	5.38 d (7.8)	C: 3 Glc5	H-3 Glc5; H-3, 5 MeGlc6
2	74.0 CH	3.94 t (7.8)	C: 1, 3 MeGlc6	
3	86.3 CH	3.61 t (8.9)	OMe; C: 2, 4 MeGlc6	OMe; H-1, 5 MeGlc6
4	69.4 CH	4.04 t (8.9)	C: 3, 5, 6 MeGlc6	
5	75.6 CH	3.97 m		H-1, 3 MeGlc6
6	*66.8* CH_2	4.94 d (10.4)	C: 4 MeGlc6	
		4.79 dd (5.2; 11.9)		
OMe	60.3 CH_3	3.73 s	C: 3 MeGlc6	

[a] Recorded at 125.67 MHz in C_5D_5N/D_2O (4/1). [b] Bold = interglycosidic positions. [c] Italic = sulfate position. [d] Recorded at 500.12 MHz in C_5D_5N/D_2O (4/1). Multiplicity by 1D TOCSY. The original spectra of **3** are provided in Figures S17–S22.

The positions of sulfate groups were established as a result of NMR spectra analyses. The typical values of chemical shifts observed due to the sulfate groups shifting effects were found. The signal of C-6 Glc3 was deshielded to δ_C 67.4, and the signal of C-5 Glc3—shielded to δ_C 74.7, indicating the sulfation of the hydroxymethylene group of this residue. At the same time, the signal of C-3 Glc3 was observed at δ_C 86.1 due to the glycosylation effect appearing because of the attachment of the terminal (Glc4) unit to this position. The ROESY correlation H-1 Glc4/H-3 Glc3 confirmed this supposition. As a result of the analysis of the isolated spin system corresponding to the Glc5 residue, two deshielded signals (in comparison with the signals of non-sulfated carbons) were assigned to sulfated carbons, C-4 Glc5 (δ_C 75.3) and C-6 Glc5 (δ_C 68.2). Another downfield shifted signal at δ_C 80.1 was attributed to position 3 of Glc5 glycosylated by the sixth monosaccharide unit. The last sulfate group was attached to C-6 of the terminal residue in the upper semi-chain—MeGlc6. A corresponding signal (C-6 MeGlc6) was observed at δ_C 66.8. The presence of a terminal non-methylated glucose unit (Glc4) was established by the shielded signal of C-3 Glc4 at δ_C 77.2, which was observed instead of the signal at δ_C ~86.5 in 3-*O*-methylated derivatives. The absence of the signal of the second OMe-group at δ_C ~60.5 in the ^{13}C NMR spectrum of **3** was an additional confirmation. Hence, chilensoside G (**3**) is a tetrasulfated hexaoside that expanded the list of the most polar glycosides of sea cucumbers found so far.

The (−)ESI-MS/MS of **3** (Figure S24) demonstrated the fragmentation of $[M_{4Na}–2Na]^{2−}$ ion with m/z 885.2 resulted in the ion-peaks at m/z 825.7 $[M_{4Na}–2Na–SO_4Na]^{2−}$, 746.7 $[M_{4Na}–2Na–MeGlcSO_3Na]^{2−}$, 695.7 $[M_{4Na}–2Na–MeGlcSO_3Na–SO_3Na]^{2−}$, and 614.7 $[M_{4Na}–2Na–MeGlcSO_3Na–Glc–SO_3Na]^{2−}$. The fragmentation of $[M_{4Na}–3Na]^{3−}$ ion at m/z 582.5 led to the presence of the ion peaks at m/z 548.8 $[M_{4Na}–3Na–SO_3Na]^{3−}$, 542.8 $[M_{4Na}–3Na–NaHSO_4]^{3−}$, and 484.1 $[M_{4Na}–3Na–MeGlcSO_3Na]^{3−}$.

These data indicate that chilensoside G (**3**) is 3*β*-*O*-{*β*-D-glucopyranosyl-(1→3)-6-*O*-sodium sulfate-*β*-D-glucopyranosyl-(1→4)-*β*-D-quinovopyranosyl-(1→2)-[6-*O*-sodium sulfate-3-*O*-methyl-*β*-D-glucopyranosyl-(1→3)-4,6-*O*-sodium disulfate-*β*-D-glucopyranosyl-(1→4)]-*β*-D-xylopyranosyl}-16-oxoholosta-9(11),24(25)-diene.

2.2. Bioactivity of the Glycosides

The cytotoxic activities of chilensosides E–G (**1**–**3**) against human cell lines, erythrocytes, and cancer cells, including neuroblastoma SH-SY5Y, adenocarcinoma HeLa, colorectal adenocarcinoma DLD-1, leukemia promyeloblast HL-60, and monocytic THP-1 have been studied. The earlier tested chitonoidoside L [3] was used as the positive control in all the tests (Table 5). The activity of the glycosides against SH-SY5Y, HeLa, and DLD-1 cells was examined using an MTT assay and against HL-60 and THP-1 cells using an MTS assay.

Table 5. The cytotoxic activities of glycosides **1**–**3** and chitonoidoside L (positive control) against human erythrocytes, SH-SY5Y, HeLa, DLD-1, HL-60, and THP-1 human cell lines.

Glycosides	ED_{50}, μM, Erythrocytes	Cytotoxicity, IC_{50} μM				
		SH-SY5Y	HeLa	DLD-1	HL-60	THP-1
Chilensoside E (1)	29.89 ± 2.67	>100.0	>100.0	>100.0	>100.0	>100.0
Chilensoside F (2)	34.31 ± 0.63	>100.0	>100.0	>100.0	>100.0	>100.0
Chilensoside G (3)	40.74 ± 1.55	>100.0	>100.0	>100.0	>100.0	>100.0
Chitonoidoside L	1.12 ± 0.10	8.06 ± 0.98	14.36 ± 1.12	9.61 ± 1.24	8.22 ± 0.65	8.32 ± 0.81

Erythrocytes, in agreement with earlier published data [15,19], exhibited an increased sensitivity to the membranolytic action of sea cucumber glycosides compared to cancer cells. The erythrocytes are a traditional and convenient model for investigating the membranolytic action of the glycosides. All compounds **1**–**3** showed moderate hemolytic activity, allowing us to suppose the cytotoxic doses of the investigated compounds will be relatively high. Chilensosides E–G (**1**–**3**) were not cytotoxic against cancer cell lines even at the maximal tested concentration (100 μM). As previously observed, the presence of four sulfate groups alone did not deplete the activity [17]. However, the decreasing membranolytic activity caused by the increasing number of sulfates was reported earlier for some of the glycosides [20]. Generally, the influence of sulfate groups on the activity of the glycosides significantly depends on the structural characteristics of their carbohydrate chains [9]. So, in the case of chilensosides E–G (**1**–**3**), the combination of sulfates quantity and positions presumably negatively affected their activity. This was in good agreement with the previous tests of cytotoxicity of tetrasulfated chilensoside D, which has been inactive in relation to three of five cancer cell lines [15]. The analysis of the structure-activity relationships of the glycosides isolated from *P. chilensis* and the comparison between the effects of trisulfated chilensoside C, tetrasulfated chilensosides D [15], and E–G (**1**–**3**) have allowed the finding of the common feature of tetrasulfated glycosides that presumably extremely decreased their activity against cancer cells—the presence of a sulfate group at C-6 Glc3 (on the bottom semi-chain). However, the much more complicated influence of the glycosides' different structural features, including aglycones' structures, carbohydrate chains, architecture and composition, and their combinations on the glycoside/membrane interactions is obvious.

2.3. Biogenesis of Chilensosides A–G

The majority of chilensosides share the same overall structures, including the same aglycones and monosaccharide composition. Consequently, biosynthesis of the glycosides of *P. chilensis* looks somewhat strictly directed. However, the combinatorial features of biosynthesis [18,21] have become evident when the character of sulfation of sugar chains of these glycosides is analyzed. The biogenetic analysis of this series of glycosides has revealed that they form a network instead of biogenetic rows due to the enzymatic introduction of sulfate groups in almost all possible positions (Figure 3). The trend has been derived from the biogenetic analysis that the upper semi-chains are sulfated before the bottom ones, and C-6 of the glucose residues attached to C-4 Xyl1 take precedence in the sulfation.

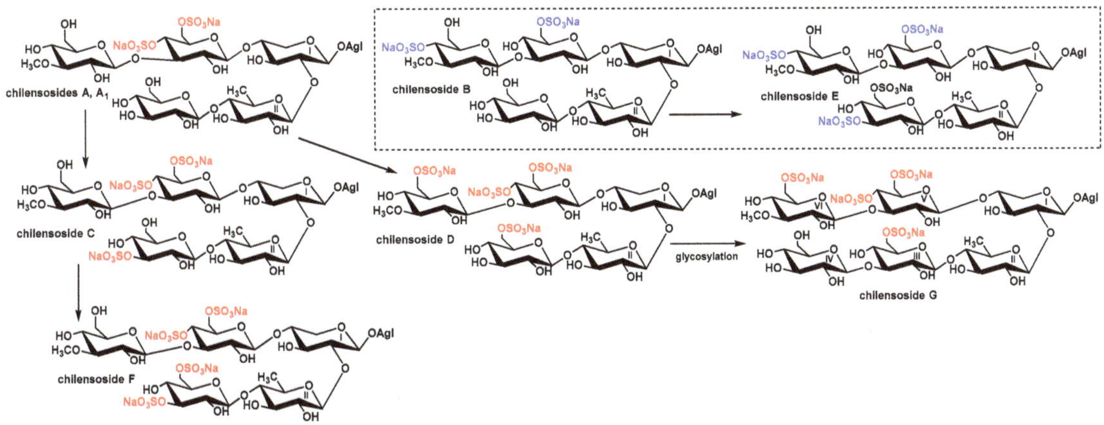

Figure 3. The directions of sulfation in the process of biosynthesis of the glycosides of *P. chilensis*. Uncombined biogenetic rows are highlighted by blue and red colors.

Noticeably, the presence of a sulfate group at C-3 Glc3—the terminal monosaccharide residue in the bottom semi-chain of compounds **1** and **2**—excludes the possibility of further sugar chain elongation. Thus, chilensosides C [15] and F (**2**) could not be the biosynthetic precursors of hexaoside chilensoside G (**3**). More likely, chilensoside G (**3**) is biosynthesized through the chilensosides A and D (Figure 2), having a free hydroxyl group available for glycosylation. Presumably, the processes of glycosylation and sulfation are concurrent biosynthetic stages. They can be shifted in time in relation to each other, a characteristic feature of the mosaic type of biosynthesis [18,21].

3. Materials and Methods

3.1. General Experimental Procedures

Specific rotation was measured on a PerkinElmer 343 Polarimeter (PerkinElmer, Waltham, MA, USA); NMR spectra were registered on a Bruker AMX 500 (Bruker BioSpin GmbH, Rheinstetten, Germany) (500.12/125.67 MHz (^{1}H/^{13}C) spectrometer; ESI MS (positive and negative ion modes) spectra were registered on an Agilent 6510 Q-TOF apparatus (Agilent Technology, Santa Clara, CA, USA), sample concentration 0.01 mg/mL; HPLC was conducted on an Agilent 1260 Infinity II equipped with a differential refractometer (Agilent Technology, Santa Clara, CA, USA); columns were used: Supelcosil LC-Si (4.6 × 150 mm, 5 µm) and Discovery Ascentis RP-Amide (10 × 250 mm, 5 µm) (Supelco, Bellefonte, PA, USA), Diasfer 110 C-8 (4.6 × 250 mm, 5 µm) (Biochemmack, Moscow, Russia).

3.2. Animals and Cells

The sea cucumber *Paracaudina chilensis* (family Caudinidae; order Molpadida) (36 specimens) was harvested in Troitsa Bay, Sea of Japan, in August 2019 by scuba diving 2–5 m in depth. The taxonomic position of the animals was determined by Boris B. Grebnev. The voucher specimen PIBOC-2019-MES-0135 is kept in G.B. Elyakov PIBOC FEB RAS, Vladivostok, Russia.

Human erythrocytes were purchased from the Station of Blood Transfusion in Vladivostok. The cells of the human adenocarcinoma line (HeLa) were provided by the N.N. Blokhin National Medicinal Research Center of Oncology of the Ministry of Health Care of the Russian Federation (Moscow, Russia). The human colorectal adenocarcinoma line DLD-1 CCL-221™ cells, human monocytic THP-1 TIB-202™ cells, human neuroblastoma line SH-SY5Y CRL-2266™, and human promyeloblast cell line HL-60 CCL-240 were received from ATCC (Manassas, VA, USA). The HeLa cell line was cultured in the medium of DMEM (Gibco Dulbecco's Modified Eagle Medium) with 1% penicillin /streptomycin sulfate (Biolot, St. Petersburg, Russia) and 10% fetal bovine serum (FBS) (Biolot, St. Petersburg, Russia). The cells of THP-1, HL-60, and DLD-1 lines were cultured in the medium of RPMI with 1% penicillin/streptomycin (Biolot, St. Petersburg, Russia) and 10% fetal bovine serum (FBS) (Biolot, St. Petersburg, Russia). The cells were incubated at 37 °C in a humidified atmosphere with 5% (*v*/*v*) CO_2. SH-SY5Y cells were cultured in MEM (Minimum Essential Medium) with 1% penicillin/streptomycin sulfate (Biolot, St. Petersburg, Russia) and with fetal bovine serum (Biolot, St. Petersburg, Russia) to a final concentration of 10%.

The study was carried out in accordance with the guidelines of the Declaration of Helsinki and approved by the Ethics Committee of G.B > Elyakov Pacific Institute of Bioorganic Chemistry (Protocol No. 0037.12.03.2021).

3.3. *Extraction and Isolation*

The ethanol extract of the sea cucumbers was purified by standard methodology [15,18], including column hydrophobic and Si gel chromatography. For the latter stage, the stepwise gradient of solvent systems CHCl3/EtOH/H2O: 100:100:17 → 100:125:25 → 100:150:50 as mobile phase was applied, followed by the additional purification of the obtained fractions with CHCl3/EtOH/H2O (100:125:25) as the mobile phase. Finally, the subfractions: I.0 (22 mg), I.1 (120 mg), II (286 mg), III.1 (66 mg), and III.2 (177 mg) were isolated [15]. HPLC of the subfraction II on the silica-based column Supelcosil LC-Si (4.6 × 150 mm, 5 μm) with $CHCl_3/MeOH/H_2O$ (55/30/4) as the mobile phase resulted in the isolation of two fractions (II.1 and II.2). The subsequent HPLC of fraction II.2 on the Supelco Ascentis RP-Amide (10 × 250 mm) column with $MeOH/H_2O/NH_4OAc$ (1 M water solution), ratio (60/38.5/1.5), as the mobile phase led to the isolation of four subfractions. The re-chromatography of two of them on a Diasfer 110 C-8 (4.6 × 250 mm) column with $MeOH/H_2O/NH_4OAc$ (1 M water solution) (50/48/2) as the mobile phase applied for the separation of each subfraction, resulted in the isolation of chilensosides E (**1**) (2.2 mg, R_t 14.12 min) and F (**2**) (2.8 mg, R_t 17.25 min). The subfraction III.2 was submitted to HPLC on a Supelco Ascentis RP-Amide (10 × 250 mm) column with $CH_3CN/H_2O/NH_4OAc$ (1 M water solution), ratio (30/68/2), as the mobile phase to give two main fractions and some minor ones. The repeated HPLC of one of the main fractions in the same conditions led to the isolation of 11.0 mg of chilensoside G (**3**) (R_t 16.67 min).

3.3.1. Chilensoside E (**1**)

Colorless powder; $[\alpha]_D^{20}$–39° (*c* 0.1, H_2O). NMR: Table S1 and Table 1, Figures S1–S8. (−)HR-ESI-MS *m*/*z*: 804.1857 (calc. 804.1874) $[M_{4Na}–2Na]^{2-}$, 528.4622 (calc. 528.4619) $[M_{4Na}–3Na]^{3-}$, 390.5999 (calc. 390.5991) $[M_{4Na}–4Na]^{4-}$); (−)ESI-MS/MS *m*/*z*: 665.7 $[M_{4Na}–2Na–C_7H_{11}O_8SNa\ (MeGlcSO_3Na)+H]^{2-}$, 456.8 $[M_{4Na}–3Na–HSO_4Na–SO_3Na+2H]^{3-}$, 430.1 $[M_{4Na}–3Na–C_7H_{12}O_9SNa\ (MeGlcSO_3Na)]^{3-}$.

3.3.2. Chilensoside F (**2**)

Colorless powder; $[\alpha]_D^{20}$–42° (*c* 0.1, H_2O). NMR: Table S2 and Table 2, Figures S9–S15. (−)HR-ESI-MS *m*/*z*: 804.1875 (calc. 804.1874) $[M_{4Na}–2Na]^{2-}$, 528.4628 (calc. 528.4619) $[M_{4Na}–3Na]^{3-}$, 390.5998 (calc. 390.5991) $[M_{4Na}–4Na]^{4-}$); (−)ESI-MS/MS *m*/*z*: 533.7 $[M_{4Na}–$

2Na–$C_7H_{13}O_5$(MeGlc)–$C_6H_9O_{11}S_2Na_2$(Glc(SO_3Na)$_2$+H]$^{2-}$, 488.8 [M_{4Na}–3Na–HSO_4Na]$^{3-}$, 430.1 [M_{4Na}–3Na–$C_7H_{13}O_5$ (MeGlc)–SO_4Na+H]$^{3-}$.

3.3.3. Chilensoside G (**3**)

Colorless powder; $[\alpha]_D{}^{20}$–53° (*c* 0.1, H_2O). NMR: Tables S3 and Table 3, Figures S17–S22. (−)HR-ESI-MS *m*/*z*: 885.2161 (calc. 885.2138) [M_{4Na}–2Na]$^{2-}$, 582.4816 (calc. 582.4795) [M_{4Na}–3Na]$^{3-}$, 431.1139 (calc. 431.1123) [M_{4Na}–4Na]$^{4-}$; (−)ESI-MS/MS *m*/*z*: 825.7 [M_{4Na}–2Na–SO_4Na]$^{2-}$, 746.7 [M_{4Na}–2Na–$C_7H_{12}O_8SNa$(MeGlcSO_3Na)]$^{2-}$, 695.7 [M_{4Na}–2Na–$C_7H_{12}O_8SNa$(MeGlcSO_3Na)–SO_3Na+H]$^{2-}$, 614.7 [M_{4Na}–2Na–$C_7H_{12}O_8SNa$ (MeGlcSO_3Na)–$C_6H_{11}O_5$(Glc)–SO_3Na]$^{2-}$, 548.8 [M_{4Na}–3Na–SO_3Na]$^{3-}$, 542.8 [M_{4Na}–3Na–$NaHSO_4$]$^{3-}$, 484.1 [M_{4Na}–3Na–$C_7H_{12}O_8SNa$ (MeGlcSO_3Na)]$^{3-}$.

3.4. Cytotoxic Activity (MTT Assay) (for SH-SY5Y, HeLa, and DLD-1 Cells)

All substances (including chitonoidoside L used as a positive control) were tested in concentrations from 0.1 µM to 100 µM using a 2-fold dilution in d-H2O. The cell suspension (180 µL) and solutions (20 µL) of tested glycosides in different concentrations were injected in wells of 96-well plates (SH-SY5Y, 1×104 cells/well, HeLa and DLD-1, 6×10^3/200 µL) and incubated at 37 °C for 24 h in the atmosphere with 5% CO_2. After the incubation, the glycosides with the medium were replaced by 100 µL of fresh medium. Then, 10 µL of MTT (3-(4,5-dimethylthiazol-2-yl)-2,5-diphenyltetrazolium bromide) (Sigma-Aldrich, St. Louis, MO, USA) stock solution (5 mg/mL) were added to each well, followed by incubation of the microplate for 4 h. After this procedure, 100 µL of SDS-HCl solution (1 g SDS/10 mL d-H2O/17 µL 6 N HCl) was added to each well and incubated for 18 h. The absorbance of the converted dye formazan was determined with a Multiskan FC microplate photometer (Thermo Fisher Scientific, Waltham, MA, USA) at 570 nm. Cytotoxic activity of the tested glycosides was calculated as a concentration that caused 50% cell metabolic activity inhibition (IC50). The experiments were conducted in triplicate, $p < 0.05$.

3.5. Cytotoxic Activity (MTS Assay) (for HL-60 and THP-1 Cells)

The cells of THP-1 (6×10^3/200 µL) and HL-60 line (10×10^3/200 µL) were placed in 96-well plates at 37 °C for 24 h in a 5% CO_2 incubator. The cells were treated with tested glycosides and chitonoidoside L as a positive control at concentrations between 0 and 100 µM for an additional 24 h. Then the cells were incubated with 10 µL MTS ([3-(4,5-dimethylthiazol-2-yl)-5-(3-carboxymethoxyphenyl)-2-(4-sulfophenyl)-2H-tetrazolium) for 4 h, and the absorbance in each well was determined at 490/630 nm with a plate reader PHERA star FS (BMG Labtech, Ortenberg, Germany). The experiments were conducted in triplicate. The results were presented as the percentage of inhibition that produced a reduction in absorbance after tested glycoside treatment compared to the non-treated cells (negative control), $p < 0.01$.

3.6. Hemolytic Activity

Erythrocytes were obtained from human blood (AB(IV) Rh+) by centrifugation with phosphate-buffered saline (PBS) (pH 7.4) at 4 °C for 5 min three times at 450 g on a centrifuge LABOFUGE 400R (Heraeus, Hanau, Germany). Then, the erythrocytes residue was resuspended in ice-cold phosphate saline buffer (pH 7.4) to a final optical density of 1.5 at 700 nm and kept on ice [22]. For the hemolytic assay, 180 µL of erythrocyte suspension was mixed with 20 µL of test compound solution (including chitonoidoside L used as a positive control) in V-bottom 96-well plates. After 1 h of incubation at 37 °C, the plates were exposed to centrifugation for 10 min at 900 g in a laboratory centrifuge LMC-3000 (Biosan, Riga, Latvia) [22]. Then, 100 µL of supernatant was carefully decanted and transferred into new flat plates. The erythrocyte lysis values were measured by measuring the supernatant's hemoglobin concentration with a microplate photometer Multiskan FC (Thermo Fisher Scientific, Waltham, MA, USA), λ = 570 nm [23]. The effective dose causing 50% hemolysis of erythrocytes (ED50) was calculated with the computer program SigmaPlot 10.0. All the experiments were carried out in triple repetitions, $p < 0.01$.

4. Conclusions

As a result of the investigation of the glycosidic composition of the sea cucumber *Paracaudina chilensis*, taking into account this paper and a previously published paper [15], the structures of eight new glycosides, chilensosides A–G have been established, and their cytotoxic activity has been studied. The structural diversity of these compounds concerned the sugar moieties, mainly the number and positions of sulfate groups. Seven glycosides shared pentasaccharides branched by C-4 Xyl1 chains with the same monosaccharide composition and sequence of residues. One glycoside from the series—chilensoside G (**3**)—contained six monosaccharide residues having additional glucose units in the bottom semi-chain. Analogously, seven glycosides have identical aglycones, and only one, chilensoside A [15], is characterized by the other structure of the aglycone side chain. Thus, the majority of chilensosides share the same overall structure. They are formed as a result of the same cascade of enzymatic reactions making this biosynthetic pathway predominant and minimizing the mosaicism (combinatorial character) [18] of the glycosides biosynthesis. However, the combinatorial features become evident in the mode of sulfation of sugar chains of these glycosides because of the attachment of sulfate groups in somewhat different positions.

Supplementary Materials: The following supporting information can be downloaded at: https://www.mdpi.com/article/10.3390/md21020114/s1; The original spectral data (Figures S1–S24 and Tables S1 and S2).

Author Contributions: Conceptualization, A.S.S. and V.I.K.; methodology, A.S.S., S.A.A. and V.I.K.; investigation, A.S.S., S.A.A., R.S.P., P.S.D., A.B.R., E.A.C. and B.B.G.; writing—original draft preparation, A.S.S. and V.I.K.; review and editing, A.S.S. and V.I.K. All authors have read and agreed to the published version of the manuscript.

Funding: The investigation was conducted with the financial support of a grant from the Russian Science Foundation No. 23-44-10009.

Institutional Review Board Statement: The study was conducted according to the guidelines of the Declaration of Helsinki and approved by the Ethics Committee of the Pacific Institute of Bioorganic Chemistry (Protocol No. 0037.12.03.2021).

Acknowledgments: The study was carried out on the equipment of the Collective Facilities Center, "The Far Eastern Center for Structural Molecular Research (NMR/MS) PIBOC FEB RAS". The authors are grateful to Valentin Stonik for reading the manuscript and providing useful comments.

Conflicts of Interest: The authors declare no conflict of interest.

References

1. Kalinin, V.I.; Silchenko, A.S.; Avilov, S.A.; Stonik, V.A. Progress in the studies of triterpene glycosides from sea cucumbers (*Holothuroidea, Echinodermata*) between 2017 and 2021. *Nat. Prod. Commun.* **2021**, *16*, 10. [CrossRef]
2. Mondol, M.A.M.; Shin, H.J.; Rahman, M.A. Sea cucumber glycosides: Chemical structures, producing species and important biological properties. *Mar. Drugs* **2017**, *15*, 317. [CrossRef] [PubMed]
3. Park, J.-I.; Bae, H.-R.; Kim, C.G.; Stonik, V.A.; Kwak, J.Y. Relationships between chemical structures and functions of triterpene glycosides isolated from sea cucumbers. *Front. Chem.* **2014**, *2*, 77. [CrossRef] [PubMed]
4. Maier, M.S. Biological activities of sulfated glycosides from Echinoderms. In *Studies in Natural Product Chemistry (Bioactive Natural Products)*; Rahman, A.U., Ed.; Elsevier: Amsterdam, The Netherlands, 2008; Volume 35, pp. 311–354.
5. Khotimchenko, Y. Pharmacological potential of sea cucumbers. *Int. J. Mol. Sci.* **2020**, *19*, 1342. [CrossRef] [PubMed]
6. Zhao, Y.-C.; Xue, C.-H.; Zhang, T.T.; Wang, Y.-M. Saponins from sea cucumber and their biological activities. *Agric. Food Chem.* **2018**, *66*, 7222–7237. [CrossRef] [PubMed]
7. Gomes, A.R.; Freitas, A.C.; Duarte, A.C.; Rocha-Santos, T.A.P. Echinoderms: A review of bioactive compounds with potential health effects. In *Studies in Natural Products Chemistry*; Rahman, A.U., Ed.; Elsevier: Amsterdam, The Netherlands, 2016; Volume 49, pp. 1–54.

8. Chludil, H.D.; Murray, A.P.; Seldes, A.M.; Maier, M.S. Biologically active triterpene glycosides from sea cucumbers (*Holothurioidea, Echinodermata*). In *Studies in Natural Products Chemistry*; Rahman, A.U., Ed.; Elsevier: Amsterdam, The Netherlands, 2003; Volume 28, pp. 587–616.
9. Zelepuga, E.A.; Silchenko, A.S.; Avilov, S.A.; Kalinin, V.I. Structure-activity relationships of holothuroid's triterpene glycosides and some in silico insights obtained by molecular dynamics study on the mechanisms of their membranolytic action. *Mar. Drugs* **2021**, *19*, 604. [CrossRef] [PubMed]
10. Honey-Escandon, M.; Arreguin-Espinosa, R.; Solis-Martin, F.A.; Samyn, Y. Biological and taxonomic perspective of triterpenoid glycosides of sea cucumbers of the family Holothuriidae (*Echinodermata, Holothuroidea*). *Comp. Biochem. Physiol.* **2015**, *180B*, 16–39. [CrossRef] [PubMed]
11. Omran, N.E.; Salem, ·H.K.; Eissa, S.H.; Kabbash, A.M.; Kandeil, ·M.A.; Salem, M.A. Chemotaxonomic study of the most abundant Egyptian sea-cucumbers using ultra-performance liquid chromatography (UPLC) coupled to high-resolution mass spectrometry (HRMS). *Chemoecology* **2020**, *30*, 35–48. [CrossRef]
12. Kalinin, V.I.; Avilov, S.A.; Silchenko, A.S.; Stonik, V.A. Triterpene glycosides of sea cucumbers (Holothuroidea, Echinodermata) as taxonomic markers. *Nat. Prod. Commun.* **2015**, *10*, 21–26. [CrossRef]
13. Smirnov, A.V. System of the class Holothuroidea. *Paleontol. J.* **2012**, *46*, 793–832. [CrossRef]
14. Miller, A.K.; Kerr, A.M.; Paulay, G.; Reich, M.; Wilson, N.G.; Carvajal, J.I.; Rouse, G.W. Molecular phylogeny of extant Holothuroidea (*Echinodermata*). *Mol. Phylogenet. Evol.* **2017**, *111*, 110–131. [CrossRef] [PubMed]
15. Silchenko, A.S.; Avilov, S.A.; Andrijaschenko, P.V.; Popov, R.S.; Chingizova, E.A.; Grebnev, B.B.; Rasin, A.B.; Kalinin, V.I. The isolation, structure elucidation and bioactivity study of chilensosides A, A_1, B, C, and D, holostane triterpene di-, tri- and tetrasulfated pentaosides from the sea cucumber *Paracaudina chilensis* (*Caudinidae, Molpadida*). *Molecules* **2022**, *27*, 7655. [CrossRef] [PubMed]
16. Silchenko, A.S.; Kalinovsky, A.I.; Avilov, S.A.; Kalinin, V.I.; Andrijaschenko, P.V.; Dmitrenok, P.S.; Popov, R.S.; Chingizova, E.A. Structures and bioactivities of psolusosides B_1, B_2, J, K, L, M, N, O, P, and Q from the sea cucumber *Psolus fabricii*. The first finding of tetrasulfated marine low molecular weight metabolites. *Mar. Drugs* **2019**, *17*, 631. [CrossRef] [PubMed]
17. Silchenko, A.S.; Avilov, S.A.; Andrijaschenko, P.V.; Popov, R.S.; Chingizova, E.A.; Dmitrenok, P.S.; Kalinovsky, A.I.; Rasin, A.B.; Kalinin, V.I. Structures and biologic activity of chitonoidosides I, J, K, K_1 and L, triterpene di-, tri- and tetrasulfated hexaosides from the sea cucumber *Psolus chitonoides*. *Mar. Drugs* **2022**, *20*, 369. [CrossRef] [PubMed]
18. Silchenko, A.S.; Avilov, S.A.; Kalinin, V.I. Separation procedures for complicated mixtures of sea cucumber triterpene glycosides with isolation of individual glycosides, their comparison with HPLC/MS metabolomic approach, and biosynthetic interpretation of the obtained structural data. In *Studies in Natural Product Chemistry*; Rahman, A.U., Ed.; Elsevier: Amsterdam, The Netherlands, 2022; Volume 72, pp. 103–146. [CrossRef]
19. Aminin, D.L.; Menchinskaya, E.S.; Pislyagin, E.A.; Silchenko, A.S.; Avilov, S.A.; Kalinin, V.I. Sea cucumber triterpene glycosides as anticancer agents. In *Studies in Natural Product Chemistry*; Rahman, A.U., Ed.; Elsevier: Amsterdam, The Netherlands, 2016; Volume 49, pp. 55–105.
20. Kalinin, V.I.; Aminin, D.L.; Avilov, S.A.; Silchenko, A.S.; Stonik, V.A. Triterpene glycosides from sea cucucmbers (*Holothurioidae, Echinodermata*), biological activities and functions. In *Studies in Natural Product Chemistry (Bioactive Natural Products)*; Rahman, A.U., Ed.; Elsevier: The Netherlands, 2008; Volume 35, pp. 135–196.
21. Kalinin, V.I.; Silchenko, A.S.; Avilov, S.A.; Stonik, V.A. Non-holostane aglycones of sea cucumber triterpene glycosides. Structure, biosynthesis, evolution. *Steroids* **2019**, *147*, 42–51. [CrossRef] [PubMed]
22. Taniyama, S.; Arakawa, O.; Terada, M.; Nishio, S.; Takatani, T.; Mahmud, Y.; Noguchi, T. *Ostreopsis* sp., a possible origin of palytoxin (PTX) in parrotfish *Scarus ovifrons*. *Toxicon* **2003**, *42*, 29–33. [CrossRef] [PubMed]
23. Malagoli, D.A. *Full-Length Protocol to Test Hemolytic Activity of Palytoxin on Human Erythrocytes*; Technical Report; Department of Animal Biology, University of Modena and Reggio Emilia: Modena, Italy, 2007; pp. 92–94.

marine drugs

Article

Anticancer Activity of the Marine Triterpene Glycoside Cucumarioside A_2-2 in Human Prostate Cancer Cells

Ekaterina S. Menchinskaya [1,2,*,†], Sergey A. Dyshlovoy [1,*,†], Simone Venz [3], Christine Jacobsen [1], Jessica Hauschild [1], Tina Rohlfing [1], Aleksandra S. Silchenko [2], Sergey A. Avilov [2], Stefan Balabanov [4], Carsten Bokemeyer [1], Dmitry L. Aminin [2,5], Gunhild von Amsberg [1,6,‡] and Friedemann Honecker [1,7,‡]

[1] Department of Oncology, Hematology and Bone Marrow Transplantation with Section Pneumology, Hubertus Wald Tumorzentrum—University Cancer Center Hamburg (UCCH), University Medical Center Hamburg-Eppendorf, 20251 Hamburg, Germany; christine.jacobsen@bnitm.de (C.J.); j.hauschild@uke.de (J.H.); t.rohlfing@uke.de (T.R.); c.bokemeyer@uke.de (C.B.); g.von-amsberg@uke.de (G.v.A.); friedemann.honecker@tbz-ost.ch (F.H.)

[2] G.B. Elyakov Pacific Institute of Bioorganic Chemistry, Far Eastern Branch of the Russian Academy of Sciences, 159 Prospect 100-letiya Vladivostoka, Vladivostok 690022, Russia; silchenko_als@piboc.dvo.ru (A.S.S.); avilov_sa@piboc.dvo.ru (S.A.A.); daminin@piboc.dvo.ru (D.L.A.)

[3] Department of Medical Biochemistry and Molecular Biology, University of Greifswald, 17475 Greifswald, Germany; simone.venz@med.uni-greifswald.de

[4] Division of Hematology, University Hospital Zurich, 8091 Zurich, Switzerland; stefan.balabanov@usz.ch

[5] Department of Biomedical Science and Environmental Biology, Kaohsiung Medical University, No. 100, Shin-Chuan 1st Road, Sanmin District, Kaohsiung City 80708, Taiwan

[6] Martini-Klinik, Prostate Cancer Center, University Hospital Hamburg-Eppendorf, 20251 Hamburg, Germany

[7] Tumor and Breast Center Eastern Switzerland, 9016 St. Gallen, Switzerland

[*] Correspondence: ekaterinamenchinskaya@gmail.com (E.S.M.); s.dyshlovoy@uke.de (S.A.D.)

[†] These authors contributed equally to this work.

[‡] These authors contributed equally to this work.

Abstract: Despite recent advances in the treatment of metastatic castration-resistant prostate cancer (CRPC), treatment is inevitably hampered by the development of drug resistance. Thus, new drugs are urgently needed. We investigated the efficacy, toxicity, and mechanism of action of the marine triterpene glycoside cucumarioside A_2-2 (CA_2-2) using an in vitro CRPC model. CA_2-2 induced a G_2/M-phase cell cycle arrest in human prostate cancer PC-3 cells and caspase-dependent apoptosis executed via an intrinsic pathway. Additionally, the drug inhibited the formation and growth of CRPC cell colonies at low micromolar concentrations. A global proteome analysis performed using the 2D-PAGE technique, followed by MALDI-MS and bioinformatical evaluation, revealed alterations in the proteins involved in cellular processes such as metastatic potential, invasion, and apoptosis. Among others, the regulation of keratin 81, CrkII, IL-1β, and cathepsin B could be identified by our proteomics approach. The effects were validated on the protein level by a 2D Western blotting analysis. Our results demonstrate the promising anticancer activity of CA_2-2 in a prostate cancer model and provide insights on the underlying mode of action.

Keywords: cucumarioside A_2-2; castration-resistant prostate cancer; PC-3 cells; apoptosis; anti-metastatic activity; proteomics

Citation: Menchinskaya, E.S.; Dyshlovoy, S.A.; Venz, S.; Jacobsen, C.; Hauschild, J.; Rohlfing, T.; Silchenko, A.S.; Avilov, S.A.; Balabanov, S.; Bokemeyer, C.; et al. Anticancer Activity of the Marine Triterpene Glycoside Cucumarioside A_2-2 in Human Prostate Cancer Cells. *Mar. Drugs* **2024**, *22*, 20. https://doi.org/10.3390/md22010020

Academic Editor: Jose A. Salas

Received: 22 November 2023
Revised: 24 December 2023
Accepted: 24 December 2023
Published: 28 December 2023

1. Introduction

Prostate cancer is the most common cancer in men aged 50+ with more than 1.3 million new cases annually, and is among the five deadliest cancers worldwide [1]. At an early stage, prostate cancer is often asymptomatic and, depending on tumor biology, can be managed by active surveillance of the patient. However, at later stages, so-called aggressive variants of prostate cancer (AVPCs) appear. AVPCs are resistant to various anticancer treatments and show an aggressive, metastases-prone phenotype [2]. Castration-resistant prostate cancer (CRPC), which is a subtype of AVPCs, is characterized by resistance to hormonal therapy, which still is the mainstay of treatment of systemic disease [3]. Despite intensive research in this area and the active search for new antitumor agents, effective ways to suppress the growth and development of CRPC tumors are still urgently needed.

Bioactive compounds derived from marine organisms are a valuable source of new anticancer agents [4]. Additionally, for many substances with established structures, previously unknown biological activities and the identification of the underlying mode of action are reported [5]. In vitro and in vivo studies of the chemo-preventive and anticancer properties of compounds isolated from fish, mollusks, starfish, and sea cucumbers indicate their great therapeutical potential [6].

Triterpene glycosides, produced by holothurians (also referred to as sea cucumbers), are a group of small molecules with a wide range of biological activities. Among others, they are known to be cytotoxic and ichthyotoxic, as well as antimicrobially, antivirally, radio-protectively, and immunomodulatorily active [7]. Using triterpene glycosides from sea cucumbers, several pharmacological preparations, biological additives, functional nutrition products, and cosmetics have already been developed and patented [8].

Of particular interest are the antitumor properties of these compounds. It has previously been described that holothurians' triterpene glycosides inhibit the proliferation of human tumor cells of various origins, including cervical and lung carcinoma, liver and stomach cancer, breast and ovarian cancer, malignant melanoma, and other types [9–16]. It is also known that triterpenoids (specifically triterpene saponins of plant origin) can effectively inhibit the division of tumor cells and suppress the growth of hormone-dependent tumors [17]. However, only little is known about the antitumor activity of marine triterpene glycosides in prostate cancer. It has been reported that the triterpene saponins cercodemasoides A–E originating from the sea cucumber *Cercodemas anceps*, stichorrenosides A–D from the sea cucumber *Stichopus horrens*, and triterpene tetraglycosides from *Stichopus herrmanni*, as well as echinoside A from *Pearsonothuria graeffei*, exhibit strong cytotoxic activity at micromolar concentrations on prostate cancer models both in vitro and in vivo [18,19]. Furthermore, it could be shown that frondoside A isolated from the sea cucumber *Cucumaria okhotensis* inhibits the proliferation and formation of colonies of metastatic castration-resistant prostate cancer cells by inhibition cell cycle progression, induction of apoptosis, and inhibition of autophagy [20,21]. Additionally, in vivo, this compound inhibited the tumor growth of PC-3 and DU145 cells with a significant reduction in lung metastasis, as well as circulating tumor cells in the peripheral blood in a mouse model [20].

Cucumarioside A_2-2 (CA_2-2) is a triterpene glycoside which was initially isolated from the sea cucumber *Cucumaria japonica*. Our group has previously shown that CA_2-2 is able to suppress the growth of mouse Ehrlich carcinoma cells in vitro and in vivo, can block DNA biosynthesis in the S-phase, induces apoptosis in a caspase-dependent manner, bypassing the activation of the p53-dependent segment, and causes apoptotic necrosis and mitoptosis [22–26]. This glycoside overcomes the multidrug resistance of tumor cells and promotes the intracellular accumulation of certain chemotherapeutic agents [27,28]. In the work presented here, we set out to elucidate the molecular mechanisms of the observed antitumor effects of CA_2-2 using human CRPC PC-3 cells. We used a global proteomics screening approach based on 2D-PAGE electrophoresis followed by MALDI-MS and subsequent bioinformatical data analysis to identify proteins showing differential regulation upon CA_2-2 treatment in these prostate cancer cells.

2. Results

2.1. CA_2-2 Inhibits Prostate Cancer Cell Viability

The anticancer activity of the triterpene glycoside CA_2-2 (Figure 1a) was assessed in metastatic castration-resistant prostate cancer PC-3 cells, known to be resistant to both anti-hormonal as well as cytotoxic agents [2]. Experiments revealed a concentration-dependent cytotoxic effect of CA_2-2 at micromolar concentrations with an IC_{50} of 2.05 μM (Figure 1b).

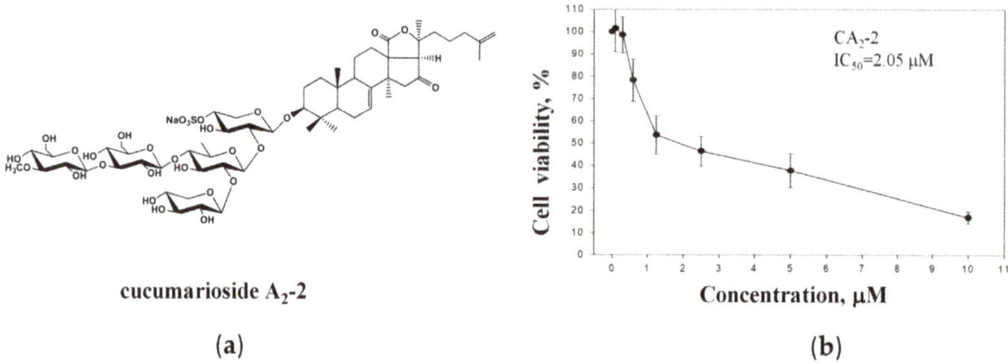

Figure 1. Chemical structure of CA$_2$-2 (**a**). Viability of PC-3 cells treated with CA$_2$-2 for 48 h and evaluated using the MTT assay (**b**).

2.2. CA$_2$-2 Induces Apoptosis and Inhibits Colony Formation in PC-3 Prostate Cancer Cells

As the next step, we analyzed the ability of CA$_2$-2 to induce apoptosis in PC-3 cells. We investigated a cleavage of poly (ADP-ribose) polymerase 1 (PARP-1) and a cleavage of caspases-3 and -9 as markers of apoptotic cell death (Figure 2a). Following 48 h of treatment, a dose-dependent cleavage of PARP-1 (89 kDa fragment) was observed in PC-3 tumor cells (Figure 2a). In addition, CA$_2$-2 activated caspase-3 and caspase-9 at concentrations of 1 and 2 μM after 48 h of treatment. Figure 2a shows the increased expression of cleaved caspase-3 and caspase-9 compared to the control, indicating the induction of these enzymes by the triterpene glycoside.

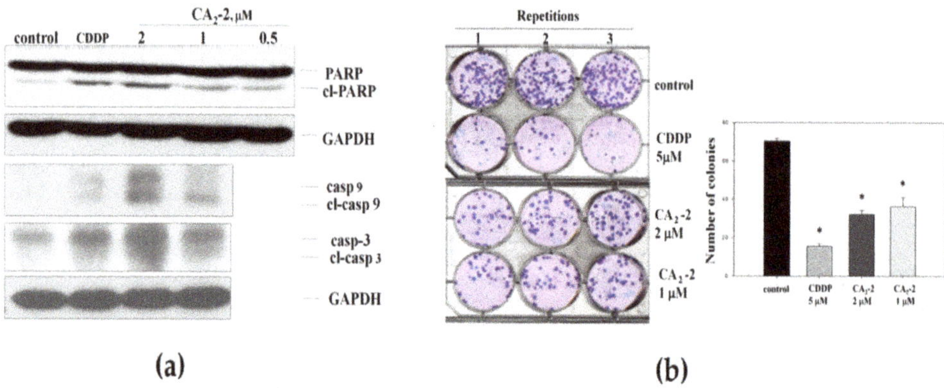

Figure 2. Cleavage of caspase-3, -9, and PARP-1 in PC-3 cells treated with CA$_2$-2 for 48 h. Cisplatin (CDDP) was used as a positive control (**a**). Effect of triterpene glycoside of CA$_2$-2 on colony formation of PC-3 tumor cells. Cisplatin (CDDP) was used as a positive control (**b**). Incubation time in either experiment was 48 h. All experiments were performed in triplicates. * $p \leq 0.05$.

The effect of CA$_2$-2 on the ability of PC-3 cells to form colonies was studied using an in vitro colony formation assay showing the ability of single cells to form colonies. Thus, this assay can be viewed as an in vitro testing of the in vivo anti-metastatic activity of a substance [29].

Figure 2b represents images of six-well plates with cancer cell colonies stained with Giemsa dye. For the experiments, we used different concentrations for CA$_2$-2, including the IC$_{50}$ of 2 μM. Cisplatin (CDDP) at a concentration of 5 μM was used as a positive control [30], strongly suppressing the growth of PC-3 cells colonies. We found that CA$_2$-2 also effectively inhibits the formation and growth of tumor cell colonies. A 30% reduction of cell colony growth was observed when CA_2-2 was applied at concentrations of 1 to 2 µM (Figure 2b).

2.3. CA_2-2 Arrests Cell Cycle of Prostate Cancer Cell

The effects of CA_2-2 on cell cycle progression were evaluated using flow cytometry. In this assay, anisomycin was used as a positive control, being able to induce apoptosis and increasing the fraction of aneuploid cells containing fragmented DNA (appearing as sub-G_0 phase) [31–34].

The incubation of PC-3 cells with the CA_2-2 led to an increased number of cells containing fragmented DNA, suggesting the induction of apoptosis. The accumulation of cells in the mitotic phase (G_2/M) indicates partial blocking of the cell cycle progression (Figure 3a, Table 1), whereas the number of cells in the G_0/G_1 phase was reduced.

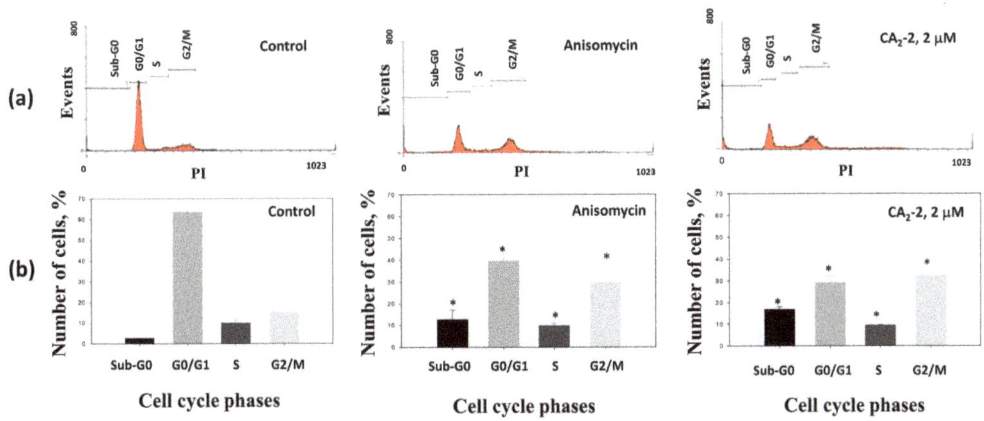

Figure 3. Distribution (**a**) and quantification (**b**) of the fractions of cells in different phases of the cell cycle depending on drug treatment for 48 h with either anisomycin (1 µM) as a positive control, or CA_2-2 (2 µM). * $p \leq 0.05$.

Table 1. Effect of CA_2-2 on cell cycle progression of PC-3 cells. Anisomycin was used as a positive control. Treatment time was 48 h.

Substance	Cell Cycle Phases (%)			
	SubG_0	G_0/G_1	S	G_2/M
Control (untreated)	2.59 ± 0.08	63.54 ± 1.01	9.85 ± 2.05	15.22 ± 1.71
CA_2-2 (1 µM)	8.78 ± 0.23 *	36.87 ± 1.53 *	9.22 ± 0.04	32.14 ± 0.30 *
CA_2-2 (2 µM)	16.92 ± 1.10 *	29.19 ± 3.11 *	9.43 ± 0.20	32.56 ± 2.72 *
Anisomycin (1 µM)	12.71 ± 4.32 *	39.63 ± 4.34 *	9.81 ± 1.09	29.62 ± 9.04

* $p \leq 0.05$ compared with control.

Thus, the triterpene glycoside CA_2-2 significantly increases the number of apoptotic cells, and simultaneously inhibits the cell cycle in the mitotic phase (G_2/M) in PC-3 cells.

2.4. CA_2-2 Regulates Expression of Proteins Involved in Growth, Migration, Invasion, and Cell Death in Prostate Cancer Cells

2.4.1. Proteomics Analysis of Proteins Using 2D-PAGE, MALDI-MS, and Bioinformatic Analysis

A global proteomics approach is a useful tool to identify drug targets on the protein level in cancer cells. The use of 2D electrophoresis, followed by mass spectrometry, helps to elucidate regulated protein spots and, thus, the effects of biologically active compounds on a molecular level.

Hence, we performed an analysis of proteins after the treatment of PC-3 cells with CA_2-2. Tumor cells were incubated with the compound for 48 h, followed by protein isolation and separation by two-dimensional gel electrophoresis. Figure 4 represents images of the 2D-PAGE gels from control versus CA_2-2-treated cells. Proteins showing both down-regulation (Figure 4a) or up-regulation (Figure 4b) upon treatment were identified. A differential image analysis revealed a total number of 946 detected protein spots. In CA_2-2-treated cells, the expression of 20 protein spots was significantly altered by the treatment ($0.5 \geq$ fold change ≥ 2; $p < 0.05$). Figure 4c shows representative magnified images of the protein spots significantly affected by the treatment. Among them, 13 protein spots were up-regulated, and 7 were down-regulated following exposure to CA_2-2 (Table 2, Figure 4d).

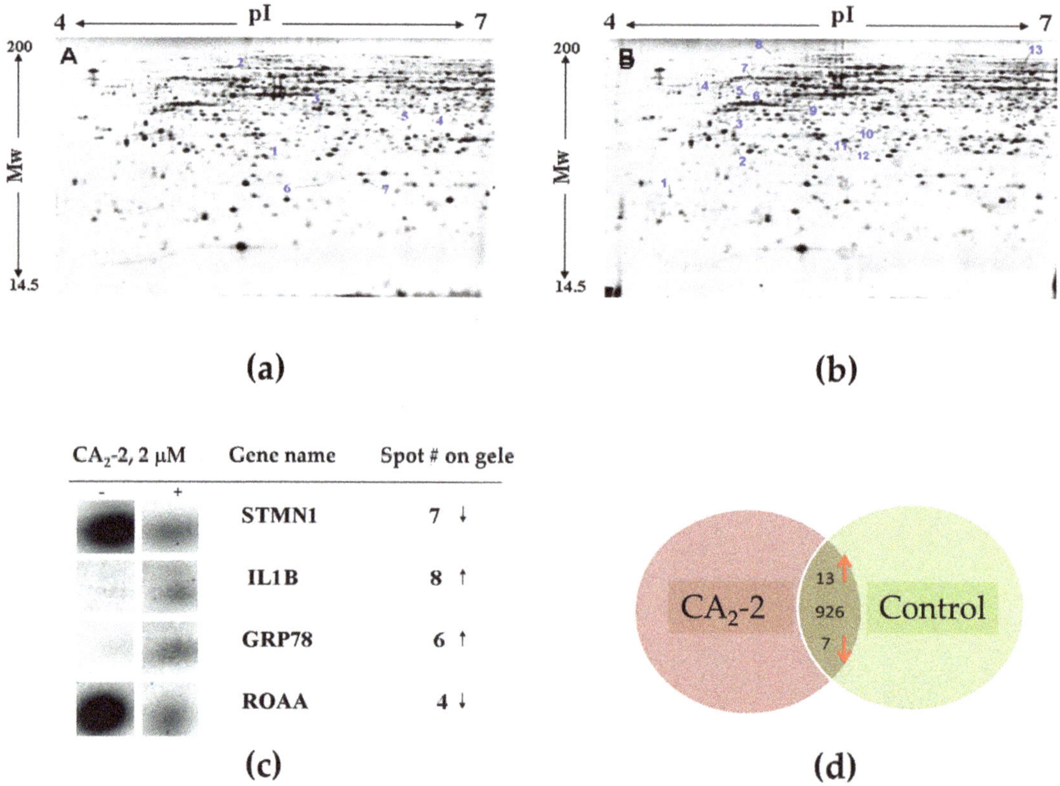

Figure 4. Images of 2D gel electrophoresis of non-treated (control) PC-3 cell lysates (**a**) versus lysates of PC-3 cells incubated with 2 μM of CA_2-2 for 48 h (**b**). Spots with down-regulation upon treatment are indicated on gel A, while up-regulated spots are indicated on gel B. Magnified images of spots on a 2D gel corresponding to representative target proteins regulated by CA_2-2 are shown in (**c**). The "−" sign in the figure denotes corresponding spots depicted for gel A (control cells), whereas "+" indicates spots depicted for gel B. Summarized information on proteins regulated in PC-3 cells is graphically shown in (**d**). Numbers indicate the total number of protein spots identified on 2D gels in the middle; arrows indicate the direction of expression of differentially regulated spots.

All significantly regulated spots were subjected to peptide mass fingerprint analysis, followed by protein identification. A typical example for the identification of a protein of interest from PC-3 cells is presented in Figure 5 by the identification of IL1B (Interleukin-1β). The expression of this protein was found to be significantly changed upon treatment with triterpene glycoside (see Table 2).

Table 2. Proteins regulated by CA_2-2 in human PC-3 prostate cancer cells. The numbers of down-regulated proteins correspond to the numbers of spots in Figure 4a; the numbers of up-regulated proteins correspond to the numbers of spots in Figure 4b.

Spot No on Gel	Gene Name	Protein	Up/Down Regulation, Fold Change	Protein Function
		Protein metabolism, enzymatic activity		
4	PDIA1	Protein disulfide-isomerase	2.44 ↑	Catalyzes the formation and destruction of disulfide bonds during protein folding
1	CATB	Cathepsin B	3.2 ↓	Takes part in apoptosis, and is a mediator of the lysosomal pathway of cell death
6	GRP78	78 kDa glucose-regulated protein	2.43 ↑	Controls the processes of invasion, apoptosis, and inflammation
		Metabolism of carbohydrates		
5	PGP	Phosphoglycolate phosphatase	2.12 ↓	Takes part in the metabolism of carbohydrates
		Cytoskeletal organization, cell motility, and division		
2	K2C1	Keratin, type II cytoskeletal 1	2.35 ↑	Participates in the formation of intermediate filaments
11	KRT81	Keratin, type II cuticular Hb1	2.96 ↑	Participates in the formation of intermediate filaments
10	KRT81	Keratin, type II cuticular Hb1	2.52 ↑	Participates in the formation of intermediate filaments
7	STMN1	Stathmin	2.16 ↓	Regulates rapid cytoskeletal remodeling in response to cell needs
5	KINH	Kinesin-1 heavy chain	2.22 ↑	Supports mitosis, meiosis, and transport of intracellular molecules
13	CALD1	Caldesmon	2,18 ↑	Binds calmodulin, and inhibits the ATPase activity of myosin
12	CRK II	Adapter molecule crk	2.93 ↑	Involved in phagocytosis of apoptotic cells, and may regulate EFNA5-EPHA3 signaling
		Immune response		
9	IL1B	Interleukin-1 beta	3.32 ↑ (precursor)	Development and regulation of the body's defense response to a pathogen
3	IL1B	Interleukin-1 beta	3.09 ↑ (mature)	Development and regulation of the body's defense response to a pathogen
		mRNA processing		
4	ROAA	Heterogeneous nuclear ribonucleoprotein A/B	2.04 ↓	Regulates the formation of telomeres and/or their stabilization, and also takes part in the control of apoptosis
8	HNRL2	Heterogeneous nuclear ribonucleoprotein U-like protein 2	2.27 ↑	Process heteronuclear RNA into mature mRNAs, and regulates of gene expression
		Response to stress		
3	NDRG1	Protein NDRG1	2.2 ↓	Participates in the formation of a response to stress and hormones, and participates in cell growth and differentiation
		Structural and functional organization of the nucleus		
6	CRABP2	Cellular retinoic acid-binding protein 2	2.16 ↓	Is an intracellular lipid-binding protein that interacts with cyclin D
1	NPM	Nucleophosmin	2.16 ↑	Takes part in the biogenesis of ribosomes, and the transport of proteins to the nucleus
2	LMNB1	Lamin-B1	4.01 ↓	Performs structural functions, and takes part in the regulation of transcription
7	SYNE1	Nesprin-1	2.15 ↑	Takes part in the nuclear organization and structural structure of the nucleus, and interacts with F-actin and with the nuclear envelope

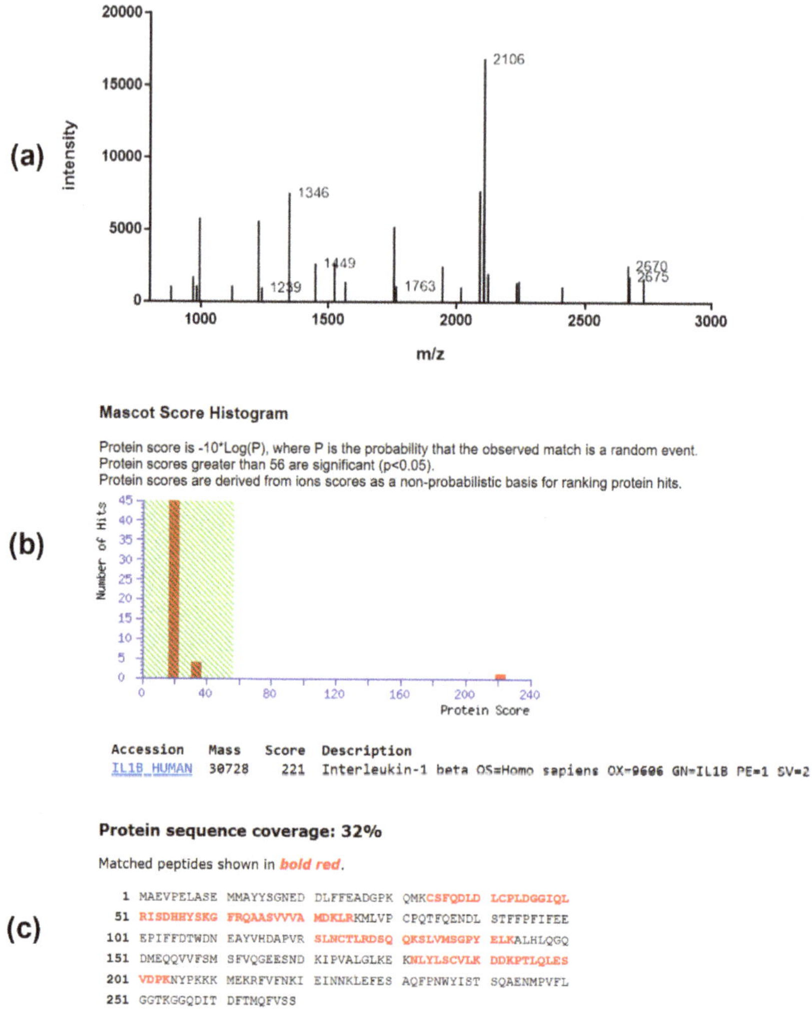

Figure 5. Peptide mass fingerprint spectrum (**a**), MOWSE score (**b**), and sequence coverage (**c**) of protein IL-1β of the gel derived from PC-3 cells. (**a**) MALDI mass spectrum obtained from peptide mixture derived from in-gel digestion of protein. (**b**) Database search result of protein revealed the presence of IL1B_HUMAN gene product in this spot. Identification scores greater than 56 (shaded area) are regarded as significant ($p < 0.05$). (**c**) The database entry for the protein sequence is depicted in the single letter code. The sequence stretches that are covered by peptide ion signals (81% sequence coverage) in the mass spectrum are shown in bold red.

The identified proteins regulated by CA_2-2 could be assigned to different groups according to their supposed biological functions (Table 2). Most of these proteins identified play an important role in the functioning of tumor cells, including some well-established tumor markers. Interestingly, the largest group of proteins regulated by CA_2-2 could be assigned to proteins involved in the structural organization of the cytoskeleton, as well as in cell movement and proliferation (32%). Another frequent finding was proteins playing an important role in the functioning of the cell nucleus (21%).

2.4.2. Verification of Protein Expression by Western Blotting and 2D Western Blotting

A proteomics analysis identified 20 proteins whose expression was significantly altered following treatment with CA_2-2. To verify the proteomics data, we selected four proteins and analyzed their expression using 1D and 2D Western blotting. The expressions of keratin-81, Crk II, interleukin-1β, and cathepsin B were examined.

Experiments on protein verification by immunoblotting showed that treatment with CA_2-2 resulted in a significant increase in the expression of both interleukin-1β isoforms (Figure 6c). However, at the same time, the expression of the keratin 81 protein remained at the control level (Figure 6a). Therefore, a 2D Western blot of keratin 81 was performed, showing a significant increase in the expression of a specific isoform of keratin 81 compared to the controls (Figure 7a,b). The expression of the CRK II protein was found to be increased when cells were incubated with CA_2-2 (Figure 6b).

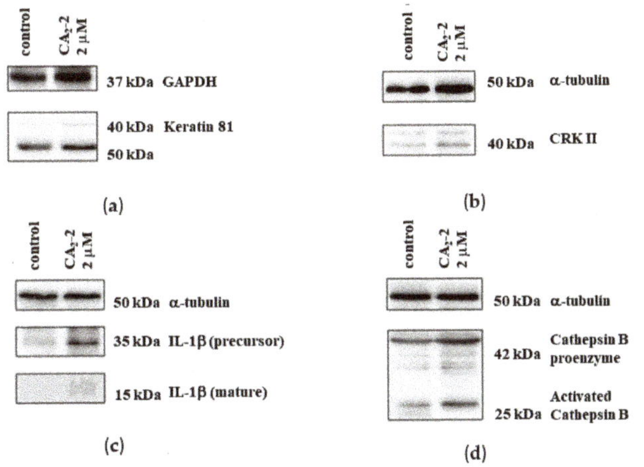

Figure 6. Immunoblots of keratin 81 (**a**), CrkII (**b**), IL-1β (**c**), and cathepsin B (**d**) upon treatment of PC-3 cells for 48 h with CA_2-2.

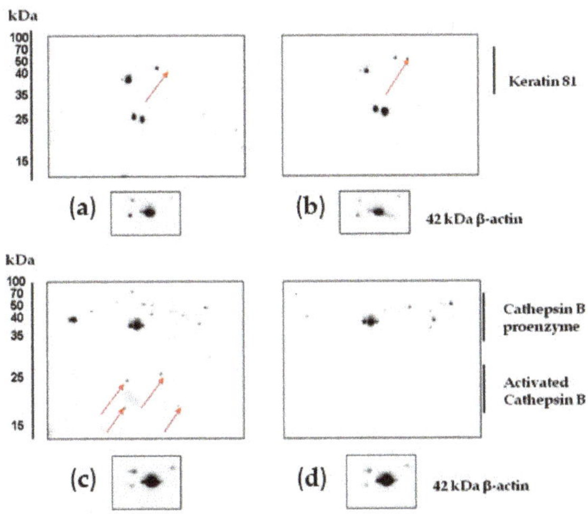

Figure 7. Results of 2D Western blot analysis of changes in the expression of keratin 81 in control (**a**) and CA_2-2-treated cells (**b**). Cathepsin B control (**c**) and CA_2-2-treated PC-3 cells (**d**). CA_2-2 was used at 2 µM for 48 h. The arrows on the gels indicate the spots corresponding to the studied proteins.

Taken together, we could demonstrate that CA_2-2 causes changes in the expression of a number of proteins that play roles in the functioning of tumor cells. These molecular/submolecular processes are involved in a broad spectrum of biological activities, including the mechanisms of cancer cell development and division, tumor invasion, and the programmed death of tumor cells. Therefore, the data generated using a global proteomics approach are in good agreement with the observation that the antitumor activity of triterpene glycosides is exerted by their ability to abolish tumor cell proliferation, block the cell cycle, and induce tumor cell apoptosis.

3. Discussion

Triterpene glycosides of holothurians are characterized by a wide spectrum of dose-dependent biological activities. Used at high concentrations, triterpene glycosides are known to exhibit membranolytic properties leading to hemolysis, cytotoxicity, and antimicrobial and embryotoxic activity [7]. On the other hand, at low nanomolar concentrations, these compounds can demonstrate immunostimulatory activity by the activation of immunocompetent cells [22], induction of apoptosis in tumor cells, and blockade of the cell cycle and proliferation [23].

Here, we present, for the first time, that the triterpene glycoside CA_2-2 induces apoptosis in drug-resistant human prostate cancer PC-3 cells. Apoptosis upon treatment with CA_2-2 is characterized by the activation of proapoptotic caspases (caspase-3 and -9) and by the appearance of a cleavage product of the substrate of activated caspases, i.e. cleaved PARP-1 protein. The effect of cucumarioside A_2-2 was accompanied by a significant increase in the number of aneuploid cells containing fragmented DNA and appearing in the sub-G_0 phase, as well as a blockade of the cell cycle in the G_2/M mitosis phase. Moreover, CA_2-2 is able to effectively abrogate the formation and growth of colonies of human tumor PC-3 cells.

Searching for the mode of action and involved pathways, we used a global proteomics screening approach. Protein groups found to be differentially expressed upon exposure of PC-3 cells to CA_2-2 can be classified as important regulators of cell metabolism, motility, and division, as well as the structural and functional organization of the cytoskeleton. The identified direct or indirect protein targets, including stathmin, kinesin, caldesmon, clathrin, CRABP2, nucleophosmin, lamin-B1, nesprin-1, CRK II, and others, play a key role in the functioning of tumor cells and are closely associated with mechanisms of the development and regulation of mitosis, tumor cell proliferation, cell migration, and tumor invasion, and control of programmed death.

It was found that CA_2-2 increases the expression of GRP78 and IL-1β. Those proteins play complex and multiple roles in the development and progression of tumors. Thus, some studies describe that the GRP78 protein promotes the growth and invasion of tumor cells [35,36]. Other reports show that the high expression of GRP78 promotes the effectiveness of chemotherapy and improves the disease-free survival of patients [37].

Speaking about the role of IL-1β, it should be noted that it plays a role in many processes in the human body. In oncogenesis, both positive and negative functions of this protein have been described, depending on the type of tumor. A high IL-1β expression in prostate cancer predicts good treatment prognosis and better progression-free survival [38]. And, when LNCaP prostate cancer cells are co-cultured with fibroblasts, the high expression of IL-1β is antiproliferative [39]. The first results of the effect of IL-1b on metastasis were shown 30 years ago. Thus, fibrosarcoma cells that overexpress IL-1β have been shown to have increased invasive potential [40]. Recently, the inhibition of IL-1β was found to reduce the metastatic potential of murine prostate cancer cells while its overexpression was increased [41]. These new insights are a first step to understanding the underlying molecular mechanisms of the observed anticancer activities hiding in holothurian glycosides.

It is well-known that some marine triterpene glycosides exhibit pronounced anticancer effects by direct interaction with tumor cells in the sub-cytotoxic concentration range [6,8]. However, we are only beginning to understand the molecular mechanisms and signaling pathways involved in their antitumoral activity. We postulate that triterpene glycosides suppress the function of tumor cells by their ability to induce caspase-dependent or -independent apoptosis, arrest the cell cycle in certain phases, and control the expression of the nuclear factor NF-κB, as well as regulate the expression of cellular receptors and enzymes involved in carcinogenesis, such as EGFR, Akt, ERK, FAK, MMP-9, and some others [5].

So far, to our knowledge, only for one triterpene glycoside, frondoside A, several target proteins were identified [20]. A global proteome analysis and bioinformatic approach revealed the regulation of proteins involved in the formation of metastases, tumor cell invasion, and apoptosis, like keratin 81, CrkII, IL-1β, and cathepsin B [20]. Comparing the results of the proteomics analyses of both frondoside A and CA_2-2, we found only five proteins that are equally regulated by these proteins. The proteins most significantly regulated by CA_2-2 were recognized as proteins encoded by the IL1B, KRT81, and CRK II genes (up-regulated), as well as proteins encoded by the CATB and ROAA genes (down-regulated). A literature search for the functional activity of these proteins in databases revealed that these proteins are directly involved in the regulation of tumor cell proliferation, tumor invasion, metastasis, and apoptosis [42–46].

Moreover, in vivo, frondoside A inhibited the tumor growth of PC-3 and DU145 cells with a notable reduction of lung metastasis, as well as circulating tumor cells in the peripheral blood [20]. The results of the current study indicate that the mechanism of action of CA_2-2 may be similar to those of frondoside A, making this class of chemicals highly interesting substances for drug development in oncology.

4. Materials and Methods

4.1. Reagents and Antibodies

The marine triterpene glycoside cucucmarioside A_2-2 was isolated from the extract of sea cucumber *Cucumaria japonica* as previously reported [47]. The purity of cucucmarioside A_2-2 was determined by ^{13}C NMR spectroscopy and ESI mass-spectrometry.

MTT solution (Sigma-Aldrich, Burlington, MA, USA), Giemsa's solution (Merck, Darmstadt, Germany), trypsin-EDTA (gibco® Life technologies™, Paisley, UK), propidium iodide solution (Sigma-Aldrich, Burlington, MA, USA), RNase (Carl Roth, Karlsruhe, Germany), CHAPS (Sigma-Aldrich, Burlington, MA, USA), 1% Pharmalyte (Sigma-Aldrich, Burlington, MA, USA), Coomassie Brilliant Blue G 250 (Thermo scientific, Rockford, IL, USA), anisomycin (Sigma-Aldrich, Burlington, MA, USA), DMEM medium (Lonza, Walkersville, MD, USA), fetal calf serum (FBS) (gibco® Life technologies™), penicillin-streptomycin (gibco® Life technologies™), polyclonal rabbit antibodies against PARP-1, titer 1:1000 (Cell signaling, Danvers, MA, USA); monoclonal rabbit antibodies against caspase-3, titer 1:1000 (Cell signaling, Danvers, MA, USA); monoclonal mouse antibodies against caspase-9, titer 1:1000 (Cell signaling, Danvers, MA, USA); and polyclonal rabbit antibodies against GAPDH (abcam, Cambridge, MA, USA) were used. Secondary goat antibodies labeled with horseradish peroxidase against rabbit, titer 1:5000 (abcam, Cambridge, MA, USA), and secondary rabbit antibodies against mouse, titer 1:10,000 (abcam, Cambridge, MA, USA) were used.

4.2. Cell Lines and Culture Conditions

The human prostate cancer androgen-independent cell line PC-3 was obtained from ATCC (CRL-1435 ™, Manassas, VA, USA). PC-3 cells were cultured as a monolayer under standard conditions (37 °C, 5% CO_2) in DMEM medium supplemented with 10% fetal bovine serum (FBS) and 1% penicillin-streptomycin.

4.3. MTT Assay

The cytotoxic activity of CA_2-2 was measured by the MTT method. 20 μL of a solution of the test compound of various concentrations was added to the wells of a 96-well plate to 180 μL of adherent PC3 cells (6×10^3 cells/well), and then incubated for 48 h at 37 °C and 5% CO_2. After incubation, 200 μL of the supernatant was replaced and 100 μL of pure medium and 10 μL of MTT solution (5 mg/mL in PBS) were added. They were further incubated for 4 h and 100 μL of SDS-HCl solution was added and incubated at 37 °C for 18 h. Absorbance was measured at 570 nm using an Infinite F200PRO plate reader (TECAN, Mannedorf, Switzerland). The cytotoxic activity of a substance was expressed as the IC_{50} concentration at which the metabolic activity of cells is inhibited by 50%.

4.4. Colony Formation Assay

PC-3 cells were seeded on small Petri dishes d = 35 mm at a concentration of 3×10^5 cells in 5 mL of DMEM medium for 24 p for adhesion at 5%, CO_2, 37 °C. Then, 50 μL of Kuk solution at various concentrations was added to the cells and placed in a CO2 incubator for 48 h. Then, the cells were treated with trypsin (0.25% trypsin in 1 mM EDTA solution) and washed with PBS by centrifugation for 5 min, 1500 rpm, and resuspended; the number of living and dead cells was counted using an automated Beckman Coulter Vi-CELL instrument (Beckman Coulter, Krefeld, Germany) and seeded into 6-well plates at a concentration of 100 live cells in 3 mL of DMEM medium. After 10 days of incubation, the medium was taken, the cells were fixed with 1 mL of 100% methanol for 25 min, washed with PBS and dried, and 1 mL of Giemsa solution was added, and then they were incubated for 25 min at room temperature and washed with dH_2O [36]. The cells were then air dried and the number of colonies was counted. Results were expressed as the number of colonies grown per well.

4.5. Cell Cycle Analysis

PC-3 cells in the amount of 2×10^6 cells/mL were plated on Petri dishes in complete growth medium and incubated for 24 h for adhesion (5% CO_2, 37 °C). After that, the medium was changed, and solutions of substance in different concentrations of 100 μL were added and placed in an incubator for 48 h. Then, the cells were treated with a solution of trypsin, the total number of cells was counted, and 1×10^6 cells were taken from each sample. The cell pellet was resuspended in 300 μL of cold phosphate-buffered saline (PBS), fixed in 700 μL of 100% ethanol, and left for 24 h at −20 °C. Then, the cells were washed twice from ethanol and resuspended in 200 μL of propidium iodide solution (PI, 5 μg/mL in PBS), RNase 20 μg/mL. The cells were then transferred to flow cytometry tubes and incubated for 30 min at room temperature in the dark. Before analysis, 500 μL of cold PBS was added to the cell suspension. The study was carried out using an FACScalibur flow cytometer (Becton Dickinson, USA) at 488 nm excitation in the FL2 channel. Histograms were evaluated using the WinMDI 2.9 Ink program (USA).

4.6. Western Blotting

For the determination of caspase-3, -9, and PARP-1 proteins by Western blotting, cells were plated on Petri dishes in the amount of 1×10^6 cells in 5 mL of complete growth medium and placed in a CO_2 incubator for 24 h for adhesion. After that, the medium was replaced with a fresh one with the addition of substance in the investigated concentrations. After 24 h or 48 h, the cells were treated with Trypsin-EDTA solution and washed with PBS three times. Next, the cell pellet was resuspended in 100 μL of lysis buffer solution (0.88% NaCl, 50 mM Tris-HCl (pH 7.6), 1% NP-40, 0.25% sodium cholate, 1 mM PFMS, 1 mM Na_3VO_4) and incubated 40 min on ice, after which it was left overnight at −20 °C. Lysed cells were pelleted by centrifugation for 10 min at 14000 rpm and 4 °C. Protein concentration was measured according to the Bradford method. Samples with proteins were heated to 95 °C for 5 min. The electrophoretic separation of proteins was carried out according to the Laemmli method [48] in 10% PAGE at a voltage of 80 V for the first 30 min and at 120 V for the remaining time.

After electrophoretic separation, proteins were transferred from the gel to a PVDF membrane (0.45 m, Millipore Corporation, Burlington, MA, USA) at 25 V for 1 h using a Trans-blot SD Semi-dry transfer cell (Bio-rad, USA). Before incubation with primary antibodies overnight at 4 °C, the membrane was blocked with 5% BSA solution in 0.05% TBS-Tween 20 solution for 1 h. Then, the membrane was washed from primary antibodies with TBS-Tween 20 solution 3 times for 5 min and incubated for 1 h with secondary antibodies labeled with horseradish peroxidase at room temperature. After that, 1 mL of Pierce ECL Western Blotting Substrate (Thermo scientific, Rockford, IL, USA) solutions were added to the membranes, a CL-XPosureTM Film 5 × 7 inches X-ray film (Thermo scientific, Rockford, IL, USA) was applied, and the signal was visualized on the X-ray film after its development using a developing machine (AGFA CP 1000, Mortsel, Belgium).

4.7. Two-Dimensional Gel Electrophoresis (2D-PAGE)

Experiments on the analysis of protein expression using the proteomics approach were carried out on PC-3 cells by two-dimensional electrophoresis (2D-PAGE) and subsequent analysis of the protein structure by mass spectrometry. The cells were seeded in flasks at a concentration of 5×10^6 cells in 30 mL of complete growth medium and placed in a CO_2 incubator for 24 h for adhesion. Then, triterpene glycoside solutions were added to the cells, and the cells were incubated for 48 h. After that, the cells were collected and washed twice by centrifugation in PBS solution. The cell pellet was resuspended in 500 µL of lysis buffer of the following composition, 9 M urea, 4% CHAPS, 1%Pharmalyte, 1% DTT, and bromophenol blue 10 µg/mL, and left overnight at −20 °C. Then, the samples were precipitated by centrifugation for 10 min, 4 °C, and 14,000 rpm, and 450 µL aliquots with 1 mg of proteins were applied to strips for isoelectric focusing (Immobiline Dry Strip pH 4–7, 24 cm, GE Healthcare Bio-Sciences AB, Uppsala, Sweden) for 24 h to rehydrate at RT. Next, the first separation was performed using a Protean IEF cell (Bio-Rad, Hercules, CA, USA). After the first separation, the strips were equilibrated in 1% DTT solution supplemented with 6 M urea, 4% SDS, 50 mM Tris-HCl pH 8.8 for 15 min, and then alkylated with 4.8% iodoacetamide solution for 15 min at RT. Then, the strips were transferred onto a 15% SDS-PAGE gel (27 cm × 21 cm × 1.5 mm) and covered with a 0.6% agarose solution. The second electrophoretic separation was carried out in SDS-buffer solution of the following composition: 1.44% glycine, 0.3% Tris base, 0.1% SDS at 20 °C for 14 h. The gels were fixed and stained with a Coomassie solution (400 mL/gel containing 0.13% Coomassie Brilliant Blue G 250, 3.6% H_2SO_4, 1.44 M NaOH, 20.3% CCl_3COOH) overnight, after which the gels were washed with dH_2O three times.

4.8. 2D-Gel Image Analysis and Protein Identification by Mass Spectrometry

The studied protein spots were excised from the gel and transferred to a 96-well plate. Spot matching, normalization of the digital images (based on total optical density), and gel image analysis were performed using Delta 2D 4.0 software (Department of Medical Biochemistry and Molecular Biology, University of Greifswald, Germany). Briefly, trypsin digestion, followed by transfer of aliquots of protein solutions to MALDI targets, was performed automatically on an Ettan Spot Handling Workstation (Amersham-Biosciences, Sweden) using a modified standard protocol. MALDI-MS measurements of sample measurements were performed on the 4800 MALDI-ToF/ToFTM Analyzer (Applied Biosystems, Waltham, MA, USA). Recording spectra in the reflex mode in the mass range from 0.8 to 4 kDa, with a mass change of 2 kDa. Additionally, for the five most intense peaks of the MS spectra, MALDI-MS/MS analysis was performed; after subtracting the peaks, there were background or trypsin fragments. Comparison of MS revenues and MS/MS data with usage efficiency was reported in databases using GPS Explorer, Ver. 3.6 (Applied Biosystems, Waltham, MA, USA). The SwissProt v56.1 database was used for identification.

4.9. Bioinformatic Analysis of Differentially Expressed Proteins

The following databases were used to search for the functions of proteins:

UniProtKB/Swiss-Prot (http://www.uniprot.org, accessed on 4 September 2023); GeneCards (http://www.genecards.org, accessed on 4 September 2023); Gene Ontology Consortium (http://www.geneontology.org, accessed on 4 September 2023); Bioinformatic

Harvester (http://www.embl.de/services/bioinformatics, accessed on 4 September 2023); PubMed (http://www.ncbi.nlm.nih.gov/pubmed, accessed on 4 September 2023).

4.10. Mini 2D Western Blotting Analysis (2D-WB)

Two-dimensional Western blotting (2D Western blotting) was used to confirm the results of 2D-PAGE electrophoresis. Aliquots of protein extracts containing 30 μg of protein were diluted using 2D-PAGE-lysis buffer to a volume of 125 μL, and then loaded onto immobilized pH-gradient (IPG) Immobiline Dry Strip (pH 4–7, 7 cm, GE Healthcare Biosciences, Sweden). The IPG strips were incubated overnight at RT for rehydration. Next, the IPG strips were subjected to isoelectric focusing (IEF) using a Protean IEF cell (Bio-Rad, Germany). Separation of proteins in the first dimension (isoelectric focusing) was carried out sequentially in 4 stages: 1–30 min at 500 V; 2–5 kV/h at 5 kV ("linear focusing"); 3–5 kV/h at 5 kV ("fast focusing"); 4–2 h at 500 V. After the IEF step, the IPG strips were stored at −80 °C.

After the first separation, the strips were equilibrated in 1% DTT solution supplemented with 6 M urea, 4% SDS, 50 mM Tris-HCl pH 8.8 for 15 min, and then alkylated with 4.8% iodoacetamide solution for 15 min at RT and loaded onto SDS-polyacrylamide gels containing 15% acrylamide and prepared as described for Western blotting, but without the concentrating gel. A marker (Spectra multicolor Broad Range Protein Ladder, Fermentas, Finland) in an amount of 10 μL was applied to 0.5 cm^2 filter paper, placed on the gel next to the IPG strip, and embedded in a 0.6% agarose solution in dH_2O containing 0.001% (*w*/*v*) bromophenol blue as a dye. Proteins were separated in a Mini-PROTEAN 3 electrophoresis chamber (Bio-Rad, Hercules, USA) in a buffer solution for protein separation in polyacrylamide gel. Electrophoresis was carried out for ~3 h at 65 V at RT to the required degree of protein separation controlled by the marker. Subsequent transfer of proteins to the membrane, blocking of the membrane, incubation with primary and secondary antibodies, and signal detection were performed in the same way as described for one-dimensional Western blotting (see Section 4.6). After the transfer step, polyacrylamide gels were stained with Coomassie Brilliant Blue colloidal solution G-250 (20 mL/gel), followed by washing with dH_2O as described for 2D-PAGE (see Section 4.8). After washing, the gels were scanned using a GS-800 Calibrated Densitometer (Bio-Rad, Hercules, USA). Digital images of the gels were used as a control for loading the same amount of protein for the 2D Western blot method. The relative optical density of protein spot signals was quantitatively analyzed using the Bio 1D 15.01 software (Vilber Lourmat, Eberhardzell, Germany).

4.11. Statistics

All data are expressed as mean ± S.E. from three or more experiments, and were statistically evaluated by Student's *t*-test using SigmaPlot 14.0 (Jandel Scientific, San Rafael, CA, USA). Differences were considered significant when $p \leq 0.05$.

5. Conclusion

In summary, the triterpene glycoside, CA_2-2, isolated from the sea cucumber *C. japonica*, is a small bioactive molecule capable of inducing apoptosis in tumor cells via the caspase-dependent intrinsic pathway. Furthermore, it leads to a blockade of the cell cycle in the G2/M phase. Among others, keratin 81, CrkII, IL-1β, and cathepsin C were identified as proteins directly or indirectly targeted by CA_2-2 in human prostate cancer cells. Given the promising anticancer activity of CA_2-2, these and other newly identified molecular targets are worth being investigated further.

Author Contributions: Cell culture, investigation, and writing, E.S.M.; methodology, validation, and writing, S.A.D.; biological experiments and data interpretation, E.S.M., S.A.D., C.J., J.H., T.R., S.B., C.B., D.L.A., G.v.A. and F.H.; CA2-2 isolation and purification, A.S.S. and S.A.A.; proteomics, S.V. and S.B.; project administration and fundraising, F.H. and G.v.A.; conceptualization, and writing—review and editing, all authors. All authors have read and agreed to the published version of the manuscript.

Funding: This research was funded by the German Academic Exchange Service (DAAD) grant #A/11/85854.

Institutional Review Board Statement: Not applicable.

Informed Consent Statement: Not applicable.

Data Availability Statement: The data that support the findings of this study are available from the corresponding author upon reasonable request.

Acknowledgments: The authors express their gratitude to Kalinin V.I. (PIBOC FEB RAS, Vladivostok, Russia) for the invaluable help in discussing the results and editing the text of the manuscript.

Conflicts of Interest: The authors declare no conflict of interest.

References

1. Rawla, P. Epidemiology of prostate cancer. *World. J. Oncol.* **2019**, *10*, 63–89. [CrossRef] [PubMed]
2. Merkens, L.; Sailer, V.; Lessel, D.; Janzen, E.; Greimeier, S.; Kirfel, J.; Perner, S.; Pantel, K.; Werner, S.; von Amsberg, G. Aggressive variants of prostate cancer: Underlying mechanisms of neuroendocrine transdifferentiation. *J. Exp. Clin. Cancer. Res.* **2022**, *41*, 46. [CrossRef] [PubMed]
3. Dyshlovoy, S.A.; Honecker, F. Marine Compounds and Cancer: Updates 2022. *Mar. Drugs* **2022**, *20*, 759. [CrossRef] [PubMed]
4. Dyshlovoy, S.A.; Honecker, F. Marine Compounds and Cancer: Updates 2020. *Mar. Drugs* **2020**, *18*, 643. [CrossRef]
5. Marta Correia-da-Silva, M.; Sousa, E.; Pinto, M.M.M.; Kijjoa, A. Anticancer and cancer preventive compounds from edible marine organisms. *Semin. Cancer Biol.* **2017**, *46*, 55–64. [CrossRef]
6. Kalinin, V.I.; Aminin, D.L.; Avilov, S.A.; Silchenko, A.S.; Stonik, V.A. Triterpene glycosides from sea cucucmbers (Holothurioidae, Echinodermata), biological activities and functions. In *Studies in Natural Product Chemistry (Bioactive Natural Products)*; Rahman, A., Ed.; Elsevier: Amsterdam, The Netherlands, 2008; Volume 35, pp. 135–196.
7. Aminin, D.L.; Menchinskaya, E.S.; Pislyagin, E.A.; Silchenko, A.S.; Avilov, S.A.; Stadnichenko, N.I.; Collin, P.D.; Kalinin, V.I. Review of patents based on triterpene glycosides of sea cucumbers. In *Studies in Natural Product Chemistry (Bioactive Natural Products)*; Rahman, A., Ed.; Elsevier: Amsterdam, The Netherlands, 2016; Volume 51, pp. 175–200.
8. Aminin, D.L.; Menchinskaya, E.S.; Pisliagin, E.A.; Silchenko, A.S.; Avilov, S.A.; Kalinin, V.I. Anticancer activity of sea cucumber triterpene glycosides. *Mar. Drugs* **2015**, *13*, 1202–1223. [CrossRef]
9. Yun, S.-H.; Sim, E.-H.; Han, S.-H.; Han, J.-Y.; Kim, S.-H.; Silchenko, A.S.; Stonik, V.A.; Park, J.-I. Holotoxin A_1 induces apoptosis by activating acid sphingomyelinase and neutral sphingomyelinase in K562 and human primary leukemia cells. *Mar. Drugs* **2018**, *16*, 123. [CrossRef]
10. Luparello, C.; Ragona, D.; Asaro, D.M.L.; Lazzara, V.; Affranchi, F.; Celi, M.; Arizza, V.; Vazzana, M. Cytotoxic potential of the coelomic fluid extracted from the sea cucumber *Holothuria tubulosa* against triple-negative MDA-MB231 breast cancer cells. *Biology* **2019**, *8*, 76. [CrossRef]
11. Liu, F.; Tang, L.; Tao, M.; Cui, C.; He, D.; Li, L.; Liao, Y.; Gao, Y.; He, J.; Sun, F.; et al. Stichoposide C exerts anticancer effects on ovarian cancer by inducing autophagy via inhibiting AKT/mTOR pathway. *Onco. Targets Ther.* **2022**, *21*, 87–101. [CrossRef]
12. Careagaa, V.P.; Buenob, C.; Muniainc, C.; Alché, L.; Maiera, M.S. Pseudocnoside A, a new cytotoxic and antiproliferative triterpene glycoside from the sea cucumber *Pseudocnus dubiosus* leoninus. *Nat. Prod. Res.* **2014**, *28*, 213–220. [CrossRef]
13. Silchenko, A.S.; Kalinovsky, A.I.; Avilov, S.A.; Popov, R.S.; Dmitrenok, P.S.; Chingizova, E.A.; Menchinskaya, E.S.; Panina, E.G.; Stepanov, V.G.; Kalinin, V.I.; et al. Djakonoviosides A, A1, A2, B1-B4—Triterpene monosulfated tetra- and pentaosides from the sea cucumber *Cucumaria djakonovi*: The first finding of a hemiketal fragment in the aglycones; activity against human breast cancer cell lines. *Int. J. Mol. Sci.* **2023**, *24*, 11128. [CrossRef] [PubMed]
14. Malyarenko, T.V.; Malyarenko, O.S.; Kicha, A.A.; Kalinovsky, A.I.; Dmitrenok, P.S.; Ivanchina, N.V. In vitro anticancer and cancer-preventive activity of new triterpene glycosides from the Far Eastern starfish *Solaster pacificus*. *Mar. Drugs* **2022**, *20*, 216. [CrossRef] [PubMed]
15. Althunibat, O.Y.; Hashim, R.B.; Taher, M.; Jamaludin, M.D.; Ikeda, M.-A.; Zali, B.I. In vitro antioxidant and antiproliferative activities of three Malaysian sea cucumber species. *Eur. J. Sci. Res.* **2009**, *37*, 376–387.
16. Fan, T.-J.; Yuan, W.-P.; Cong, R.-S.; Yang, X.-X.; Wang, W.-W.; Jing, Z. Studies on the purification of water-soluble holothurian glycosides from *Apostichopus japonicus* and their tumor suppressing activity. *Yaoxue Xuebao* **2009**, *44*, 25–31.
17. Cuong, N.X.; Vien, L.T.; Hanh, T.T.H.; Thao, N.P.; Thao, D.T.; Thanh, N.V.; Nam, N.Y.; Thung, D.C.; Kiem, P.V.; Minh, C.V. Cytotoxic triterpene saponins from *Cercodemas anceps*. *Bioorg. Med. Chem. Lett.* **2015**, *25*, 3151–3156. [CrossRef] [PubMed]
18. Vien, L.T.; Hanh, T.T.H.; Quang, T.H.; Thanh, N.V.; Thao, D.T.; Cuong, N.X.; Nam, N.H.; Thung, D.C.; Kiem, P.V. Triterpene tetraglycosides from *Stichopus Herrmanni* Semper, 1868. *Nat. Prod. Commun.* **2022**, *17*, 1–5. [CrossRef]
19. Li, M.; Miao, Z.H.; Chen, Z.; Chen, Q.; Gui, M.; Lin, L.P.; Sun, P.; Yi, Y.H.; Ding, J. Echinoside A, a new marine-derived anticancer saponin, targets topoisomerase 2α by unique interference with its DNA binding and catalytic cycle. *Ann. Oncol.* **2010**, *21*, 597–607. [CrossRef]

20. Dyshlovoy, S.A.; Menchinskaya, E.S.; Venz, S.; Rast, S.; Amann, K.; Hauschild, J.; Otte, K.; Kalinin, V.I.; Silchenko, A.S.; Avilov, S.A.; et al. The marine triterpene glycoside frondoside A exhibits activity in vitro and in vivo in prostate cancer. *Int. J. Cancer* **2016**, *138*, 2450–2465. [CrossRef]
21. Dyshlovoy, S.A.; Madanchi, R.; Hauschild, J.; Otte, K.; Alsdorf, W.H.; Schumacher, U.; Kalinin, V.I.; Silchenko, A.S.; Avilov, S.A.; Honecker, F.; et al. The marine triterpene glycoside frondoside A induces p53-independent apoptosis and inhibits autophagy in urothelial carcinoma cells. *BMC Cancer* **2017**, *17*, 93. [CrossRef]
22. Aminin, D.L.; Chaykina, E.L.; Agafonova, I.G.; Avilov, S.A.; Kalinin, V.I.; Stonik, V.A. Antitumor activity of the immunomodulatory lead Cumaside. *Int. Immunopharmacol.* **2010**, *10*, 648–654. [CrossRef]
23. Menchinskaya, E.S.; Pislyagin, E.A.; Kovalchyk, S.N.; Davydova, V.N.; Silchenko, A.S.; Avilov, S.A.; Kalinin, V.I.; Aminin, D.L. Antitumor activity of cucumarioside A_2-2. *Chemotherapy* **2013**, *59*, 181–191. [CrossRef]
24. Reunov, A.A.; Reunov, A.V.; Pimenova, E.A.; Reunova, Y.A.; Menchinskaya, E.S.; Lapshina, L.A.; Aminin, D.L. Cucumarioside A_2-2 stimulates apoptotic necrosis in Ehrlich ascites carcinoma cells. *Dokl. Biol. Sci.* **2015**, *462*, 161–163. [CrossRef]
25. Reunov, A.; Reunov, A.; Pimenova, E.; Reunova, Y.; Menchinskaya, E.; Lapshina, L.; Aminin, D. The study of the calpain and caspase-1 expression in ultrastructural dynamics of Ehrlich ascites carcinoma necrosis. *Gene* **2018**, *658*, 1–9. [CrossRef]
26. Pimenova, E.A.; Reunova, Y.A.; Menchinskaya, E.S.; Reunov, A.A.; Aminin, D.L. An unusual pathway of mitoptosis found in Ehrlich Carcinoma Cells. *Dokl. Biol. Sci.* **2020**, *494*, 240–243. [CrossRef]
27. Menchinskaya, E.S.; Aminin, D.L.; Avilov, S.A.; Silchenko, A.S.; Andryjashchenko, P.V.; Kalinin, V.I.; Stonik, V.A. Inhibition of tumor cells multidrug resistance by cucumarioside A_2–2, frondoside A and their complexes with cholesterol. *Nat. Prod. Commun.* **2013**, *8*, 1377–1380. [CrossRef]
28. Menchinskaya, E.; Gorpenchenko, T.; Silchenko, A.; Avilov, S.; Aminin, D. Modulation of doxorubicin intracellular accumulation and anticancer activity by triterpene glycoside cucumarioside A_2-2. *Mar. Drugs* **2019**, *17*, 597. [CrossRef]
29. Franken, N.A.; Rodermond, H.M.; Stap, J.; Haveman, J.; van Bree, C. Clonogenic assay of cells in vitro. *Nat. Protoc.* **2006**, *1*, 2315–2319. [CrossRef]
30. Wang, Y.; Xu, L.; Shi, S.; Wu, S.; Meng, R.; Chen, H.; Jiang, Z. Deficiency of NEIL3 enhances the chemotherapy resistance of prostate cancer. *Int. J. Mol. Sci.* **2021**, *22*, 4098. [CrossRef]
31. Hazzalin, C.A.; Le Panse, R.; Cano, E.; Mahadevan, L.C. Anisomycin selectively desensitizes signalling components involved in stress kinase activation and fos and jun induction. *J. Mol. Cell. Biol.* **1998**, *18*, 1844–1854. [CrossRef]
32. Kochi, S.K.; Collier, R.J. DNA fragmentation and cytolysis in U937 cells treated with diphtheria toxin or other inhibitors of protein synthesis. *Exp. Cell Res.* **1993**, *208*, 296–302. [CrossRef]
33. Torocsik, B.; Szeberenyi, J. Anisomycin affects both pro- and antiapoptotic mechanisms in PC12 cells. *Biochem. Biophys. Res. Commun.* **2000**, *278*, 550–556. [CrossRef]
34. Curtin, J.F.; Cotter, T.G. Anisomycin activates JNK and sensitises DU 145 prostate carcinoma cells to Fas mediated apoptosis. *Br. J. Cancer* **2002**, *87*, 1188–1194. [CrossRef]
35. Niu, Z.; Wang, M.; Zhou, L.; Yao, L.; Liao, Q.; Zhao, Y. Elevated GRP78 expression is associated with poor prognosis in patients with pancreatic cancer. *Sci Rep.* **2015**, *4*, 16067. [CrossRef]
36. Ulianich, L.; Insabato, L. Endoplasmic reticulum stress in endometrial cancer. *Front. Med.* **2014**, *1*, 55. [CrossRef]
37. Yerushalmi, R.; Raiter, A.; Nalbandyan, K.; Hardy, B. Cell surface GRP78: A potential marker of good prognosis and response to chemotherapy in breast cancer. *Oncol. Lett.* **2015**, *10*, 2149–2155. [CrossRef]
38. Rodríguez-Berriguete, G.; Sánchez-Espiridión, B.; Cansino, J.R.; Olmedilla, G.; Martínez-Onsurbe, P.; Sánchez-Chapado, M.; Paniagua, R.; Fraile, B.; Royuela, M. Clinical significance of both tumor and stromal expression of components of the IL-1 and TNF-α signaling pathways in prostate cancer. *Cytokine* **2013**, *64*, 555–563. [CrossRef]
39. Kawada, M.; Ishizuka, M.; Takeuchi, T. Enhancement of antiproliferative effects of interleukin-1beta and tumor necrosis factor-alpha on human prostate cancer LNCaP cells by coculture with normal fibroblasts through secreted interleukin-6. *Jpn J. Cancer Res.* **1999**, *90*, 546–554. [CrossRef]
40. Song, X.; Voronov, E.; Dvorkin, T.; Fima, E.; Cagnano, E.; Benharroch, D.; Shendler, Y.; Bjorkdahl, O.; Segal, S.; Dinarello, C.A.; et al. Differential effects of IL-1 alpha and IL-1 beta on tumorigenicity patterns and invasiveness. *J. Immunol.* **2003**, *15*, 6448–6456. [CrossRef]
41. Liu, Q.; Russell, M.R.; Shahriari, K.; Jernigan, D.L.; Lioni, M.I.; Garcia, F.U.; Fatatis, A. Interleukin-1β promotes skeletal colonization and progression of metastatic prostate cancer cells with neuroendocrine features. *Cancer Res.* **2013**, *73*, 3297–3305. [CrossRef]
42. Kaler, P.; Galea, V.; Augenlicht, L.; Klampfer, L. Tumor associated macrophages protect colon cancer cells from TRAIL-Induced apoptosis through IL-1β- Dependent stabilization of snail in tumor cells. *PLoS ONE* **2010**, *5*, e11700. [CrossRef]
43. Karantza, V. Keratins in health and cancer: More than mere epithelial cell markers. *Oncogene* **2011**, *30*, 27–138. [CrossRef]
44. Cirman, T.; Oresić, K.; Mazovec, G.D.; Turk, V.; Reed, J.C.; Myers, R.M.; Salvesen, G.S.; Turk, B. Selective disruption of lysosomes in HeLa cells triggers apoptosis mediated by cleavage of Bid by multiple papain-like lysosomal cathepsins. *J. Biol. Chem.* **2004**, *279*, 3578–3587. [CrossRef]
45. Yaowu, H.; Brown, M.A.; Rothnagel, J.A.; Saunders, N.A.; Smith, R. Roles of heterogeneous nuclear ribonucleoproteins A and B in cell proliferation. *J. Cell Sci.* **2005**, *118*, 3173–3183. [CrossRef]

46. Rodrigues, S.P.; Fathers, K.E.; Chan, G.; Zuo, D.; Halwani, F.; Meterissian, S.; Park, M. CrkI and CrkII function as key signaling integrators for migration and invasion of cancer cells. *Mol. Cancer Res.* **2005**, *3*, 183–194. [CrossRef]
47. Avilov, S.A.; Stonik, V.A.; Kalinovskii, A.I. Structures of four new triterpene glycosides from the holothurians *Cucumaria japonica*. *Chem. Nat. Compd.* **1990**, *26*, 670–675. [CrossRef]
48. Laemmli, U.K. Cleavage of structural proteins during the assembly of the head of bacteriophage T4. *Nature* **1970**, *227*, 680–685. [CrossRef]

marine drugs

MDPI

Article

New Ceramides and Cerebrosides from the Deep-Sea Far Eastern Starfish *Ceramaster patagonicus*

Timofey V. Malyarenko [1,2,*], Viktor M. Zakharenko [1,2], Alla A. Kicha [1], Alexandra S. Kuzmich [1], Olesya S. Malyarenko [1], Anatoly I. Kalinovsky [1], Roman S. Popov [1], Vasily I. Svetashev [3] and Natalia V. Ivanchina [1,*]

[1] G.B. Elyakov Pacific Institute of Bioorganic Chemistry, Far Eastern Branch, Russian Academy of Sciences, Pr. 100-let Vladivostoku 159, 690022 Vladivostok, Russia

[2] Department of Bioorganic Chemistry and Biotechnology, School of Natural Sciences, Far Eastern Federal University, Russky Island, Ajax Bay 10, 690922 Vladivostok, Russia

[3] A.V. Zhirmunsky National Scientific Center of Marine Biology, Far Eastern Branch, Russian Academy of Sciences, ul. Palchevskogo 17, 690041 Vladivostok, Russia

* Correspondence: malyarenko-tv@mail.ru (T.V.M.); ivanchina@piboc.dvo.ru (N.V.I.); Tel.: +7-423-2312-360 (N.V.I.); Fax: +7-423-2314-050 (N.V.I.)

Abstract: Three new ceramides (**1–3**) and three new cerebrosides (**4**, **8**, and **9**), along with three previously known cerebrosides (ophidiocerebrosides C (**5**), D (**6**), and CE-3-2 (**7**)), were isolated from a deep-sea starfish species, the orange cookie starfish *Ceramaster patagonicus*. The structures of **1–4**, **8**, and **9** were determined by the NMR and ESIMS techniques and also through chemical transformations. Ceramides **1–3** contain *iso*-C_{21} or C_{23} Δ^9-phytosphingosine as a long-chain base and have C_{16} or C_{17} (2*R*)-2-hydroxy-fatty acids of the normal type. Cerebroside **4** contains C_{22} Δ^9-sphingosine *anteiso*-type as a long-chain base and (2*R*)-2-hydroxyheptadecanoic acid of the normal type, while compounds **8** and **9** contain saturated C-17 phytosphingosine *anteiso*-type as a long-chain base and differ from each other in the length of the polymethylene chain of (2*R*)-2-hydroxy-fatty acids of the normal type: C_{23} in **8** and C_{24} in **9**. All the new cerebrosides (**4**, **8**, and **9**) have β-D-glucopyranose as a monosaccharide residue. The composition of neutral sphingolipids from *C. patagonicus* was described for the first time. The investigated compounds **1–3**, **5–7**, and **9** exhibit slight to moderate cytotoxic activity against human cancer cells (HT-29, SK-MEL-28, and MDA-MB-231) and normal embryonic kidney cells HEK293. Compounds **2**, **5**, and **6** at a concentration of 20 μM inhibit colony formation of MDA-MB-231 cells by 68%, 54%, and 68%, respectively. The colony-inhibiting activity of compounds **2**, **5**, and **6** is comparable to the effect of doxorubicin, which reduces the number of colonies by 70% at the same concentration.

Keywords: ceramides; cerebrosides; NMR spectra; fatty acids; long-chain bases; starfish; *Ceramaster patagonicus*; cytotoxic activity; inhibition of colony formation

Citation: Malyarenko, T.V.; Zakharenko, V.M.; Kicha, A.A.; Kuzmich, A.S.; Malyarenko, O.S.; Kalinovsky, A.I.; Popov, R.S.; Svetashev, V.I.; Ivanchina, N.V. New Ceramides and Cerebrosides from the Deep-Sea Far Eastern Starfish *Ceramaster patagonicus*. *Mar. Drugs* **2022**, *20*, 641. https://doi.org/10.3390/md20100641

Academic Editor: Ricardo Calado

Received: 9 September 2022
Accepted: 12 October 2022
Published: 14 October 2022

1. Introduction

Starfish (also called sea stars) are found throughout the world's oceans at a wide range of depths: from intertidal to deep-sea habitats. Their ecological characteristics and life-history features make them a rich source of various low-molecular-weight compounds. The best-studied starfish-derived substances are polar steroidal compounds that were found in almost all of the species analyzed [1–9]. In addition, anthraquinoid pigments, triterpene glycosides, carotenoids, and sphingolipids were found in starfish [5,9–15].

Sphingolipids are a group of heterogeneous lipids including those present in the plasma membranes that, along with phospholipids and sterols, play a fundamental role in important phenomena such as cell–cell recognition and antigenic specificity [16,17]. Sphingolipids can be divided into several structural groups: ceramides, cerebrosides, gangliosides, and sphingophospholipids.

Ceramides are hydrophobic molecules consisting of a long-chain base (LCB) and an amide-linked fatty acid (FA) residue. Ceramides are biosynthesized during the reaction of S-acyl-coenzyme A (usually C_{16}-CoA) with serine, which is catalyzed by serine palmitoyl transcriptase or related enzymes, followed by the reduction of the carbonyl group by ketosphinganine reductase and the N-acylation by ceramide synthase. Surprisingly, the LCB hydroxylation, which leads to the production of so-called phytosphinganine derivatives, also occurs in plants and many echinoderms. When hydroxylases act on FA in these invertebrates, an additional hydroxyl group is introduced also into the α–position of FA [17]. Additionally, both bases and FA moieties in this type of natural product may contain normal chains, as well as those with *iso-* and/or *anteiso*-branching. Unfortunately, the biological activity of starfish-derived ceramides has not been sufficiently studied. It was previously reported that asteriaceramide A from *Asterias amurensis* showed it could actively stimulate the root growth of *Brassica campestris* [18].

Cerebrosides are glycosylceramides containing, as a rule, glucose and galactose or other rare monosaccharide residues in their carbohydrate moieties. These compounds are synthesized by special enzymes, glycosyl-transferases, that attach monosaccharide residues to C-1 of ceramide [17]. On the basis of the chemical structure of cerebrosides, these can be divided into three groups: monoglycosides, biglycosides (mainly lactosides), and oligoglycosides. In addition to glucose and galactose residues, this class of glycosylated lipids can contain an aminosugar residue (globosides) in their carbohydrate moieties or be sulfated. The interest in sphingolipids and their derivatives is mainly associated with their wide range of biological activities. Some studies have shown that sphingolipids can inhibit the growth of microalgae, fungi, and bacteria [19]. It was previously reported that starfish cerebrosides showed neuritogenic activity against the rat pheochromocytoma PC12 cells in the presence of nerve growth factor (NGF), an anti-inflammatory effect, in vitro cytotoxic activity against Caco-2 colon cancer cells, improvement of the barrier function of the skin, and other properties [14].

Thus, the study of starfish sphingolipids is an interesting and relevant scientific issue. It is also worth noting that previously, sphingolipids were studied from starfish that live at shallow depths. To date, there have been no reports on sphingolipids derived from starfish dwelling at depths greater than 150 m.

Recently, we found that conjugates of polyhydroxysteroids with long-chain FAs from the same starfish species exhibited potent anticancer activity in vitro [20]. In the present report, continuing the search for anticancer compounds from marine organisms, we provide the results of our studies on the structures of ceramides and cerebrosides derived from the starfish *Ceramaster patagonicus*, and also their effects on the viability of human normal and cancer cells and the colony formation of cancer cells.

2. Results and Discussion

2.1. The Isolation and Determination of the Structures of Compounds **1–9** from C. patagonicus

The concentrated methanol–chloroform–ethanolic extract of *C. patagonicus* was separated between H_2O and AcOEt/BuOH, with the organic layer dried and washed with cold acetone. The acetone-soluble fraction was separated by chromatography on a silica gel column, followed by HPLC on semi-preparative Diasfer-110-C18, Discovery C18, and Discovery HS C18-10 columns. As a result, we obtained three new ceramides (**1–3**) and three new cerebrosides (**4**, **8**, and **9**) along with three known cerebrosides: ophidiacerebrosides C (**5**) and D (**6**) that had been previously isolated from the purple starfish *Ophidiaster ophidianus* [21] and CE-3-2 (**7**) from the sea cucumber *Cucumaria echinata* [22] (Figure 1).

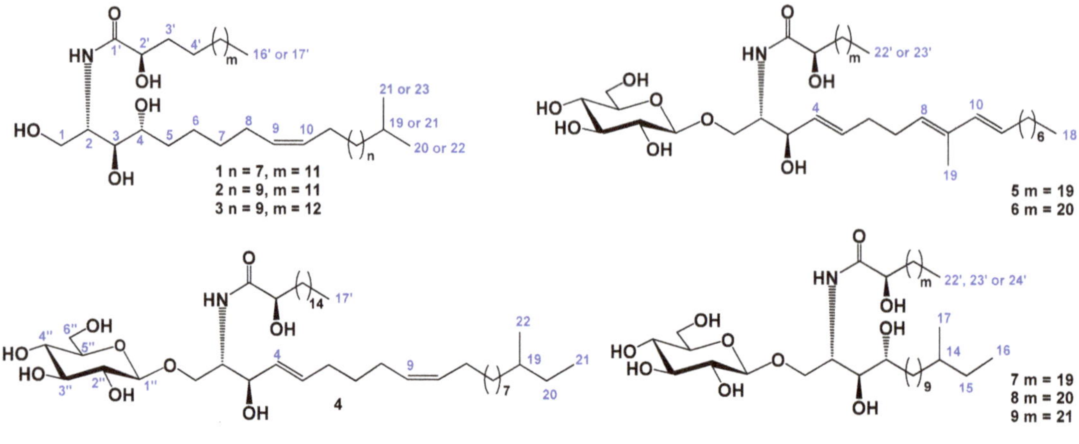

Figure 1. The structures of compounds **1–9** isolated from *C. patagonicus*.

The IR spectrum of compound **1** showed the presence of hydroxyl (3402 cm^{-1}) and amide (1656, 1523 cm^{-1}) groups. The molecular formula of compound **1** was determined as $C_{37}H_{73}NO_5$ from the $[M + Na]^+$ sodium adduct ion peak at m/z 634.5376 in the (+)HRESIMS and the $[M - H]^-$ deprotonated molecular ion peak at m/z 610.5412 in the (–)HRESIMS (Figures S1 and S2). The ^{1}H- and ^{13}C-NMR spectroscopic data of **1** showed the resonances of protons and carbons of three terminal methyls CH_3-20, CH_3-21, and CH_3-16′ [δ_H 2 × 0.88 d (6.5), 0.89 t (7.2); δ_C 2 × 22.5, 14.0], four oxygenated groups CH_2-1 [δ_{Ha} 4.51 dt (10.9, 4.2), δ_{Hb} 4.43 m; δ_C 61.8], CH-3 (δ_H 4.34 m; δ_C 76.7), CH-4 (δ_H 4.28 m; δ_C 72.8), and CH-2′ (δ_H 4.62 m; δ_C 72.3), one amide group NH-CO [δ_H 8.55 d (9.0); δ_C 175.0], the 9(10)-double bond (δ_H 5.50 m, 5.48 m; δ_C 130.1, 129.9), and one characteristic methine CH-2 [δ_H 5.10 sext (4.5); δ_C 52.8] attached at the nitrogen atom (Table 1, Figures S3 and S4). Thus, the ^{1}H- and ^{13}C-NMR spectra of **1** exhibited the characteristic signals of an unsaturated phytosphingosine-type ceramide with a 2-hydroxy fatty acid (Figure 1). Moreover, ceramide **1** has normal and *iso*-types of side chains; in the terminal methyl groups, the carbon atom signals were observed at δ_C 14.0 (normal form) and 2 × 22.5 (*iso*-form) in the ^{13}C-NMR spectrum (Table 1, Figure S4). The ^{1}H-^{1}H COSY and HSQC correlations of **1** revealed the corresponding sequences of protons at C-1 to C-11; C-20 to C-21 through C-19; C-2 to NH; C-2′ to C-4′, and C-16′ to C-14′ (Table 1, Figures 2A, S5 and S6). The key HMBC cross-peaks, such as Hb-1/C-3; H-2/C-1, C-3, C-1′; H-3/C-2, C-4; Hb-5/C-4, C-6, C-7; H-8/C-6, C-7, C-9, C-10; H-9 and H-10/C-8, C-11; H-11/C-9, C-10; H-19/C-20, C-21; H_3-20/C-19, C-21; H_3-21/C-19, C-20; NH/C-2, C-1′; H-2′/C-1′; Ha-3′/C-1′, C-4′; Hb-3′/C-2′; H_2-15′/C-14′; and H_3-16′/C-14′, C-15′ confirmed the overall structure of ceramide **1** (Figures 2A and S7).

The polymethylene chain length of LCB and FA and the absolute configuration of the ceramide **1** were determined as follows. When **1** was methanolyzed with methanolic hydrochloric acid, fatty acid methyl ester (FAME) was obtained together with LCB. A gas chromatography–mass spectrometry (GC–MS) analysis of FAME showed the existence of one component that was characterized as saturated methyl 2-hydroxyhexadecanoate of normal type (FAME-1). The normal type of FAME-1 was also confirmed by ^{1}H-NMR spectra, which consisted only of one triplet terminal methyl group at δ_H 0.89. The optical rotation of FAME-1 ($[\alpha]_D^{25}$ –3.5° (c = 1.0, $CHCl_3$)) is consistent with the data $[\alpha]_D^{25}$–3.21° reported in the literature [23]; therefore, the absolute stereochemistry at C-2′ is suggested to be *R*. Based on this suggestion, as well as on NMR and mass spectrometric data, we assumed LCB of ceramide **1** to have 21 carbon atoms and *iso*-type of unsaturated polymethylene chain. The geometry of the double bond in LCB can be determined on the basis of the ^{13}C-NMR chemical shift of the methylene carbon adjacent to the olefinic carbon ($\delta_C \approx 27$ for (*Z*) isomers and $\delta_C \approx 32$ for (*E*) isomers [24]). The ^{13}C-NMR spectrum of compound **1** indicated the presence of two characteristic allyl carbons, C-8 (δ_C 27.5) and C-11 (δ_C 27.3). Thus, the olefinic group in **1** was determined to have a *cis* (*Z*) geometry. The location of the double bond in the LCB moiety at position C-9 was determined through ^{1}H-^{1}H COSY, HMBC, and 2D TOCSY NMR experiments (Figures 2A, S5, S7 and S8). The absolute configuration of LCB of ceramide **1** was assumed to be *D-ribo*-(2*S*,3*S*,4*R*) on the basis of similarities in optical rotation ($[\alpha]_D^{25}$ +28.0° (c = 1.0, $CHCl_3$)) with synthetic *D-ribo*-(2*S*,3*S*,4*R*)-phytosphingosine ($[\alpha]_D^{25}$ +26.8° (c = 1.1, $CHCl_3$)) [25].

Table 1. ^{1}H- (700.13 MHz) and ^{13}C- (176.04 MHz) NMR chemical shifts of ceramides **1**–**3** in C_5D_5N, at 30 °C, δ in ppm, J values in Hz.

Position	δ_H	δ_C	Position	δ_H	δ_C
1a 1b	4.51 dt (10.9, 4.2) 4.43 m	61.8	11	2.10 q (6.7)	27.3
1-OH	6.60 m		19 or 21	1.51 m	27.9
2	5.10 sext (4.5)	52.8	20 or 22	0.88 d (6.5)	22.5
3	4.34 m	76.7	21 or 23	0.88 d (6.5)	22.5
3-OH	6.59 d (6.8)		NH	8.55 d (9.0)	
4	4.28 m	72.8	1′		175.0
4-OH	6.14 d (6.5)		2′	4.62 m	72.3
5a 5b	2.28 m 1.95 m	33.9	2′-OH	7.53 d (5.1)	
6a 6b	1.74 m 1.39 m	26.1	3′a 3′b	2.24 m 2.05 m	35.5
7a 7b	1.58 m 1.53 m	30.2	4′a 4′b	1.80 m 1.73 m	25.6
8	2.17 q (7.0)	27.5	14′ or 15′	1.25 m	31.9
9	5.50 m	130.1	15′ or 16′	1.27 m	22.7
10	5.48 m	129.9	16′ or 17′	0.89 t (7.2)	14.0

Based on all above-mentioned data, we determined the structure of **1** to be (2*S*,3*S*,4*R*,9*Z*)-2-[(2*R*)-2-hydroxyhexadecanoylamino]-19-methyl-9-icosen-1,3,4-triol. As far as we know, a ceramide with such a chemical structure was isolated for the first time.

The IR spectrum of compound **2** showed the presence of hydroxyl (3404 cm^{-1}) and amide (1657, 1522 cm^{-1}) groups. The molecular formula of compound **2** was determined as $C_{39}H_{77}NO_5$ from the $[M + Na]^+$ sodium adduct ion peak at m/z 662.5324 in the (+)HRESIMS and the $[M - H]^-$ deprotonated molecular ion peak at m/z 638.5364 in the (–)HRESIMS (Figures S9 and S10). A comparison of the ^{1}H-, ^{13}C-NMR spectra and an extensive 2D NMR analysis of compounds **1**, **2**, and **3** revealed that the unsaturated phytosphingosine-type ceramide with a 2-hydroxy fatty acid of **2** and **3** is identical to that of compound **1**, while the polymethylene chain lengths of LCB and/or FA of **1**–**3** differ from each other (Figures 1 and S11–S15, Table 1). A comparison of the molecular weights (MWs) of **1** and **2** showed that they differed by 28 amu.

The FA unit in **2** was identified by GC analysis and the mass spectra of the FAME-2 derivative were measured by GC–MS similarly to compound **1**. The GC–MS analysis showed that FAME-2 was identical to FAME-1. Moreover, the normal type of FAME-2 was also confirmed by the ^{1}H-NMR spectrum, which consisted of only one triplet terminal methyl group at δ_H 0.89. Thus, the FA of ceramide **2** was determined to be (2*R*)-2-hydroxyhexadecanoic acid. Based on this finding, as well as on the NMR and mass spectrometry data, we suggested that LCB of ceramide **2** has 23 carbon atoms and an *iso*-type of unsaturated polymethylene chain. Accordingly, the structure of **2** was determined to be (2*S*,3*S*,4*R*,9*Z*)-2-[(2*R*)-2-hydroxyhexadecanoylamino]-21-methyl-9-docosen-1,3,4-triol.

Compound **3** was characterized from a mixture with compound **2** at a ratio of 2:1 on the basis of the evaluation of the ion peak intensities in ESI mass-spectra. The IR spectrum of compound **3** showed the presence of hydroxyl (3407 cm^{-1}) and amide (1655, 1522 cm^{-1}) groups. The positive HRESI mass spectrum of this mixture showed two $[M + Na]^+$ ion peaks at m/z 662.5324 corresponding to compound **2** and at m/z 676.5463 corresponding to compound **3**. Therefore, the molecular formula of compound **3** was determined as $C_{40}H_{79}NO_5$ from the $[M + Na]^+$ sodium adduct ion peak at *m/z* 676.5463 in the (+)HRESIMS and the $[M - H]^-$ deprotonated molecular ion peak at *m/z* 652.5520 in the (–)HRESIMS (Figures S16 and S17). The NMR spectra of compounds **3** and **2** were almost identical (Figures S18–S22), but the MWs of **3** and **2** differed by 14 amu. A GC–MS analysis and mass spectra of fatty acid methyl esters obtained from the mixture of **3** and **2** showed the presence of FAME-2 containing saturated methyl 2-hydroxyhexadecanoate of the normal type and FAME-3 containing methyl 2-hydroxyheptadecanoate of the normal type. Thus, compounds **2** and **3** differed from each other by FA residues, C_{16} in **2** and C_{17} in **3**, and had an identical C_{23} unsaturated phytosphingosine-type LCB. Thus, the structure of **3** was determined to be (2*S*,3*S*,4*R*,9*Z*)-2-[(2*R*)-2-hydroxyheptadecanoylamino]-21-methyl-9-docosen-1,3,4-triol.

The IR spectrum of compound **4** showed the presence of hydroxyl (3383 cm^{-1}) and amide (1649, 1538 cm^{-1}) groups. The molecular formula of compound **4** was determined as $C_{45}H_{85}NO_9$ from the $[M + Na]^+$ sodium adduct ion peak at *m/z* 806.6110 in the (+)HRESIMS and the $[M - H]^-$ deprotonated molecular ion peak at *m/z* 782.6154 in the (–)HRESIMS (Figures S23 and S24). The ^1H- and ^{13}C-NMR spectroscopic data of the ceramide moiety of **4** showed the resonances of protons and carbons of three terminal methyls CH_3-21, CH_3-22, and CH_3-17′ [δ_H 0.87 t (7.3), 0.87 d (6.2), 0.89 t (7.0); δ_C 11.3, 19.1, 14.0], three oxygenated groups CH_2-1 [δ_{Ha} 4.70 dd (10.6, 5.6), δ_{Hb} 4.26 dd (10.6, 4.1); δ_C 69.9], CH-3 (δ_H 4.77 m; δ_C 72.1), and CH-2′ (δ_H 4.58 m; δ_C 72.3), one amide group NH-CO [δ_H 8.32 d (9.0); δ_C 175.4], the 4(5)-double bond [δ_H 5.99 dd (15.8, 6.2), 5.93 dt (15.8, 6.2); δ_C 131.7, 132.2], the 9(10)-double bond (δ_H 5.50 m, 5.49 m; δ_C 130.3, 129.6), and one characteristic methine CH-2 attached to nitrogen atom [δ_H 4.81 m; δ_C 54.4] (Table 2, Figures S25 and S26). Thus, the presence of characteristic signals of an unsaturated sphingosine-type ceramide including the 2-hydroxy FA residue in the ^1H- and ^{13}C-NMR spectra of the ceramide part of **4** was shown (Figure 1). Moreover, the ceramide moiety of **4** had normal and *anteiso*-types of side chains because the carbon atom signals of the terminal methyl groups were observed at δ_C 14.0 (normal form) and 11.3 and 19.1 (*anteiso*-form) in the ^{13}C-NMR spectrum (Table 2). The ^1H-^1H COSY and HSQC correlations in the NMR spectra of **4** indicated the corresponding sequences of protons at C-1 to C-11; C-21 to C-22 through C-19 and C-20; C-2 to NH; C-2′ to C-4′, and C-16′ to C-14′ (Table 2, Figures 2B, S27 and S28). The key HMBC cross-peaks such as Hb-1/C-2, C-3; H-2/C-1′; H-3/C-2, C-4; H-4/C-5, C-6; H-5/C-6, C-7; H-8/C-6, C-7, C-10; H-9/C-8, C-10, C-11; H-11/C-9; H_3-21/C-19, C-20; H_3-22/C-19; NH/C-2, C-1′; H-2′/C-1′; Ha-3′/C-1′, C-2′, C-4′; H_2-16′/C-15′; and H_3-17′/C-15′, C-16′ confirmed the overall structure of the ceramide part of **4** (Figures 2B and S29).

A GC–MS analysis of FAME-4 showed the existence of one component belonging to saturated methyl 2-hydroxyheptadecanoate of the normal type. Based on this finding, as well as on the NMR and mass spectrometry data, we assumed that the LCB of the ceramide part of **4** has 22 carbon atoms and an *anteiso*-type of unsaturated polymethylene chain. The *E*-configuration of the 4(5)-double bond in LCB was determined on the basis of the coupling constant between H-4 and H-5 (15.8 Hz) in the ^1H-NMR spectrum of **4** (Table 2). The geometry of the 9(10)-double bond in LCB was characterized as *Z* on the basis of the ^{13}C-NMR chemical shifts of methylene carbons at δ_C 27.2 (C-8) and δ_C 27.0 (C-11) [19]. The location of the double bond in the LCB moiety was determined through ^1H-^1H COSY and HMBC NMR experiments (Table 2, Figures 2B, S27 and S29).

The absolute configuration of C-2 and C-3 in LCB of the ceramide part of **4** is suggested to be (2*S*,3*R*) according to the similarities of the ^1H-NMR data with the previously known asteriacerebroside G with a (2*S*,3*R*)-configuration of asymmetric centers [18].

Table 2. ^{1}H- (700.13 MHz) and ^{13}C- (176.04 MHz) NMR chemical shifts of cerebrosides **4** and **8**, **9** in C_5D_5N, at 30 °C, δ in ppm, *J* values in Hz.

Position	**4**		**8, 9**	
	δ_H	δ_C	δ_H	δ_C
1a 1b	4.70 dd (10.6, 5.6) 4.26 dd (10.6, 4.1)	69.9	4.71 dd (10.6, 6.5) 4.53 dd (10.6, 4.4)	70.3
2	4.81 m	54.4	5.26 m	51.5
3	4.77 m	72.1	4.32 dd (12.3, 5.5)	75.7
3-OH	6.75 d (4.7)		6.69 d (6.0)	
4	5.99 dd (15.8, 6.2)	131.7	4.21 m	72.4
5a 5b	5.93 dt (15.8, 6.2)	132.2	2.24 m 1.91 m	33.9
6	2.12 m	32.1		
7	1.50 m	29.4		
8	2.11 m	27.2		
9	5.50 m	130.3		
10	5.49 m	129.6		
11	2.12 m	27.0		
19 or 14	1.30 m	34.4	1.30 m	34.5
20a or 15a 20b or 15b	1.30 m 1.12 m	36.7	1.30 m 1.10 m	36.8
21 or 16	0.87 t (7.3)	11.3	0.86 t (7.5)	11.3
22 or 17	0.87 d (6.2)	19.1	0.86 d (6.4)	19.2
NH	8.32 d (9.0)		8.54 d (9.2)	
1′		175.4		175.4
2′	4.58 m	72.3	4.58 m	72.3
2′-OH	7.53 d (5.0)		7.55 m	
3′a 3′b	2.21 m 2.02 m	35.4	2.20m 2.01 m	35.4
4′a 4′b	1.82 m 1.73 m	25.7	1.77 m 1.70 m	25.0
Terminal CH_3	0.89 t (7.0)	14.0	0.88 t (7.0)	14.0
1″	4.92 d (7.8)	105.4	4.95 d (7.9)	105.5
2″	4.03 m	74.9	4.00 m	74.9
2″-OH	7.11 brs			
3″	4.20 m	78.2	4.17 m	78.2
4″	4.20 m	71.4	4.17 m	71.4
5″	3.91 m	78.3	3.89 m	78.3
6″a 6″b	4.51 brd (12.5) 4.35 m	62.5	4.48 brd (12.0) 4.33 m	62.5
6″-OH	6.26 brs		6.27 brs	

In addition to the above-mentioned signals, the ^{1}H-NMR spectrum of **4** exhibited one resonance in the de-shielded region due to the anomeric proton of the monosaccharide unit at δ_H 4.92 that correlated in the HSQC experiment with a carbon signal at δ_C 105.4 (Table 2). The ^{1}H-^{1}H COSY, HSQC, HMBC, and ROESY experiments led to the assignment of all the proton and carbon signals to the carbohydrate residue of **4** (Table 2, Figures 2B and S25–S29). The coupling constant (7.8 Hz) of the anomeric proton was indicative of a β-configuration of the glycosidic bond. The NMR spectroscopic data of the monosaccharide moiety strictly coincided with those of a β-glucopyranosyl residue of the known asteriacerebroside G from *A. amurensis* [18]. The attachment of the monosaccharide to the ceramide part of **4** was deduced from the long-range correlations in the HMBC spectrum. There were cross-peaks between H-1″ of Glc*p* and C-1 of aglycon, as well as between H-1 of the ceramide part and C-1″ of Glc*p* (Figure 2B). Acid hydrolysis of cerebroside **4** with 2M TFA was carried out to confirm the identification of its monosaccharide unit as glucose. An alcoholysis of sugar by ©-(−)-2-octanol followed by acetylation, a GC analysis, and a comparison with the corresponding derivatives of standard monosaccharides allowed us to identify the D-configuration for the β-glucopyranosyl residue of **4**.

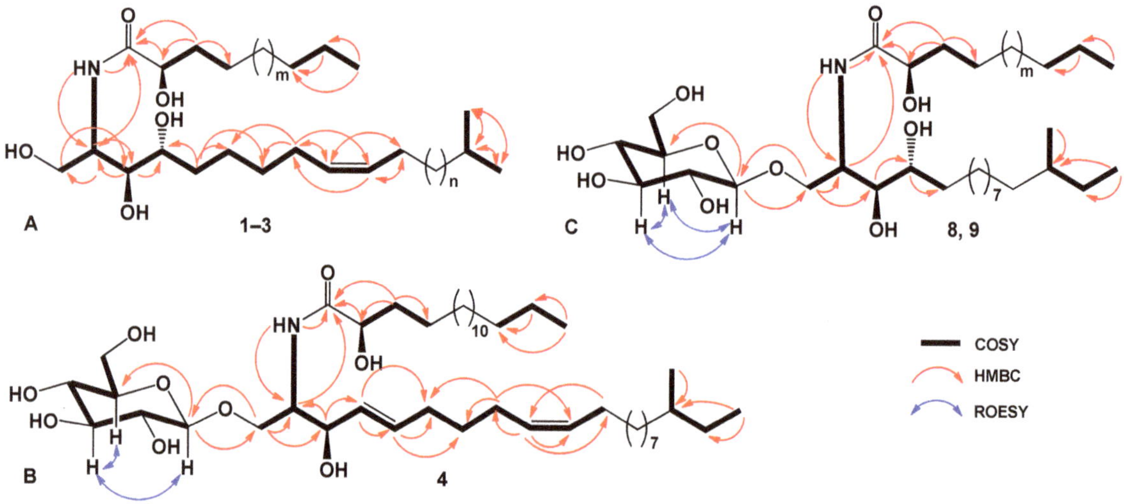

Figure 2. ^{1}H-^{1}H COSY and key HMBC correlations for compounds **1–3** (**A**) and ^{1}H-^{1}H COSY, key HMBC, and ROESY correlations for compounds **4** (**B**), **8**, and **9** (**C**).

The presence of a monosaccharide unit and the structure of the ceramide part of cerebroside **4** were confirmed by ESIMS/MS data. In fact, the (−)ESIMS/MS spectrum of the molecular anion peak –M − H]$^-$ at *m/z* 782 showed fragmentary peaks obtained through the loss of a sugar unit at *m/z* 602 [–M − H)−180]$^-$ (Z_0-ion) and 179 [hexo–e − H]$^-$ (C_2-ion), and also a few characteristic fragmentary peaks due to the cleavage of the ceramide part of **4**: *m/z* 326 [–M − H)−456]$^-$, the loss of a monosaccharide residue, and the cleavage of the bond between C-2 and C-3 (S-ion); *m/z* 310 [–M − H)−472]$^-$, the loss of monosaccharide residue, and the cleavage of the bond between C-2 and C-3 (T-ion); *m/z* 293 [–M − H)−489]$^-$, the cleavage of the bond between C-2 and C-3 (G-ion); *m/z* 283/284 [–M − H)−499/500]$^-$, the cleavage of the bond between C-2 and NH (U-ion); *m/z* 267 [–M − H)−515]$^-$, the cleavage of the amide bond (V-ion), and *m/z* 239 [–M − H)−543]$^-$, and the cleavage of the bond between C-1′ and C-2′ (W-ion) (Figure 3).

On the basis of all the above-mentioned data, we determined the structure of cerebroside **4** to be (2*S*,3*R*,4*E*,9*Z*)-1-*O*-(β-D-glucopyranosyl)-2-[(2*R*)-2-hydroxyheptadecanoylamino]-19-methyl-4,9-henicosadien-3-ol.

After carrying out an extensive 2D NMR and MS analysis of cerebrosides **4**, **8**, and **9**, we suggested that the monosaccharide moiety of **4** is identical to those of glycosides **8** and **9**.

The IR spectrum of compound **8** showed the presence of hydroxyl (3401 cm^{-1}) and amide (1634, 1541 cm^{-1}) groups. The molecular formula of compound **8** was determined as $C_{46}H_{91}NO_{10}$ from the [M + Na]$^+$ sodium adduct ion peak at *m/z* 840.6336 in the (+)HRESIMS and the [M − H]$^-$ deprotonated molecular ion peak at *m/z* 816.6574 in the (−)HRESIMS (Figures S30 and S31). The ^{1}H- and ^{13}C-NMR spectroscopic data of the ceramide part of **8** showed the resonances of protons and carbons of three terminal methyls CH_3-16, CH_3-17, and CH_3-22′ [δ_H 0.86 t (7.5), 0.86 d (6.4), 0.88 t (7.0); δ_C 11.3, 19.2, 14.0], four oxygenated groups CH_2-1 [δ_{Ha} 4.71 dd (10.6, 6.5), δ_{Hb} 4.53 dd (10.6, 4.4); δ_C 70.3], CH-3 [δ_H 4.32 dd (12.3, 5.5); δ_C 75.7], CH-4 (δ_H 4.21 m; δ_C 72.4), and CH-2′ (δ_H 4.58 m; δ_C 72.3), one amide group NH-CO [δ_H 8.54 d (9.2); δ_C 175.4], and one characteristic methine CH-2 [δ_H 5.26 m; δ_C 51.5] attached at the nitrogen atom (Table 2, Figures S32 and S33). Thus, the ^{1}H- and ^{13}C-NMR spectra of **8** exhibited the characteristic signals of a saturated phytosphingosine-type ceramide residue containing a 2-hydroxy FA (Figure 1). Moreover, the ceramide part of **8** had the normal and *anteiso*-types of side chains because the carbon atom signals of the terminal methyl groups were observed at δ_C 14.0 (normal form) and 11.3 and 19.2 (*anteiso*-form) in the ^{13}C-NMR spectrum (Table 2). The ^{1}H-^{1}H COSY and HSQC correlations in the NMR spectra of **8** indicated the corresponding sequences of protons at C-1 to C-5; C-16 to C-17 through C-15 and C-14; C-2 to NH; C-2′ to C-4′, and C-22′ to C-20′ (Table 2, Figures 2C, S34 and S35). The key HMBC cross-peaks such as Hb-1/C-2, C-3; H-2/C-1′; H-3/C-4; H-4/C-5; H_3-16/C-14, C-15; H_3-17/C-14; NH/C-2, C-1′; H-2′/C-1′; Ha-3′/C-1′, C-2′, C-4′; H_2-21′/C-20′; H_3-22′/C-21′ confirmed the common structure of the ceramide part of **8** (Figures 2C and S36).

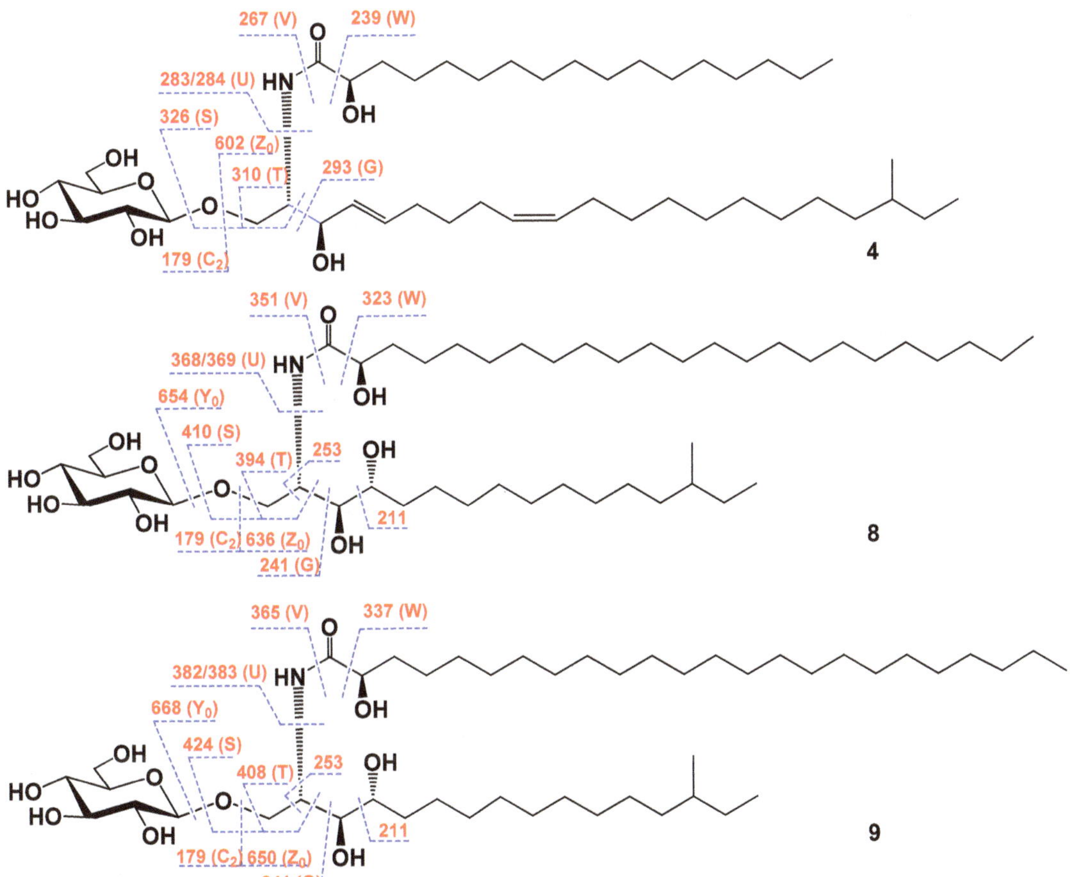

Figure 3. The key fragmentary peaks in (–)ESIMS/MS spectra of compounds **4**, **8**, and **9**.

A GC–MS analysis of FAME-5 showed the existence of one component that belonged to a saturated methyl 2-hydroxytricosanoate of the normal type. Based on this finding, as well as on the NMR and mass spectrometric data, we assumed that the LCB of the ceramide part of **8** has 17 carbon atoms and the *anteiso*-type of saturated polymethylene chain. The absolute configuration of LCB of the ceramide moiety of **8** is suggested to be D-*ribo*-(2*S*,3*S*,4*R*) on the basis of the similarity of its ^{1}H-NMR spectroscopic data with those of the LCB of ceramide **1**. The attachment of the monosaccharide to the ceramide part of **8** was deduced from long-range correlations in the HMBC spectrum. There were cross-peaks between H-1″ of Glc*p* and C-1 of aglycon, as well as between H-1 of the ceramide part and C-1″ of Glc*p* (Figures 2C and S36).

The presence of a monosaccharide unit and the structure of the ceramide part of cerebroside **8** were confirmed by ESIMS/MS data. In fact, the (−)ESIMS/MS spectrum of the molecular anion peak -M − H]$^-$ at *m*/*z* 816 showed fragmentary peaks obtained through the loss of a sugar unit at *m*/*z* 654 [-M − H)−162]$^-$ (Y_0-ion), *m*/*z* 636 [-M − H)−180]$^-$ (Z_0-ion) and 179 [hexo–e − H]$^-$ (C_2-ion), and also few characteristic fragmentary peaks due to the cleavage of the ceramide part of **8**: *m*/*z* 410 [-M − H)−406]$^-$, the loss of a monosaccharide residue and the cleavage of the bond between C-2 and C-3 (S-ion); *m*/*z* 394 [-M − H)−422]$^-$, the loss of a monosaccharide residue and the cleavage of the bond between C-2 and C-3 (T-ion); *m*/*z* 368/369 [-M - H)−448/449]$^-$, the cleavage of the bond between C-2 and NH (U-ion); *m*/*z* 351 [-M - H)−465]$^-$, the cleavage of the amide bond (V-ion); *m*/*z* 323 [-M - H)−493]$^-$, the cleavage of the bond between C-1′ and C-2′ (W-ion); *m*/*z* 253 [-M - H)−563]$^-$, the cleavage of the bonds between C-1 and C-2 and simultaneously between C-2 and NH; *m*/*z* 241 [-M - H)−575]$^-$, the cleavage of the bond between C-2 and C-3 (G-ion); and 211 [-M - H)−605]$^-$, the cleavage of the bond between C-3 and C-4 (Figure 3).

Hence, we determined the structure of cerebroside **8** to be (2*S*,3*S*,4*R*)-1-*O*-(β-D-glucopy ranosyl)-2-[(2*R*)-2-hydroxytricosanoylamino]-14-methylhexadecan-3,4-diol.

The IR spectrum of compound **9** showed the presence of hydroxyl (3401 cm^{-1}) and amide (1626, 1540 cm^{-1}) groups. The molecular formula of compound **9** was determined as $C_{47}H_{93}NO_{10}$ from the [M + Na]$^+$ sodium adduct ion peak at *m*/*z* 854.6687 in the (+)HRESIMS and the [M - H]$^-$ deprotonated molecular ion peak at *m*/*z* 830.6730 in the (-)HRESIMS (Figures S37 and S38). Based on a thorough 2D NMR analysis of cerebrosides **9** and **8**, we suggested that the ceramide part of **9** is almost identical to those of compound **8** (Figure S39–S43). However, a comparison of the molecular weights of **8** and **9** showed that they differ in MW by 14 amu. A GC–MS analysis of FAME-6 showed the presence of one component that was characterized as saturated methyl 2-hydroxytetracosanoate normal type.

The presence of a monosaccharide unit and the structure of the ceramide part of cerebroside **9** were confirmed by ESIMS/MS data. In fact, the (−)ESIMS/MS spectrum of the molecular anion peak -M - H]$^-$ at *m*/*z* 830 showed fragmentary peaks obtained through the loss of a sugar unit at *m*/*z* 668 [-M - H)−162]$^-$ (Y_0-ion), *m*/*z* 650 [-M - H)−180]$^-$ (Z_0-ion), and 179 [hexo–e - H]$^-$ (C_2-ion) and also few characteristic fragmentary peaks due to the cleavage of the ceramide part of **9**: *m*/*z* 424 [-M - H)−406]$^-$, the loss of a monosaccharide residue and the cleavage of the bond between C-2 and C-3 (S-ion); *m*/*z* 408 [-M - H)−422]$^-$, the loss of a monosaccharide residue and the cleavage of the bond between C-2 and C-3 (T-ion); *m*/*z* 382/383 [-M - H)−448/449]$^-$, the cleavage of the bond between C-2 and NH (U-ion); *m*/*z* 365 [-M - H)−465]$^-$, the cleavage of the amide bond (V-ion); *m*/*z* 337 [-M - H)−493]$^-$, the cleavage of the bond between C-1′ and C-2′ (W-ion); *m*/*z* 253 [-M - H)−577]$^-$, the cleavage of the bonds between C-1 and C-2 and simultaneously between C-2 and NH; *m*/*z* 241 [-M - H)−589]$^-$, the cleavage of the bond between C-2 and C-3 (G-ion); and 211 [-M - H)−619]$^-$, and the cleavage of the bond between C-3 and C-4 (Figure 3).

Thus, we determined the structure of cerebroside **9** to be (2*S*,3*S*,4*R*)-1-*O*-(β-D-glucopyr anosyl)-2-[(2*R*)-2-hydroxytetracosanoylamino]-14-methyl-hexadecan-3,4-diol.

2.2. *The Cytotoxic Activity of Compounds* **1–3**, **5–7**, *and* **9** *against Normal and Cancer Cells and their Effect on Colony Formation and Growth of Human Cancer Cells*

In the present study, the cytotoxic activity of compounds **1–3**, **5–7**, and **9** against human normal embryonic kidney cells HEK293 and a panel of human cancer cells HT-29, SK-MEL-28, and MDA-MB-231 was measured by the MTS assay after 24 h of exposure. Different concentrations of doxorubicin (10, 50, and 100 μM), used as a positive control, and compounds **1–3**, **5–7**, and **9** (1, 10, and 50 μM) were studied. As a result, it was found that compounds **1–3**, **5–7**, and **9** had moderate cytotoxic activity against HEK293, HT-29, and SK-MEL-28 (Table 3). These compounds slightly inhibited the viability of HEK293, HT-29, and SK-MEL-28 cells at concentrations of up to 50 μM (with a percentage of inhibition lower than 15%). Compounds **1**, **7**, and **9** also possessed slight cytotoxic activity against MDA-MB-231 cells at concentrations of up to 50 μM. The half maximal inhibitory concentration (IC_{50}) of compounds **2**, **3**, **5**, and **6** that caused inhibition of 50% cell viability was recorded only for breast carcinoma cells MDA-MB-231 and was comparable among the compounds under study (Table 3). The IC_{50} of doxorubicin (Doxo) was 35.7 μM, 21.8 μM, 40.0 μM, and 22.3 μM for HEK293, HT-29, SK-MEL-28, and MDA-MB-231 cell lines, respectively (Table 3).

Table 3. Cytotoxic activities of compounds **1–3**, **5–7**, and **9** against normal and cancer cells.

Compounds	Half Maximal Inhibitory Concentration (IC_{50}), μM			
	HEK293	HT-29	SK-MEL-28	MDA-MB-231
Doxorubicin	35.7 ± 1.2	21.8 ± 3.2	40.0 ± 5.0	22.3 ± 0.2
1	>50.0	>50.0	>50.0	>50.0
2	>50.0	>50.0	>50.0	48.7 ± 2.4
3	>50.0	>50.0	>50.0	49.4 ± 1.6
5	>50.0	>50.0	>50.0	48.7 ± 1.8
6	>50.0	>50.0	>50.0	40.5 ± 0.5
7	>50.0	>50.0	>50.0	>50.0
9	>50.0	>50.0	>50.0	>50.0

IC_{50} is the concentration of compounds that caused a 50% reduction in cell viability of tested normal and cancer cells. Values are mean ± standard deviation.

Since compounds **1–3**, **5–7**, and **9** inhibited the viability of breast cancer cells MDA-MB-231, we then tested their ability to inhibit colony formation of MDA-MB-231 cells using the soft agar assay. The colony formation assay, also referred to as soft agar assay, allows for screening the therapeutic efficacy of compounds for anchorage-independent cell growth, which is one of the hallmark characteristics of cellular transformation and uncontrolled growth of cancer cells [26].

As a result, we found that compounds **1**, **3**, **7**, and **9** at a concentration of 20 μM had a comparable effect on MDA-MB-231 colony formation and decreased the number of colonies by 46%, 48%, 44%, and 50%, respectively. Compounds **2**, **5**, and **6** (20 μM) inhibited colony formation of MDA-MB-231 cells by 68%, 54%, and 68%, respectively. The colony-inhibiting activity of compounds **2**, **5**, and **6** (20 μM) was comparable with the effect of doxorubicin that reduced the number of colonies by 70% at a concentration of 20 μM (Figure 4).

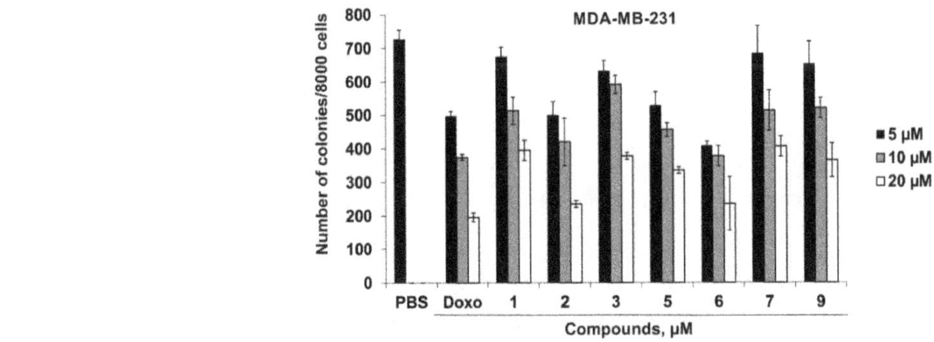

Figure 4. The effect of compounds **1–3**, **5–7**, and **9** on colony formation of human breast cancer cells MDA-MB-231 (2.4 × 10^4) that were exposed to PBS (control), Doxo (5, 10, and 20 μM), or the compounds under study (5, 10, and 20 μM) and placed on dishes with 0.3% Basal Medium Eagle (BME) agar containing 10% fetal bovine serum FBS, 2 mM L-glutamine, and 25 μg/mL gentamicin. After 14 days of incubation, the number of colonies was counted under a microscope using the ImageJ software program.

3. Materials and Methods

3.1. General Procedures

Optical rotations were determined on a PerkinElmer 343 polarimeter (Waltham, MA, USA). UV spectra were recorded on a Shimadzu UV-1601 PC spectrophotometer (Shimadzu, Kyoto, Japan). IR spectra were recorded using a Bruker Equinox 55 spectrophotometer in $CDCl_3$ (Bruker, Göttingen, Germany). The ^1H- and ^{13}C-NMR spectra were obtained on Bruker Avance III 700 spectrometer (Bruker BioSpin, Bremen, Germany) at 700.13 and 176.04 MHz, respectively; chemical shifts were referenced to the corresponding residual solvent signals (δ_H 7.21/δ_C 123.5 for C_5D_5N). The HRESIMS spectra were recorded on a Bruker Impact II Q-TOF mass spectrometer (Bruker, Bremen, Germany); the samples were dissolved in MeOH (c 0.001 mg/mL). HPLC separations were carried out on an Agilent 1100 Series chromatograph (Agilent Technologies, Santa Clara, CA, USA) equipped with a differential refractometer and with the following columns used: Diasfer-110-C18 (10 μm, 250 × 15 mm, Biochemmack, Moscow, Russia), Discovery HS C18-10 (10 μm, 250 × 21.2 mm, Supelco, North Harrison, PA, USA), and Discovery C18 (5 μm, 250 × 4 mm, Supelco, North Harrison, PA, USA). GC and GC–MS analyses were performed on a GC 2010 chromatograph equipped with a flame ionization detector and a gas chromatograph–mass spectrometer GCMS-QP5050, both Shimadzu (Kioto, Japan), and with fused silica capillary columns Supelcowax 10 and MDN-5S (both columns 30 m, 0.25 mm ID, 0.25 lm film, Supelco, USA). Low-pressure liquid column chromatography was carried out using Polychrom-1 (powdered Teflon, 0.25–0.50 mm; Biolar, Olaine, Latvia), Si gel KSK (50–160 μm, Sorbpolimer, Krasnodar, Russia), and Florisil (60–100 μm, Sigma Aldrich, St. Louis, MO, USA). Sorbfil Si gel plates (4.5 × 6.0 cm, 5–17 μm, Sorbpolimer, Krasnodar, Russia) were used for thin-layer chromatography.

3.2. Animal Material

Specimens of *Ceramaster patagonicus* Sladen, 1889 (order Valvatida, family Goniasteridae) were collected at a depth of 150–300 m in the Sea of Okhotsk, off Iturup Island, during the 42nd research cruise aboard the R/V *Akademik Oparin* in August 2012. The species was identified by B.B. Grebnev (G.B. Elyakov Pacific Institute of Bioorganic Chemistry FEB RAS, Vladivostok, Russia). A voucher specimen [no. 042-67] is deposited at the marine specimen collection of the G.B. Elyakov Pacific Institute of Bioorganic Chemistry FEB RAS, Vladivostok, Russia.

3.3. Extraction and Isolation

The fresh *C. patagonicus* specimens (3 kg wet weight) were cut into small pieces and extracted with $CHCl_3$:MeOH (2:1) followed by further extraction with $CHCl_3$:MeOH (1:1) and EtOH. The combined extracts were concentrated in vacuo to a residue of 159.5 g. This residue was separated between H_2O (1.5 L) and AcOEt:BuOH (2:1) (4.5 L), and the organic layer was concentrated *in vacuo* to obtain a less polar fraction (51.5 g), which was washed with cold acetone (1 L). The acetone-soluble fraction (28.5 g) was chromatographed on a Si gel column (19 × 4.5 cm) using $CHCl_3$, $CHCl_3$:MeOH (97:3), and $CHCl_3$:MeOH (9:1) to yield four fractions: 1 (932 mg), 2 (486 mg), 3 (735 mg), and 4 (1.04 g). Fractions 1–4 were further chromatographed on a Si gel column (10 × 4 cm) using *n*-hexane:AcOEt:MeOH (stepwise gradient, 6:3:0.1→6:3:0.7, v/v/v) to yield six subfractions: 21 (123 mg), 31 (475 mg), 32 (231 mg), 41 (55 mg), 42 (212 mg), and 43 (570 mg), which were then analyzed by TLC in the eluent system $CHCl_3$:MeOH:H_2O (8:1:0.1, v/v/v). Subfractions 21–43 mainly contained ceramides, cerebrosides, admixtures of pigments, and other concomitant lipids. A HPLC separation of subfraction 31 (475 mg) on a Diasfer-110-C18 column (2.5 mL/min) with MeOH as an eluent yielded pure **2** (12.0 mg, R_t 68.2 min) and seventeen subfractions 31-3–31-15 and 31-17–31-20. A HPLC separation of subfractions 31-14 and 31-18 on a Discovery C18 column (2.5 mL/min) with MeOH as an eluent yielded pure **1** (2.0 mg, R_t 15.1 min) and unseparated mixture of **3** and **2** (1.5 mg, R_t 17.8 min). HPLC separation of subfraction 42 (212 mg) on a Discovery HS C18-10 column (4.0 mL/min) with MeOH as an eluent yielded pure **5** (3.0 mg, R_t 133.2 min), **6** (4.5 mg, R_t 153.2 min), and thirteen subfractions 42-4–42-15 and 42-17. A HPLC separation of subfractions 42-12, 42-13, 42-15, and 42-17 on a Discovery C18 column (2.5 mL/min) with MeOH as an eluent yielded pure **4** (1.0 mg, R_t 15.9 min), **7** (2.5 mg, R_t 18.9 min), **8** (0.5 mg, R_t 21.7 min), and **9** (1.5 mg, R_t 26.9 min).

3.4. Compounds Characterization Data

(2S,3S,4R,9Z)-2-[(2R)-2-hydroxyhexadecanoylamino]-19-methyl-9-icosen-1,3,4-triol (**1**): Amorphous powder; $[\alpha]_D^{25}$: +15.4 (*c* 0.1, MeOH); IR ($CDCl_3$) ν_{max} 3402, 2934, 2920, 2855, 1722, 1656, 1625, 1523, 1493, 1462, 1365, 1274, 1186, 1130, 1080 cm^{-1}; (+)HRESIMS *m/z* 634.5376 [M + Na]$^+$ (calcd for $[C_{37}H_{73}NO_5Na]^+$, 634.5381); (–)HRESIMS *m/z* 610.5412 [M – H]$^-$ (calcd for $[C_{37}H_{72}NO_5]^-$, 610.5416); ^{1}H- and ^{13}C-NMR data, see Table 1.

(2S,3S,4R,9Z)-2-[(2R)-2-hydroxyhexadecanoylamino]-21-methyl-9-docosen-1,3,4-triol (**2**): Amorphous powder; $[\alpha]_D^{25}$: +10.8 (*c* 0.1, MeOH); IR ($CDCl_3$) ν_{max} 3404, 2937, 2922, 2853, 1724, 1657, 1625, 1522, 1495, 1460, 1365, 1275, 1187, 1131, 1080 cm^{-1}; (+)HRESIMS *m/z* 662.5324 [M + Na]$^+$ (calcd for $[C_{39}H_{77}NO_5Na]^+$, 662.5694); (–)HRESIMS *m/z* 638.5364 [M – H]$^-$ (calcd for $[C_{39}H_{76}NO_5]^-$, 638.5729); ^{1}H- and ^{13}C-NMR data, see Table 1.

(2S,3S,4R,9Z)-2-[(2R)-2-hydroxyheptadecanoylamino]-21-methyl-9-docosen-1,3,4-triol (**3**) as mixed with **2** (ratio 2:1): Amorphous powder; $[\alpha]_D^{25}$: +10.0 (*c* 0.1, MeOH); IR ($CDCl_3$) ν_{max} 3407, 2932, 2927, 2855, 1728, 1655, 1624, 1522, 1495, 1464, 1366, 1278, 1183, 1130, 1081 cm^{-1}; (+)HRESIMS *m/z* 676.5463 [M + Na]$^+$ (calcd for $[C_{40}H_{79}NO_5Na]^+$, 676.5850); (–)HRESIMS *m/z* 652.5520 [M – H]$^-$ (calcd for $[C_{40}H_{78}NO_5]^-$, 652.5885); ^{1}H- and ^{13}C-NMR data, see Table 1.

(2S,3R,4E,9Z)-1-O-(β-D-glucopyranosyl)-2-[(2R)-2-hydroxyheptadecanoylamino]-19-methyl-4,9-henicosadien-3-ol (**4**): Amorphous powder; $[\alpha]_D^{25}$: –9.7 (*c* 0.05, MeOH); IR ($CDCl_3$) ν_{max} 3383, 2927, 2854, 1718, 1649, 1602, 1538, 1461, 1366, 1261, 1098, 1078, 1034 cm^{-1}; (+)HRESIMS *m/z* 806.6110 [M + Na]$^+$ (calcd for $C_{45}H_{85}NO_9Na$, 806.6117); (–)HRESIMS *m/z* 782.6154 [M – H]$^-$ (calcd for $C_{45}H_{84}NO_9$, 782.6152); (–)ESIMS/MS of the ion at *m/z* 782: *m/z* 602 [–M – H)–180]$^-$, 326 [–M – H)–456]$^-$, 310 [–M – H)–472]$^-$, 293 [–M – H)–489]$^-$, 283/284 [–M – H)–499/500]$^-$, 267 [–M – H)–515]$^-$, 239 [–M – H)–543]$^-$, 179 [hexo–e – H]$^-$; ^{1}H- and ^{13}C-NMR data, see Table 2.

(2S,3S,4R)-1-O-(β-D-glucopyranosyl)-2-[(2R)-2-hydroxytricosanoylamino]-14-methylhexadecan-3,4-diol (**8**): Amorphous powder; $[\alpha]_D^{25}$: –14.0 (*c* 0.1, MeOH); IR ($CDCl_3$) ν_{max} 3401, 2925, 2854, 1732, 1634, 1602, 1541, 1457, 1363, 1261, 1078, 1018 cm^{-1}; (+)HRESIMS *m/z* 840.6336 [M + Na]$^+$ (calcd for $C_{46}H_{91}NO_{10}Na$, 840.6335); (–)HRESIMS *m/z* 816.6574 [M – H]$^-$ (calcd for $C_{46}H_{90}NO_{10}$, 816.6570); (–)ESIMS/MS of the ion at *m/z* 816: *m/z* 654 [–M – H)–162]$^-$, 636 [–M – H)–180]$^-$, 410 [–M – H)–406]$^-$, 394 [–M – H)–422]$^-$, 368/369 [–M – H)–448/449]$^-$, 351 [–M – H)–465]$^-$, 323 [–M – H)–493]$^-$, 253 [–M – H)–563]$^-$, 241 [–M – H)–575]$^-$, 211 [–M – H)–605]$^-$, 179 [hexo–e – H]$^-$; ^{1}H- and ^{13}C-NMR data, see Table 2.

(2S,3S,4R)-1-O-(β-D-glucopyranosyl)-2-[(2R)-2-hydroxytetracosanoylamino]-14-methylhexadecan-3,4-diol (**9**): Amorphous powder; $[\alpha]_D^{25}$: –17.5 (*c* 0.1, MeOH); IR ($CDCl_3$) ν_{max} 3401, 2925, 2854, 1731, 1626, 1602, 1540, 1455, 1297, 1103, 1078, 1029 cm^{-1}; (+)HRESIMS *m/z* 854.6687 [M + Na]$^+$ (calcd for $C_{47}H_{93}NO_{10}Na$, 854.6692); (–)HRESIMS *m/z* 830.6730 [M – H]$^-$ (calcd for $C_{47}H_{92}NO_{10}$, 830.6727); (–)ESIMS/MS of the ion at *m/z* 830: *m/z* 668 [–M – H)–162]$^-$, 650 [–M – H)–180]$^-$, 424 [–M – H)–406]$^-$, 408 [–M – H)–422]$^-$, 382/383 [–M – H)–448/449]$^-$, 365 [–M – H)–465]$^-$, 337 [–M – H)–493]$^-$, 253 [–M – H)–577]$^-$, 241 [–M – H)–589]$^-$, 211 [–M – H)–619]$^-$, 179 [hexo–e – H]$^-$; ^{1}H- and ^{13}C-NMR data, see Table 2.

3.5. Methanolysis of Compounds 1–9 and Analysis of FAMEs

Compounds **1–9** (1 mg) were heated with 1 N HCl in 80% aqus. MeOH (1.0 mL) at 80°C for 4 h. The reaction mixtures were then extracted with *n*-hexane and the extracts were concentrated in vacuo to yield FAME-1–FAME-9. The FAMEs were analyzed on Supelcowax 10 columns at 200 °C. Helium was used as the carrier gas at a linear velocity of 30 cm/s. Mass spectra were recorded at 70 eV. The obtained mass spectra were compared with the NIST library and a FA mass spectra archive accessible online.

3.6. Acid Hydrolysis and Determination of Absolute Configurations of Monosaccharides

The acid hydrolysis of **4** (0.5 mg) was carried out in a solution of 2 M trifluoroacetic acid (TFA) (1 mL) in a sealed vial on an H_2O bath at 100 °C for 2 h. The H_2O layer was washed with $CHCl_3$ (3 × 1.0 mL) and concentrated in vacuo. One drop of concentrated TFA and 0.5 mL of *R*-(−)-2-octanol (Sigma Aldrich) were added to the sugar fraction, and the sealed vial was heated in a glycerol bath at 130 °C for 6 h. The solution was evaporated in vacuo and exposed to a mixture of pyridine/acetic anhydride (1:1, 0.5 mL) for 24 h at room temperature. The acetylated 2-octylglycosides were analyzed by GC using the corresponding authentic samples prepared by the same procedure. The following peaks were detected in the hydrolysate of **4**: D-glucose (t_R 24.24, 24.84, 25.08, and 25.38 min). The retention times of the authentic samples were as follows: D-glucose (t_R 24.23, 24.83, 25.06, and 25.37 min), L-glucose (t_R 24.39, 24.63, 24.83, and 25.06 min).

3.7. Bioactivity Assay

3.7.1. Reagents

The McCoy's 5A Modified Medium (McCoy's 5A), the Dulbecco's Modified Eagle's Medium (DMEM), Basal Medium Eagle (BME), phosphate-buffered saline (PBS), L-glutamine, penicillin–streptomycin solution (10 000 U/mL, 10 μg/mL) and trypsin were all purchased from Sigma-Aldrich (St. Louis, MO, USA). The MTS reagent was purchased from Promega (Madison, WI, USA). Fetal bovine serum (FBS), agar and gentamicin were purchased from ThermoFisher Scientific (Waltham, Massachusetts, USA).

3.7.2. Cell Lines

The human embryonic kidney cells HEK293 (ATCC® CRL-1573™), the colorectal adenocarcinoma cell line HT-29 (ATCC® HTB-38), the melanoma SK-MEL-28 (ATCC® HTB-72™), and breast adenocarcinoma MDA-MB-231 (ATCC® HTB-26™) cell lines were obtained from the American Type Culture Collection (Manassas, VA, USA).

3.7.3. Cells Culture Conditions

HT 29 cells were cultured in McCoy's 5A medium; HEK293, SK-MEL-28, and MDA-MB-231 cell lines were maintained in the DMEM medium. The culture media were supplemented with 10% FBS and a 1% penicillin/streptomycin solution. The cell cultures were maintained at 37 °C in a humidified atmosphere containing 5% CO_2.

3.7.4. Preparation of Compounds

Compounds **1–3**, **5–7**, and **9** were dissolved in a sterile dimethyl sulfoxide solution (DMSO) to prepare stock concentrations of 20 mM. Cells were exposed to serially diluted **1–3**, **5–7**, and **9** (1–100 μM) (with the culture medium used as a diluent) (the final concentration of DMSO was less than 0.5%).

Doxorubicin (Doxo) (Teva Pharmaceutical Industries, Ltd. (Israel)) was dissolved in sterile PBS to prepare stock concentrations of 10 mM. Cells were exposed to serially diluted Doxo (10–100 μM) (with the culture medium used as a diluent).

The vehicle control is the cells exposed to the equivalent volume of DMSO (the final concentration was less than 0.5%) for all of the experiments conducted.

3.7.5. Cell Viability Assay

The effect of compounds **1–3**, **5–7**, and **9** on the viability of tested cell lines was evaluated by the MTS assay. HEK293 (1.0 × 10^4/200 μL), HT-29 (1.0 × 10^4/200 μL), SK-MEL-28 (0.8 × 10^4/200 μL), and MDA-MB-231 (1.0 × 10^4/200 μL) cells were seeded on a 96-well plate and incubated for 24 h at 37°C in a CO_2 incubator. The cells were exposed to either DMSO (vehicle control) or Doxo at concentrations of 1, 10, 50 μM (positive control) or **1–3**, **5–7**, and **9** at concentrations of 1, 10, and 50 μM for 24 h. The cells were subsequently incubated with 15 μL MTS reagent for 3 h, and the absorbance of each well was measured at 490/630 nm on a Power Wave XS microplate reader (BioTek, Winooski, VT, USA). The concentration at which a compound exerted half of its maximal inhibitory effect on cell viability (IC_{50}) was calculated using the AAT-Bioquest® online calculator [27].

3.7.6. Colony Formation Assay

MDA-MB-231 cells (2.4×10^4/mL) were exposed to either DMSO (vehicle control) or Doxo (positive control) at concentrations of 5, 10, and 20 μM, and to **1–3**, **5–7**, and **9** at concentrations of 5, 10, 20 μM. Then, the cells were applied on dishes with 0.3% BME agar containing 10% FBS, 2 mM L-glutamine, and 25 μg/mL gentamicin. The cultures were maintained at 37°C in a 5% CO_2 incubator for 14 days. The number and size of the colonies were estimated under a Motic microscope AE 20 and using the ImageJ software bundled with 64-bit Java 1.8.0_112 (NIH, Bethesda, Maryland, USA).

3.7.7. Statistical Analysis

All assays were performed in triplicate. Results are presented as mean ± standard deviation (SD).

4. Conclusions

Three new ceramides (**1–3**) and three new cerebrosides (**4**, **8**, and **9**) along with three previously known cerebrosides—ophidiacerebrosides C (**5**), D (**6**), and CE-3-2 (**7**)—were isolated from a deep-sea starfish species, the orange cookie star *Ceramaster patagonicus*. Ceramides **1–3** contain *iso*-C_{21} or C_{23} Δ^9-phytosphingosine as LCB and have C_{16} or C_{17} (2*R*)-2-hydroxy-fatty acids of the normal type. As far as we know, ceramides with the *iso*-type of LCB were isolated from this starfish for the first time. It is also worth noting that starfish-derived ceramides are the least studied class of sphingolipids. This may be due to the challenge of isolating certain components from complex mixtures of ceramides and other lipids. However, new data on the structures of starfish ceramides allow a better understanding of the biosynthetic features of these animals. Furthermore, these ceramides can also be used as chemotaxonomic markers.

Cerebroside **4** contains C_{22} Δ^9-sphingosine of *anteiso*-type and (2*R*)-2-hydroxyheptadecanoic acid of the normal type. The C_{22} Δ^9-sphingosine of *anteiso*-type was found in starfish cerebrosides for the first time. New compounds **8** and **9** contain saturated C-17 phytosphingosine of *anteiso*-type as LCB and differ from each other in the length of the polymethylene chain of (2*R*)-2-hydroxy-fatty acids: C_{23} at **8** and C_{24} at **9**. All the isolated cerebrosides have β-D-glucopyranose as a monosaccharide residue. It is also worth noting that all the isolated ceramides and cerebrosides have only (2*R*)-2-hydroxy-fatty acids of the normal type. As far as we know, the composition of neutral sphingolipids of the starfish *C. patagonicus* was described for the first time.

Compounds **1–3**, **5–7**, and **9** under study exhibit slight or moderate cytotoxic activity against HEK293, HT-29, SK-MEL-28, and MDA-MB-231 cells. On the other hand, compounds **1**, **3**, **7**, and **9** at a non-toxic concentration of 20 μM significantly decreased the number of colonies of MDA-MB-231 cells. The colony-inhibiting activity of compounds **2**, **5**, and **6** is comparable to the anticancer effect of doxorubicin. Ophidiacerebrosides C (**5**) and D (**6**) with (4*E*,8*E*,10*E*)-9-methylsphinga-4,8,10-trienine as LCB showed the highest cytotoxic and colony formation inhibitory effects among the cerebrosides analyzed. These data agree well with the results of published studies where these cerebrosides showed strong cytotoxic activity at a concentration of 2 μM against murine leukemic cells L1210 and moderate cytotoxic activity against a number of human cancer cell lines at concentrations of 15–34 μM [21,28]. The inhibition of cancer cell colony formation by starfish sphingolipids was shown for the first time.

Supplementary Materials: The following supporting information can be downloaded at: https://www.mdpi.com/article/10.3390/md20100641/s1, copies of the HRESIMS (Figures S1, S2, S9, S10, S16, S17, S23, S24, S30, S31, S37, and S38), 1H-NMR (Figures S3, S11, S18, S25, S32, and S39), 13C-NMR (Figures S4, S12, S19, S26, S33, and S40), 1H-1H-COSY (Figures S5, S13, S20, S27, S34, and S41), HSQC (Figures S6, S14, S21, S28, S35, and S42), and HMBC (Figures S7, S15, S22, S29, S36, and S43) spectra of compounds 1, 2, 3, 4, 8, and 9, respectively. This material is available free online.

Author Contributions: T.V.M.: Conceptualization, methodology (chemical), writing—original draft. V.M.Z.: Methodology (chemical). A.A.K.: writing—review and editing. A.S.K.: Methodology (biological). O.S.M.: Conceptualization, methodology (biological), writing—review and editing. A.I.K.: Methodology (NMR). R.S.P.: Methodology (mass spectrometry). V.I.S.: Methodology (chemical). N.V.I.: Writing—review and editing, supervision. All authors have read and agreed to the published version of the manuscript.

Funding: This study was supported by the Grant from the Ministry of Science and Higher Education of the Russian Federation No. 13.1902.21.0012 (contract No. 075-15-2020-796).

Institutional Review Board Statement: Not applicable.

Data Availability Statement: The data presented in this study are available on request from the corresponding authors.

Acknowledgments: The study was carried out using the equipment of the Collective Facilities Center "The Far Eastern Center for Structural Molecular Research (NMR/MS) of PIBOC FEB RAS". We are grateful to B.B. Grebnev (G.B. Elyakov Pacific Institute of Bioorganic Chemistry FEB RAS, Vladivostok, Russia) for the identification of the starfish species.

Conflicts of Interest: The authors declare no conflict of interest.

References

1. Minale, L.; Riccio, R.; Zollo, F. Steroidal oligoglycosides and polyhydroxysteroids from Echinoderms. *Fortschr. Chem. Org. Nat.* **1993**, *62*, 75–308. [CrossRef]
2. Stonik, V.A. Marine polar steroids. *Russ. Chem. Rev.* **2001**, *70*, 673–715. [CrossRef]
3. Iorizzi, M.; De Marino, S.; Zollo, F. Steroidal oligoglycosides from the Asteroidea. *Curr. Org. Chem.* **2001**, *5*, 951–973. [CrossRef]
4. Stonik, V.A.; Ivanchina, N.V.; Kicha, A.A. New polar steroids from starfish. *Nat. Prod. Commun.* **2008**, *3*, 1587–1610. [CrossRef]
5. Dong, G.; Xu, T.H.; Yang, B.; Lin, X.P.; Zhou, X.F.; Yang, X.W.; Liu, Y.H. Chemical constituents and bioactivities of starfish. *Chem. Biodivers.* **2011**, *8*, 740–791. [CrossRef]
6. Ivanchina, N.V.; Kicha, A.A.; Stonik, V.A. Steroid glycosides from marine organisms. *Steroids* **2011**, *76*, 425–454. [CrossRef] [PubMed]
7. Ivanchina, N.V.; Kicha, A.A.; Malyarenko, T.V.; Stonik, V.A. *Advances in Natural Products Discovery*; Gomes, A.R., Rocha-Santos, T., Duarte, A., Eds.; Nova Science Publishers: Hauppauge, NY, USA, 2017; Volume 6, pp. 191–224.
8. Stonik, V.A.; Kicha, A.A.; Malyarenko, T.V.; Ivanchina, N.V. Asterosaponins: Structures, taxonomic distribution, biogenesis and biological activities. *Mar. Drugs* **2020**, *18*, 584. [CrossRef]
9. Xia, J.M.; Miao, Z.; Xie, C.L.; Zhang, J.W.; Yang, X.W. Chemical constituents and bioactivities of starfish: An update. *Chem. Biodivers.* **2020**, *17*, e1900638. [CrossRef] [PubMed]
10. Muralidhar, P.; Radhika, P.; Krishna, N.; Rao, D.V.; Rao, C.B. Sphingolipids from marine organisms: A review. *Nat. Prod. Sci.* **2003**, *9*, 117–142.
11. Higuchi, R.; Inagaki, M.; Yamada, K.; Miyamoto, T. Biologically active gangliosides from echinoderms. *J. Nat. Med.* **2007**, *61*, 367–370. [CrossRef]
12. Yamada, K. Chemo-pharmaceutical studies on the glycosphingolipid constituents from Echinoderm, sea cucumbers, as the medicinal materials. *Yakugaku Zasshi* **2002**, *122*, 1133–1143. [CrossRef] [PubMed]
13. Inagaki, M. Structure and biological activity of glycosphingolipids from starfish and feather star. *Yakugaku Zasshi* **2008**, *128*, 1187–1194. [CrossRef] [PubMed]
14. Malyarenko, T.V.; Kicha, A.A.; Stonik, V.A.; Ivanchina, N.V. Sphingolipids of Asteroidea and Holothuroidea: Structures and biological activities. *Mar. Drugs* **2021**, *19*, 330. [CrossRef]
15. Degroote, S.; Wolthoorn, J.; van Meer, G. The cell biology of glycosphingolipids. *Semin. Cell Dev. Biol.* **2004**, *15*, 375–387. [CrossRef]
16. Hakomori, S.; Igarashi, Y. Functional role of glycosphingolipids in cell recognition and signaling. *J. Biochem.* **1995**, *118*, 1091–1103. [CrossRef]
17. Sperling, P.; Heinz, E. Plant sphingolipids: Structural diversity, biosynthesis, first genes and functions. *Biochem. Biophys. Acta* **2003**, *1632*, 1–15. [CrossRef]

18. Ishii, T.; Okino, T.; Mino, Y. A ceramide and cerebroside from the starfish *Asterias amurensis* Lütken and their plant-growth promotion activities. *J. Nat. Prod.* **2006**, *69*, 1080–1082. [CrossRef]
19. Fischer, C.L.; Drake, D.R.; Dawson, D.V.; Blanchette, D.R.; Brogden, K.A.; Wertz, P.W. Antibacterial activity of sphingoid bases and fatty acids against gram-positive and gram-negative bacteria. *Antimicrob. Agents Chemother.* **2012**, *56*, 1157–1161. [CrossRef]
20. Malyarenko, T.V.; Kicha, A.A.; Malyarenko, O.S.; Zakharenko, V.M.; Kotlyarov, I.P.; Kalinovsky, A.I.; Popov, R.S.; Svetashev, V.I.; Ivanchina, N.V. New conjugates of polyhydroxysteroids with long-chain fatty acids from the deep-water Far Eastern starfish *Ceramaster patagonicus* and their anticancer activity. *Mar. Drugs* **2020**, *18*, 260. [CrossRef]
21. Jin, W.; Rinehart, K.L.; Jares-Erijman, E.A. Ophidiacerebrosides: Cytotoxic glycosphingolipids containing a novel sphingosine from a sea star. *J. Org. Chem.* **1994**, *59*, 144–147. [CrossRef]
22. Yamada, K.; Hara, E.; Miyamoto, T.; Higuchi, R.; Isobe, R.; Honda, S. Constituents of Holothuroidea, 6-Isolation and structure of biologically active glycosphingolipids from the sea cucumber *Cucumaria echinata*. *Eur. J. Org. Chem.* **1998**, 371–378. [CrossRef]
23. Higuchi, R.; Natori, T.; Komori, T. Biologically active glycosides from Asteroidea, XX. Glycosphingolipids from the starfish *Asterina pectinifera*, 1. Isolation and characterization of acanthacerebroside B and structure elucidation of related, nearly homogeneous cerebrosides. *Liebigs Ann. Chem.* **1990**, *1*, 51–55. [CrossRef]
24. De Haan, J.W.; van de Ven, L.J.M. Configurations and Conformations in Acyclic, Unsaturated Hydrocarbons. A ^{13}C NMR Study. *Magn. Reson Chem.* **1973**, *5*, 147–153. [CrossRef]
25. Sugiyama, S.; Honda, M.; Komori, T. Biologically active glycosides from Asteroidea, XXIV 1). Stereochemistry of the four diastereomers of phytosphingosine. *Liebigs Ann. Chem.* **1990**, *11*, 1069–1078. [CrossRef]
26. Horibata, S.; Vo, T.V.; Subramanian, V.; Thompson, P.R.; Coonrod, S.A. Utilization of the soft agar colony formation assay to identify inhibitors of tumorigenicity in breast cancer cells. *J. Vis. Exp.* **2015**, *99*, e52727. [CrossRef]
27. AAT Bioquest. Available online: https://www.aatbio.com/tools/ic50-calculator (accessed on 10 November 2020).
28. Kawatake, S.; Nakamura, K.; Inagaki, M.; Higuchi, R. Isolation and structure determination of six glucocerebrosides from the starfish *Luidia maculata*. *Chem. Pharm. Bull.* **2002**, *50*, 1091–1096. [CrossRef]

marine drugs

MDPI

Article

Carotenoids from Starfish *Patiria pectinifera*: Therapeutic Activity in Models of Inflammatory Diseases

Aleksandr M. Popov [1,*,†], Emma P. Kozlovskaya [1,*,†], Anna A. Klimovich [1], Tatyana A. Rutckova [1], Aleksey I. Vakhrushev [1], Dmitry M. Hushpulian [2,3], Irina G. Gazaryan [2,4,5], Vyacheslav V. Makhankov [1], Oksana M. Son [6,7] and Liudmila A. Tekutyeva [6,7]

1 G.B. Elyakov Pacific Institute of Bioorganic Chemistry, Far Eastern Branch of the Russian Academy of Science, 159 Prospect 100-letiya Vladivostoka, Vladivostok 690022, Russia; anka_zaraza13@mail.ru (A.A.K.); tanya1119@yandex.ru (T.A.R.); aivahr@mail.ru (A.I.V.); mvvslav@mail.ru (V.V.M.)
2 Faculty of Biology and Biotechnology, National Research University Higher School of Economics, 13/4 Myasnitskaya str., Moscow 117997, Russia; Hushpulian@gmail.com (D.M.H.); igazaryan@gmail.com (I.G.G.)
3 Bach Institute of Biochemistry, Federal Research Centre "Fundamentals of Biotechnology" of the Russian Academy of Sciences, Leninski prospect 33, Moscow 1190721, Russia
4 Department of Chemical Enzymology, M.V. Lomonosov Moscow State University, Moscow 119991, Russia
5 Department of Chemistry and Physical Sciences, Dyson College of Art and Sciences, Pace University, 861 Bedford Road, Pleasantville, NY 10570, USA
6 School of Advanced Engineering Studies, Institute of Biotechnology, Bioengineering and Food Systems, Far Eastern Federal University, p. Ajax 10, Russky Island, Vladivostok 690922, Russia; oksana_son@bk.ru (O.M.S.); lat7777@mail.ru (L.A.T.)
7 ARNIKA, Territory of PDA Nadezhdinskaya, Volno-Nadezhdinskoye 692481, Russia
* Correspondence: popovam@piboc.dvo.ru (A.M.P.); kozempa@mail.ru (E.P.K.); Tel.: +7-950-298-4684 (A.M.P.); +7-914-677-4202 (E.P.K.)
† These authors contributed equally to this work.

Citation: Popov, A.M.; Kozlovskaya, E.P.; Klimovich, A.A.; Rutckova, T.A.; Vakhrushev, A.I.; Hushpulian, D.M.; Gazaryan, I.G.; Makhankov, V.V.; Son, O.M.; Tekutyeva, L.A. Carotenoids from Starfish *Patiria pectinifera*: Therapeutic Activity in Models of Inflammatory Diseases. *Mar. Drugs* **2023**, *21*, 470. https://doi.org/10.3390/md21090470

Academic Editors: Jose Maria Vega, Vladimir I. Kalinin, Pavel S. Dmitrenok and Natalia V. Ivanchina

Received: 14 July 2023
Revised: 20 August 2023
Accepted: 25 August 2023
Published: 27 August 2023

Abstract: The carotenoids mixture (MC) isolated from the starfish *Patiria. pectinifera* contains more than 50% astaxanthin, 4–6% each zeaxanthine and lutein, and less pharmacologically active components such as free fatty acids and their glycerides. Astaxanthin, the major component of MC, belongs to the xanthophyll class of carotenoids, and is well known for its antioxidant properties. In this work, in vitro and in vivo studies on the biological activity of MC were carried out. The complex was shown to exhibit anti-inflammatory, anti-allergic and cancer-preventive activity, without any toxicity at a dose of 500 mg/kg. MC effectively improves the clinical picture of the disease progressing, as well as normalizing the cytokine profile and the antioxidant defense system in the in vivo animal models of inflammatory diseases, namely: skin carcinogenesis, allergic contact dermatitis (ACD) and systemic inflammation (SI). In the skin carcinogenesis induced by 7,12-dimethylbenzanthracene, the incidence of papillomas was decreased 1.5 times; 1% MC ointment form in allergic contact dermatitis showed an 80% reduced severity of pathomorphological skin manifestations. Obtained results show that MC from starfish *P. pectinifera* is an effective remedy for the treatment and prevention of inflammatory processes.

Keywords: carotenoids mixture (MC); starfish *Patiria pectinifera*; astaxanthin; therapeutic activity; inflammatory diseases

1. Introduction

Currently, natural biologically active substances (BASs), which have beneficial properties for human health, are of great interest to the consumer. A special place among these BASs is occupied by carotenoids (CRs), which are characterized by a high potential for protecting the body from a wide range of diseases, such as chronic inflammatory [1–4], neurodegenerative [5–9], cardiovascular [10–13], cancer [14–17], metabolic syndrome and diabetes [18–21], liver [22,23], skin [24–26] and eye [27] diseases, exercise-induced fatigue [21], infertility [28,29] and many other pathologies. Carotenoids are of particular interest as an adjuvant drug to reduce the cytokine storm, including those associated with COVID-19 [30,31]. It has been established that astaxanthin blocks oxidative damage to DNA, reduces the level of C-reactive protein and other biomarkers of inflammation [1,30–32]. Eating a diet rich in carotenoids has been shown to combat inflammation by increasing plasma concentrations of IFN-α2 and decreasing MIP-1β and TNF-α in human trials [1].

The CRs family is represented by more than 850 natural fat-soluble pigments, which are synthesized by algae, phytoplankton, plants, and some varieties of fungi and bacteria [33–35]. CRs are responsible for the coloration of various photosynthetic organisms [36]. Animals and humans are not able to synthesize CRs de novo; they obtain them with food [37–39] and actively use the metabolic capabilities of CRs [40]. For example, β-carotene is converted in the body to vitamin A, whereas lutein and zeaxanthin are macular pigments protecting the retina from damage by UV light [27,41].

As suggested [42,43], the antioxidant properties of CRs are the main mechanism by which they maintain and improve human health. Marine CRs (astaxanthin, fucoxanthin, β-carotene, lutein, as well as the rare siphonaxanthin, sioxanthin, and myxol) have recently demonstrated antioxidant properties in reducing markers of oxidative stress [1,44,45]. However, the interpretation of some results remains conflicting. It is difficult to explain CRs' physiological effects only by their antioxidant activity; therefore more research is necessary to clarify the relationship between CRs' action and ROS-mediated disorders. Over the past few years, genomic studies have focused on the unusual ability of CRs to regulate the expression of specific genes involved in cell metabolism [1,44,45].

To date, a large body of data has been collected in a number of epidemiological, interventional, and clinical studies, mainly from experiments with β-carotene, lycopene, astaxanthin, lutein, and zeaxanthin, which generally support the observation that an adequate intake of CR-rich fruits and vegetables or CR supplements can significantly reduce the risk of certain chronic diseases. The World Health Organization estimates that low fruit and vegetable intake is responsible for 31% of coronary heart disease, 11% of strokes, and 19% of gastrointestinal cancers worldwide.

Preparations based on echinoderm extracts have not yet been widely used in modern medicine, but are actively used in Chinese medicine for the prevention and treatment of a wide range of diseases. Increasingly, scientific works began to appear showing the effectiveness of the use of echinoderm extracts in the treatment of various diseases. In this regard, the research on the composition of metabolites of this type of animal is of great scientific interest [46,47].

Patiria (=Asterina) pectinifera [48] is an unpretentious species of starfish, widely distributed along the Western coast of the Pacific Ocean from the Sakhalin Island to the Yellow Sea. It is the most numerous species in the Peter the Great Bay (Sea of Japan).

The tissues of *P. pectinifera* contain a large amount of CRs, with the major component being astaxanthin [49–51]. Takashi Maoka et al. [51] almost completely, with the exception of 2%, identified carotenoids of the starfishes *A. pectinifera* and *Asterias amurensis*. They showed that starfish carotenoids are mainly astaxanthin derivatives. Our carotenoids mixture (MC) preparation contained 56% astaxanthin, 4% zeaxanthin, 6% lutein, and unidentified carotenoids (Figure 1). They can be used as raw materials for the production of new medicinal, cosmetic, and food products.

The goal of this work was to test the biological activity of a mixture of carotenoid pigments enriched in astaxanthin obtained from the starfish *P. pectinifera* according to the Russian patent [49] in a series of models of skin damage and systemic inflammatory diseases in vivo.

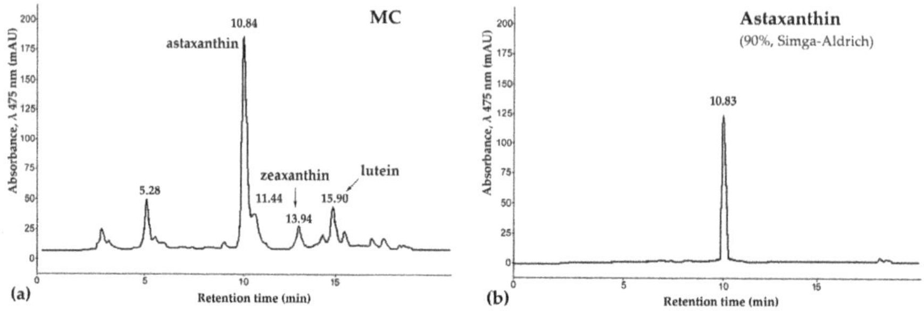

Figure 1. The high-performance liquid chromatography (HPLC) chromatogram of carotenoids mixture of starfish *P. pectinifera* (**a**) and sample of astaxanthin (90%, Sigma-Aldrich, St. Louis, MO, USA) (**b**). Chromatographic conditions: HPLC was carried out on a chromatograph LaChrom 2000 (Hitachi/Merck, Darmstadt, Germany) equipped with UV detector L-7400, pump L-7100, thermostat L-7300, integrator D-7500; column, Agilent Technologies Zorbax Eclipse XDB-C18, 3.5 µm (75 mm × 4.6 mm) with guard column Hypersil ODS, 5 µm (4.0 mm × 4.0 mm); detection, UV 475 nm; solvent, acetonitrile/water (85/15, *v*/*v*) with 1% glacial acetic acid; flow rate, 0.2 mL/min; thermostated at 30 °C. Abbreviations: MC—carotenoid mixture of starfish *P. pectinifera*.

2. Results

2.1. Evaluation of Safety of Carotenoid mixture of Starfish P. pectinifera (MC)

To access the toxicity of the preparation, 12 healthy mice (6♂and 6♀) of the intact group of animals were treated for 14 days. The changes in the body weight were in the normal mode, characteristic of healthy animals (Table 1). The general condition and behavior of the intact group met the criteria of the norm: they actively moved around the cage, drank water, and ate food. Food and water intake as well as hairline and mucous membranes remained normal.

Table 1. Weight (g) of healthy intact mice, mice after a single intraperitoneal or oral administration of a water–alcohol solution (solvent) and MC.

No., Gender	Intact Animals		Water-Alcohol Solution (Solvent) in a Volume of 0.5 mL		MC at a Dose of 500 mg/kg.	
	First Day	On the 14th Day	Before Administration	On the 14th Day	Before Administration	On the 14th Day
			IP injection			
1♂	23.5	30.2	22.6	30.6	23.0	31.1
2♂	23.4	31.1	23.4	30.1	23.8	31.8
3♂	22.8	31.6	22.8	31.1	22.3	31.2
1♀	21.3	30.8	22.1	30.9	22.4	29.8
2♀	23.4	31.1	23.2	32.1	23.0	31.1
3♀	23.2	31.6	22.4	31.3	22.6	30.3
			Oral administration			
1♂	22.4	30.9	22.2	31.3	23.6	29.9
2♂	22.9	31.1	23.4	31.1	23.2	30.4
3♂	23.3	31.1	23.8	31.8	22.3	30.1
1♀	22.9	30.4	22.3	30.4	23.3	30.4
2♀	23.5	31.8	23.8	32.1	23.8	32.1
3♀	22.4	33.1	23.4	32.3	22.8	31.3

Abbreviations: MC—carotenoids mixture of starfish *P. pectinifera;* IP—Intraperitoneally administration; ♂—male; ♀—female.

The evaluation of the product's safety in mice established that a single dose of the carotenoid complex orally and intraperitoneally had no toxic effect on the body weight compared to intact animals. The general condition of this group of experimental animals remained the same as the groups of healthy intact animals. Their behavior throughout the experiment did not differ from the behavior of healthy animals, i.e., they actively moved around the cage, drank water, and ate food. Feed and water intake as well as the condition of the hairline and mucous membranes remained normal after oral and intraperitoneal administration, with no local irritation observed, leading us to conclude on the safety of the MC use for therapeutic and prophylactic purposes.

2.2. The MC Effectiveness in Treatment of Carcinogenic and Allergic Skin Pathologies, and Systemic Inflammation (SI)

2.2.1. Evaluation of Cancer-Preventive Activity of MC

The study of the cancer-preventive activity of MC was carried out in a model of skin carcinogenesis induced by 7,12-dimethylbenzanthracene (DMBA). It was shown that during the period of active development of the oncological process, after frequent applications of DMBA, the groups of animals that received prophylactic oral administration of MC (10 mg/kg three times a week for 8 weeks) saw a decrease in the incidence of new tumor foci and in the growth intensity (Figure 2). It should be emphasized that MC was more effective than the reference substance, rosmarinic acid (RA), in inhibiting the formation of new papillomas (Table 2). So, in the MC C group, by the 11th week of the experiment, the formation of new tumor foci almost completely stopped, and by the 15th week, the number and size of papillomas were more than 1.5 times smaller than in the C(−) group (Figure 2b).

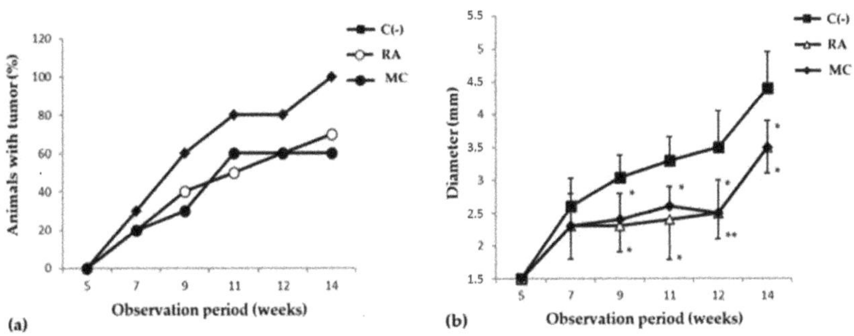

Figure 2. Dynamics of skin cancer incidence (**a**) and changes in the size of papillomas (**b**) (y-axis) in the experimental groups of animals. The data are presented from the moment the first changes appeared on the DMBA-treated animal skin areas (5 weeks after the start of the experiment). The results are presented as m ± SD (standard deviation) (n = 10), * $p \leq 0.05$, ** $p \leq 0.01$, compared with C(−) group. Abbreviations: C(−)—negative control; RA—rosmarinic acid (positive control); MC—carotenoids mixture of starfish *P. pectinifera*.

Table 2. Chemopreventive effect of MC in the experimental model of skin carcinogenesis.

Group	Number of Animals with Tumor, %	Number of Tumors	Average Tumor Diameter, mm	Latent Period, Week
Intact	-	-	-	-
C(−)	100	27	4.4 ± 1.6	4.2 ± 2.9
RA	70	15	3.5 ± 0.9	4.6 ± 1.1
MC	60	13	3.5 ± 1.01	4.3 ± 2.7

Abbreviations: C(−)—negative control; RA—rosmarinic acid (positive control); MC—carotenoids mixture of starfish *P. pectinifera*.

Oxygenated carotenoids (astaxantin, zeaxanthin and lutein), the components of MC, are supposed to exert protective activity in oncological, allergic, and inflammatory processes mainly due to their ability to modulate the immune response [3–5,21,23,31,50,52]. When evaluating the functional activity of the immune system of experimental animals, it was shown that at the stage of progressive tumor growth in the C(−) group, onco-dependent immunosuppression was observed, which was expressed as a decrease in the level of IL-1, IL-17, IFN-γ, TNF-α, IL- 10, both systemically (blood serum) and locally (skin homogenate) (Figure 3a,b). Prophylactic oral intake of MC contributed to the stimulation of immunological surveillance, through an increase in the production of all of the above cytokines to the level close to the intact group, with the exception of IL-4. The data obtained indicate the ability of MC to modulate immunological reactions in skin lesions caused by DMBA and enhance the antitumor immune response.

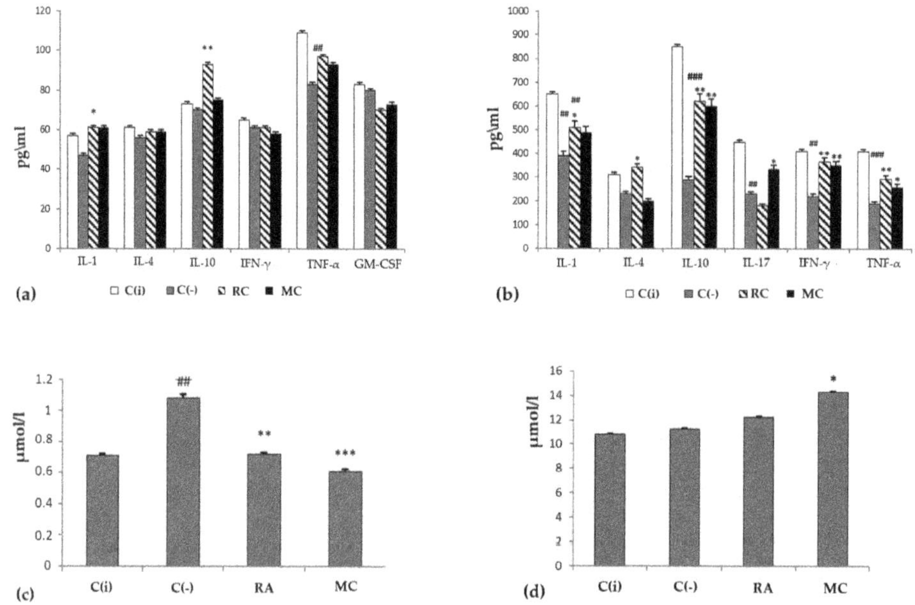

Figure 3. The profile of cytokines (*y*-axis, pg/mL) in blood serum (**a**) and skin homogenate (**b**), the level of malonic dialdehyde (MDA) (**c**) and total bilirubin (**d**) in the blood serum of animals (*y*-axis, μmol/L) at experimental modeling of skin carcinogenesis. The results are presented as m ± SD (standard deviation) (n = 10), * $p \leq 0.05$, ** $p \leq 0.01$, *** $p \leq 0.01$, compared with the C(−) group; ## $p \leq 0.01$, ### $p \leq 0.01$, C(−) group compared to C(i) group. Abbreviations: C(−)—negative control; C(i)–intact control; RA—rosmarinic acid (positive control); MC—carotenoids mixture of starfish *P. pectinifera;* IL-1,4,10,17—interleukins-1,4,10,17; IFN-γ—interferon-gamma; TNF-α—tumor necrosis factor; GM-CSF—granulocyte–macrophage colony-stimulating factor.

The state of the antioxidant system in the experimental animals was assessed by the concentration of malonic dialdehyde (MDA) (Figure 3c) in the blood serum. It was shown that in the C(−) group, the level of MDA was increased by 1.5 times compared to the group of intact animals. At the same time, MC turned out to be an effective corrector of this parameter. The general condition of the experimental animals was assessed by the level of bilirubin. The analysis showed that total bilirubin in the MC and RA groups was within the normal range (Figure 4d). Consequently, long-term administration of RA and MC did not exert a noticeable toxic effect on the animal organism.

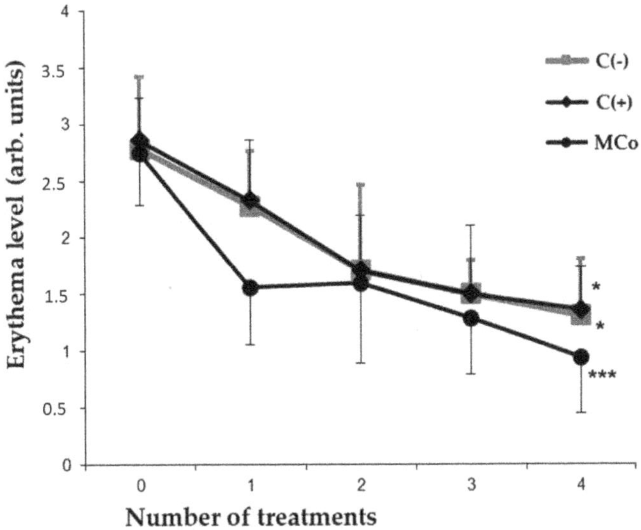

Figure 4. Indicators of the level of erythema (y-axis) caused by DNFB in different groups of animals after each day of treatment (x-axis). The results are presented as m ± SD (standard deviation) (n = 8), * $p \leq 0.05$, *** $p \leq 0.01$ (paired Student's t-test). Abbreviations: DNFB—2,4-dinitrofluorobenzene; C(−)—negative control; C(+)—positive control ointment "Fucidin"; Mco—ointment, containing 1% carotenoids mixture of starfish *P. pectinifera*.

Thus, in the groups of experimental animals treated with MC, a pronounced cancer-preventive effect was observed. MC is effective similarly to RA and improves the clinical picture of the disease progression, while normalizing the cytokine profile and the antioxidant defense system in the experimental animals.

2.2.2. Evaluation of the MC Ointment Form in Allergic Contact Dermatitis (ACD)

In the case of administering a 1% MC ointment as an active compound, a significant restoration of the initial parameters of the skin and a decrease in the severity of external pathomorphological manifestations of ACD were observed. By the 4th day of treatment of experimental animals in the MC group, the index of reducing the severity of pathomorphological skin manifestations was 80% (Table 3). At the same time, the level of erythema decreased by approximately two times as compared to the C(−) group (Figure 4). It should be noted that the studied MC-based ointment was superior to the commercial anti-inflammatory drug "Fucidin" in terms of the effectiveness of the pharmacological action. MC acts as an effective corrector of the inflammatory process in ACD, the severity of which was assessed by the level of pro-inflammatory and anti-inflammatory cytokines in the blood serum by enzyme-linked immunosorbent assay (ELISA) (Figure 5).

Table 3. Effect of MC ointment on the process of restoration of skin integuments affected by DNFB.

Group Animals	1 Days	2 Days	4 Days
	Healing, %, (m ± σ)		
C(−)	-	-	-
Fucidin	8 ± 2.5	20 ± 3.5	24 ± 2.3
MC, 1%	43 ± 2.9	40 ± 2.6	80 ± 1.3

Abbreviations: C(−)—negative control without treatment; Fucidin—positive control treatment ointment "Fucidin"; MCo—ointment, containing 1% carotenoids mixture of starfish *P. pectinifera*.

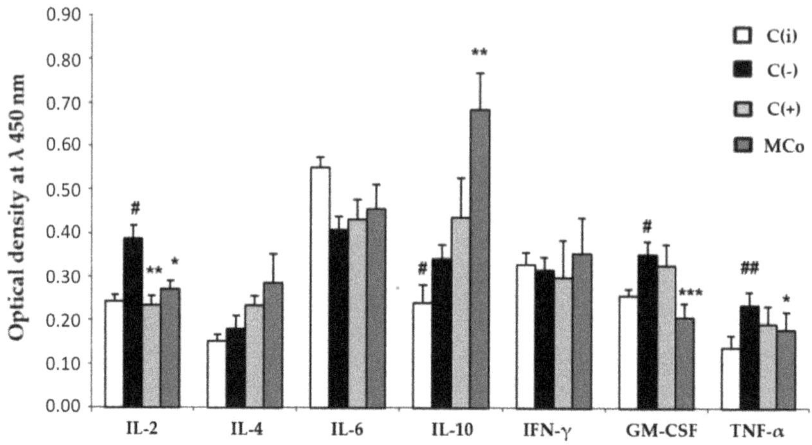

Figure 5. The level of cytokines in the blood serum (*y*-axis, optical density at λ 450 nm) in the experimental model of allergic contact dermatitis (ACD) induced by DNFB. Blood samples were obtained 24 h after the last application of the ointments. The results are presented as m ± SD (standard deviation) (n = 8), * $p \leq 0.05$, ** $p \leq 0.01$, *** $p \leq 0.01$, compared with the C(−) group; # $p \leq 0.05$, ## $p \leq 0.01$, C(−) group versus C(i) group. Abbreviations: DNFB—2,4-dinitrofluorobenzene; C(−)—negative control; C(i)–intact control; C(+)—positive control ointment "Fucidin"; Mco—ointment containing 1% carotenoids mixture of starfish *P. pectinifera*; IL-2,4,6,10—interleukins–2,4,6,10; IFN-γ—interferon-gamma; TNF-α—tumor necrosis factor; GM-CSF—granulocyte–macrophage colony-stimulating factor.

It was found that the use of MC led to a decrease in the level of the main pro-inflammatory cytokines, involved in the pathogenesis of ACD, IL-2, TNF-α and GM-CSF, with a sharp increase in the main anti-inflammatory cytokines, IL-10 and IL-4, which may indicate the ability of MC components to enhance the anti-inflammatory response in ACD.

2.2.3. Anti-Inflammatory Activity of MC in SI Model

When lipopolysaccharide (LPS) was administered to experimental animals, an increase in the serum content of pro-inflammatory cytokines, e.g., IL-1, -6, IFN-γ, and TNF-α, in the C (−) group was recorded (Figure 6a).

It is typical that the effect of MC and the reference drug "Dexamethasone" on the level of cytokines largely depends on the dose and route of administration of the drug. When administered orally, MC and Dexamethasone contributed to a sharp increase in the level of all tested cytokines compared to the C(i) and C(−) groups. However, with intraperitoneal administration, MC was more effective than Oftan-Dexamethasone in suppressing the production of key pro-inflammatory cytokines, IL-1 and TNF-α, bringing their level to that of the intact animals.

Thus, the ointment containing 1% MC of starfish *P. pectinifera* has high pharmacological prospects in the treatment of ACD and can be further used in various preclinical studies.

It should be noted that under conditions of SI, regardless of the dose and route of administration, the studied MC preparation had a protective effect on redox homeostasis, functional activity of the liver, biliary tract, and hematopoiesis, analyzed by the content of MDA and total bilirubin in blood serum in different experimental groups (Figure 6b,c). The conducted experimental studies in the SI model showed that MC has pronounced antioxidant and anti-inflammatory protective effects on the organism of experimental animals.

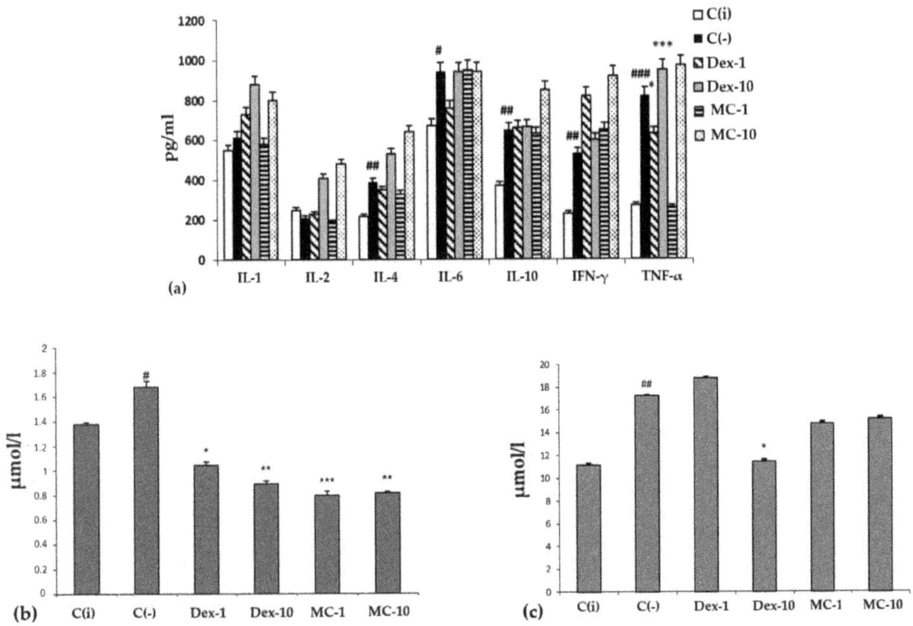

Figure 6. The content of cytokines (**a**) (y-axis, pg/mL), malonic dialdehyde (MDA) (**b**) and total bilirubin (**c**) (y-axis, in µmol/L) in the blood serum. The results are presented as m $\pm$ SD (standard deviation) (n = 6), * $p \leq 0.05$, ** $p \leq 0.01$, *** $p \leq 0.01$, compared with the C(−) group; # $p \leq 0.05$, ## $p \leq 0.01$, ### $p \leq 0.01$, C(−) group versus C(i) group. Abbreviations: C(−)—negative control, C(i) —intact control; Dex-1, Dex-10—"Dexamethasone" in doses of 1 and 10 mg/kg, respectively; MC-1, MC-10—oxygenated carotenoid complex of starfish *P. pectinifera* at doses of 1 and 10 mg/kg, respectively; MDA—malonic dialdehyde; IL-2,4,6,10—interleukins–2,4,6,10; IFN-γ—interferon-gamma; TNF-α—tumor necrosis factor.

3. Discussion

One of the most promising strategies in the control of the incidence of oncological and inflammatory pathologies is chemoprophylaxis, which consists in the systematic use of prophylactic agents and functional nutrients that have a tumor-protective effect. At the same time, special attention is paid to compounds that have the ability to maintain redox homeostasis, exhibit an immunomodulatory effect, and have a positive effect on the functioning of various cell-signaling pathways with no side effects on the body of experimental animals. The results of our studies indicate that there is a direct correlation between the therapeutic use of MC of starfish *P. pectinifera* and a reduction in the risk of developing skin cancer (Figure 3, 1.5-fold reduction), and MC is more effective than rosmarinic acid (Table 2).

The Inhibition of the formation and growth of skin neoplasms observed with the use of MC may be the result not only from the antioxidant and immunomodulatory activity of carotenoids but also their ability to neutralize the carcinogenic effect of DMBA on epidermal cells, through the activation of the aryl hydrocarbon receptor (AhR). By stimulating the activity of AhR, they contribute to the stimulation of gene expression of biotransformation enzymes and the detoxification and elimination of xenobiotics, reducing the risk of accumulation of carcinogens in epithelial cells and thereby inhibiting the initial stages of carcinogenesis. It has been shown that AhR agonists are able to restrain neoplastic processes in the cell induced by chemical carcinogens, including DMBA [44,53–55]. The use of MC as prophylactic agent and functional nutrient can prevent the adverse effects of carcinogenic factors, maintain the optimal immunological status of the body and, thereby, reduce the risk of cancer. In addition to preventive action, MC has been shown to enhance the antitumor effect of the well-known cytostatic doxorubicin, which is widely used in chemotherapy, when they are used together [50]. In this regard, MC, after conducting preclinical trials, can expand the arsenal of complementary therapies in the treatment of oncological diseases.

Since carotenoids are considered not only as functional components of food, but also as active ingredients of cosmetic and cosmeceutical products, an important task is to evaluate their therapeutic effect in skin diseases, particularly in allergies. Allergic dermatoses are common inflammatory skin diseases with multifactorial etiology and complex pathogenesis. The most well-known diseases are psoriasis, contact dermatitis, eczema, chronic urticaria, and others. Pathogenesis includes primarily damage to the epidermal barrier as a result of an excessive inflammatory response from the immune system, in response to the action of various irritants (mechanical damage to the skin, contact with chemical, food, household, and other allergens). Currently widely used antiallergic drugs (glucocorticoids, antihistamines, mast cell stabilizers, and immunosuppressants) have a strong antiallergic effect. However, due to the narrow range of pharmacological action, none of these drugs are universal for the treatment of all types of skin allergies, and often they require long-term use, which can lead to a number of side effects: central nervous system depression, impaired carbohydrate and lipid metabolism, myasthenia gravis, osteoporosis, decreased resistance to infections [55]. In this regard, it is necessary to develop universal effective drugs that, along with antiallergic activity, can also inhibit inflammatory reactions and stimulate reparative processes in the epidermis. The therapeutic effect of the MC—based ointment was seen in a noticeable improvement (80% skin restoration effect, Table 3) in pathomorphological and biochemical parameters of the disease, i.e., a significant decrease in the level of erythema and correction of the cytokine profile. Previously, Meephansan et al. studied the effect of astaxanthin on skin wound healing [56]. Full-thickness skin wounds were created in 36 healthy female mice, which were divided into a control group and a group treated with 78.9 mg locally of astaxanthin twice a day for 15 days. Astaxanthin-treated wounds showed marked narrowing by day 3 of treatment and complete wound closure by day 9, while wounds in control mice showed only partial epithelialization and still had scabs. Biological markers of wound healing, including Col1A1 and bFGF, were significantly elevated in the astaxanthin group from day 1. The results show that astaxanthin is an effective wound-healing compound. Chow et al. showed that the astaxanthin-enriched extract (EAE) from H. pluvialis increased the expression of certain proteins that promote cell proliferation, had a higher cell growth capacity than doxycycline, and was able to proliferate more collagen than the control group over seven days [57]. This indicates that it is the best alternative for collagen production.

Immunological blood testing (Figure 6) showed that MC acts as an effective corrector of the inflammatory process in ACD, its use led to a decrease in the level of the main pro-inflammatory cytokines involved in the pathogenesis of ACD: IL-2, TNF-α and GM-CSF. It should be noted that under the action of MC, the content of the main anti-inflammatory cytokines sharply increased: IL-10 and IL-4. This fact may indicate the ability of MC to enhance the anti-inflammatory response in ACD. It can be assumed that these carotenoids have an anti-inflammatory effect, preventing the synthesis of endogenous inhibitors of matrix metalloproteinases. The latter play a central role in the metabolism of connective tissue proteins, as they promote tissue renewal and are essential for skin repair and removal of inflammatory agents [58]. In 2010, Park et al. conducted the first comprehensive study investigating the effects of dietary ASX on modulating immune response, oxidative status, and inflammation in young healthy adult female subjects [59]. After eight weeks of supplementation, ASX enhanced both cell-mediated and humoral immune responses, including T cell and B cell mitogen-induced lymphocyte proliferation, NK cell cytotoxic activity, and IL-6 production. ASX did not affect plasma C-reactive protein concentrations, but levels of 8-hydroxy-2′-deoxyguanosine (8-oHdG) (a biomarker of DNA damage) were significantly lower in the higher-dose ASX group.

The study on the effect of MC on the immunological status of animals in the model of skin carcinogenesis, ACD and SI, indicates that it is an anti-inflammatory agent when administered intraperitoneally, but not orally. Altogether, the obtained results show that MC from sea star *P. pectinifera* is an effective remedy for the treatment and prevention of inflammatory processes.

4. Materials and Methods

4.1. Preparations

The method for obtaining an MC from starfish *P. pectinifera* includes the extraction of raw materials with 96% alcohol with the addition of food acid and ascorbic acid as a food antioxidant, column chromatography on a hydrophobic sorbent Polychrome-1, balanced with 20–30% alcohol. The column with adsorbed carotenoids is washed with alcohol in a gradient of 30→50%. The mixture of carotenoids is eluted with 60–65% alcohol. The eluate is evaporated in vacuum at 40–60 °C. Additional purification of the MC from phospholipids carried out by dissolving the concentrate in 96% alcohol, settling solution for 24–48 h at (−18 °C), centrifugation and subsequent evaporation in vacuum at 40–60 °C [49].

The MC composition was determined by high-performance liquid chromatography, UV and IR spectroscopy (Figure 1).

For intraperitoneal and oral injections, MC were weighed according to the dose used and prepared in aqueous, water–alcohol (100:1, v/v) or aqueous solutions with DMSO (Sigma-Aldrich, St. Louis, MO, USA) (10:1, v/v). Ointment for external use was prepared on a lanolin–Vaseline basis (1:3 by weight): 1 g of MC (1%) was added to 100 g of the ointment base.

4.2. Animals

CBA, CD-1, and BALB/c mice (20 ± 2 g) were used as a test system for studying the activity of natural substances. The experiments were performed on animals purchased from the laboratory animal nursery "Pushchino" and bred in the vivarium of the Pacific Institute of Bioorganic Chemistry of the Far Eastern Branch of the Russian Academy of Sciences (PIBOC FEB RAS) (certificate available).

The animals were kept in accordance with the GOST 33216-2014 "Guidelines for accommodation and care of animals. Species-specific provisions for laboratory rodents and rabbits". After the experiment was completed, they were subjected to euthanasia. Studies using experimental animals were carried out in accordance with the GOST 33044-2014 "Principles of good laboratory practice" and "Guidelines for conducting preclinical studies of drugs", edited by Mironov et al. [60]. All experiments were approved by the Ethical Committee for Animal Research of the PIBOC FEB RAS, protocol code 08/19, date of approval 27 March 2019.

4.3. Safety Assessment of MC

The safety assessment of MC was carried out in accordance with the guidelines for studying the toxic effects of pharmacological substances [61]. The experiments were performed on female and male mice of the CD-1 line (24 ± 2 g). Experimental groups of animals: 36 mice were used in the experiment, including 18 females and 18 males. Animals were randomized into 3 groups. The experimental group included 12 animals, 6 males and 6 females, who were administered MC orally and intraperitoneally in a water–alcohol solution in a volume of 0.5 mL. The control group of intact animals, 6 females and 6 males, were injected with distilled water. The control negative group, 6 females and 6 males, consisted of the animals that were injected with a water–alcohol solution (solvent).

4.4. Murine Model of Skin Carcinogenesis

The study of the cancer-preventive activity of MC was carried out on a model of skin carcinogenesis induced by 7,12-dimethylbenzanthracene (DMBA) (Sigma-Aldrich, St. Louis, MO, USA), which was dissolved in benzene and applied to the shaved interscapular region of experimental animals at a dose of 10 μg per mouse, 3 times a week for 8 weeks. The cancer-protective effect of the studied substances was assessed by their effect on pathomorphological changes in tissues: changes in the number of animals with a tumor, size, number of tumor formations (diameter, mm) and latent period (the period between the appearance of the first signs of oncogenesis until the appearance in the group of 50% of animals with tumor). Rosmarinic acid (RA) (Sigma-Aldrich, St. Louis, MO, USA) was used as a positive control.

4.5. Murine model of Experimental Dermatitis

Experimental ACD was reproduced using the obligate allergen 2,4-dinitrofluorobenzene (DNFB) (Sigma-Aldrich, St. Louis, MO, USA) in the form of a 0.5% oil-acetone mixture (acetone: olive oil, 4:1, by volume) for sensitization (applied once on shaved area of the peritoneum) and a 0.2% mixture to obtain an extensive disease (double application to the inner and outer surface of the ear). The effect of ointment preparations on the degree of healing of the affected areas and the level of erythema (redness and induration of the area, inflamed tissues with hyperemia, lichenification of the skin and the formation of a superficial hemorrhagic crust and areas of necrosis) was assessed. Commercial ointment "Fucidin" (Leo Laboratoris Limited, Dublin, Ireland) was used as a reference drug.

4.6. Murine Model of Inflammation

SI was induced by LPS from *Escherichia coli* (Sigma-Aldrich, St. Louis, MO, USA) at a dose of 0.1 mg/kg. The anti-inflammatory commercial drug Oftan Dexamethasone (Santen AO, Tampere, Finland) was used as a positive control. The test substances were administered to the animals 1 h before LPS induction. An hour and a half after SW induction, blood samples were taken for immunological and biochemical analyses.

4.7. Immunological and Biochemical Parameters Studies

The functional state of the immune system was assessed by determining the level of cytokines by ELISA using diagnostic kits (BD Bioscience OptEIA, Bergen, NJ, USA).

The severity of free-radical processes in pathological processes was determined by the content of TBA-reactive products (MDA) in blood plasma, by reaction with thiobarbituric acid (TBA) (Sigma-Aldrich, St. Louis, MO, USA).

To assess the effect of the studied dietary supplements on the general condition of the body, an analysis was made for total bilirubin using the Novogluk-KM kit (Vector-Best, Novosibirsk, Russia). The optical density of the samples was measured using an ELS 808 iu plate reader (BioTek, Winooski, VT, USA) at a wavelength of 450 nm.

4.8. Statistics

Statistical and graphical processing of experimental data was carried out using the statistical package Microsoft Excel Office version number 12.0. The obtained values were expressed as mean ± SD (standard deviation). The significance of differences was determined using a parametric Student's t-test. Differences at $p < 0.05$ were taken as significant.

5. Conclusions

Evaluation of the functional activity of the antioxidant defense system and the general state of the body of experimental animals, when modeling the SI, showed that MC in all studied doses and routes had a protective effect on redox homeostasis and the functional activity of the liver, biliary tract, and hematopoiesis, and had no toxic side effects.

The results of the study of the effect of MC on the immunological status of animals, when modeling skin carcinogenesis, ACD, and SI, testify in favor of the conclusion that MC manifests itself as a local, and oral, anti-inflammatory agent. Namely, MC reduces the excessive immune response by direct action on the focus of inflammation (ip injection in SI, applications on the inflamed skin area in ACD), and at the same time has an immunostimulatory effect when administered enterally (oral administration for skin carcinogenesis and SI). Therefore, the selection of the route of administration of the oxygenated carotenoid products largely determines their physiological effect on the body. Based on the results obtained, it can be concluded that the *P. pectinifera* starfish MC is a promising candidate for the treatment and prevention of inflammatory processes. This finding creates the background for a comprehensive study of MC biopharmaceutical properties and its optimization for medical and cosmetic practice.

Author Contributions: Conceptualization, E.P.K. and A.M.P.; methodology, A.A.K. and A.M.P.; formal analysis, A.A.K.; investigation, A.A.K., V.V.M. and A.I.V.; resources, L.A.T.; writing—original draft preparation, A.A.K. and A.M.P.; writing—review and editing, E.P.K., I.G.G., D.M.H.; visualization, A.A.K., T.A.R. and V.V.M.; supervision, E.P.K. and A.M.P.; project administration, T.A.R. and O.M.S. All authors have read and agreed to the published version of the manuscript.

Funding: The work was financially supported by the Ministry of Education and Science of the Russian Federation as part of the implementation of support programs for the development of advanced engineering schools, in accordance with Decree of the Government of the Russian Federation of 4 August 2022 No. 619, agreement No. 075-15-2022-1143 of 7 July 2022. D.M.H. and I.G.G. acknowledge the support from Russian Scientific Foundation (project 20-15-00207).

Institutional Review Board Statement: All experiments were approved by the Ethical Committee for Animal Research of the Pacific Institute Bioorganic Chemistry of the Far Eastern Branch of the Russian Academy of Sciences (PIBOC FEB RAS) (protocol code 08/19, date of approval 27 March 2019).

Data Availability Statement: Not applicable.

Conflicts of Interest: The authors declare no conflict of interest.

References

1. Jahns, L.; Conrad, Z.; Johnson, L.K.; Whigham, L.D.; Wu, D.; Claycombe-Larson, K.J. A diet high in carotenoid-rich vegetables and fruits favorably impacts inflammation status by increasing plasma concentrations of IFN-α2 and decreasing MIP-1β and TNF-α in healthy individuals during a controlled feeding trial. *Nutr. Res.* **2018**, *52*, 98–104. [CrossRef]
2. Kaulmann, A.; Bohn, T. Carotenoids, inflammation, and oxidative stress-implications of cellular signaling pathways and relation to chronic disease prevention. *Nutr. Res.* **2014**, *34*, 907–929. [CrossRef]
3. Kohandel, Z.; Farkhondeh, T.; Aschner, M.; Pourbagher-Shahri, A.M.; Samarghandian, S. Anti-inflammatory action of astaxanthin and its use in the treatment of various diseases. *Biomed. Pharmacother.* **2022**, *145*, 112179. [CrossRef]
4. Chang, M.X.; Xiong, F. Astaxanthin and its Effects in Inflammatory Responses and Inflammation-Associated Diseases: Recent Advances and Future Directions. *Molecules* **2020**, *25*, 5342. [CrossRef] [PubMed]
5. Kowsalya, K.; Vidya, N.; Vijayalakshmi, V.; Arun, M. Super nutritive marine astaxanthin, an effectual dietary carotenoid for neurodegenerative diseases. *Int. Res. J. Multidiscip. Technovation* **2019**, *1*, 115–124. [CrossRef]
6. Cho, K.S.; Shin, M.; Kim, S.; Lee, S.B. Role of natural antioxidants from functional foods in neurodegenerative and metabolic disorders. *Oxidative Med. Cell. Longev.* **2018**, *2018*, 4120458. [CrossRef]
7. Mullan, K.; Williams, M.A.; Cardwell, C.R.; McGuinness, B.; Passmore, P.; Silvestri, G.; Woodside, J.V.; McKay, G.J. Serum concentrations of vitamin E and carotenoids are altered in Alzheimer's disease: A case-control study. *Alzheimers Dement.* **2017**, *3*, 432–439. [CrossRef]
8. Guest, J.; Grant, R. Carotenoids and neurobiological health. *Adv. Neurobiol.* **2016**, *12*, 199–228. [CrossRef] [PubMed]
9. Honarvar, N.M.; Saedisomeolia, A.; Abdolahi, M.; Shayeganrad, A.; Sangsari, G.T.; Rad, B.H.; Muench, G. Molecular anti-inflammatory mechanisms of retinoids and carotenoids in Alzheimer's disease: A review of current Evidence. *J. Mol. Neurosci.* **2017**, *61*, 289–304. [CrossRef]
10. Giordano, P.; Scicchitano, P.; Locorotondo, M.; Mandurino, C.; Ricci, G.; Carbonara, S.; Gesualdo, M.; Zito, A.; Dachille, A.; Caputo, P.; et al. Carotenoids and cardiovascular risk. *Curr. Pharm. Des.* **2012**, *18*, 5577–5589. [CrossRef]
11. Kulczyński, B.; Gramza-Michałowska, A.; Kobus-Cisowska, J.; Kmiecik, D. The role of carotenoids in the prevention and treatment of cardiovascular disease. Current state of knowledge. *J. Funct. Foods* **2017**, *38*, 45–65. [CrossRef]
12. Wolak, T.; Paran, E. Can carotenoids attenuate vascular aging? *Vascul. Pharmacol.* **2013**, *59*, 63–66. [CrossRef] [PubMed]
13. Gammone, M.A.; Pluchinotta, F.R.; Bergante, S.; Tettamanti, G.; d'Orazio, N. Prevention of cardiovascular diseases with carotenoids. *Front. Biosci.* **2017**, *9*, 165–171. [CrossRef]
14. Saini, R.K.; Keum, Y.S.; Daglia, M.; Rengasamy, K.R.R. Dietary carotenoids in cancer chemoprevention and chemotherapy: A review of emerging evidence. *Pharmacol. Res.* **2020**, *157*, 104830. [CrossRef] [PubMed]
15. Huang, J.; Lu, M.S.; Fang, Y.J.; Xu, M.; Huang, W.Q.; Pan, Z.Z.; Chen, Y.M.; Zhang, C.X. Serum carotenoids and colorectal cancer risk: A case-control study in Guangdong, China. *Mol. Nutr. Food Res.* **2017**, *61*, 1700267. [CrossRef]

16. Butalla, A.C.; Crane, T.F.; Patil, B.; Wertheim, B.C.; Thomson, P.; Thomson, C.A. Effect of a carrot juice intervention on plasma carotenoids, oxidative stress, and inflammation in overweigth breast cancer survivors. *Nutr. Cancer.* **2012**, *64*, 331–341. [CrossRef] [PubMed]
17. Koklesova, L.; Liskova, A.; Samec, M.; Zhai, K.; Abotaleb, M.; Ashrafizadeh, M.; Brockmueller, A.; Shakibaei, M.; Biringer, K.; Bugos, O.; et al. Carotenoids in cancer metastasis—Status quo and outlook. *Biomolecules* **2020**, *10*, 1653. [CrossRef]
18. Roohbakhsh, A.; Karimi, G.; Iranshahi, M. Carotenoids in the treatment of diabetes mellitus and its complications: A mechanistic review. *Biomed. Pharmacother.* **2017**, *91*, 31–42. [CrossRef]
19. McClinton, K.J.; Aliani, M.; Kuny, S.; Sauvé, Y.; Suh, M. Differential effect of a carotenoid-rich diet on retina function in non-diabetic and diabetic rats. *Nutr. Neurosci.* **2020**, *23*, 838–848. [CrossRef]
20. Zhu, R.; Chen, B.; Bai, Y.; Miao, T.; Rui, L.; Zhang, H.; Xia, B.; Li, Y.; Gao, S.; Wang, X.D. Lycopene in protection against obesity and diabetes: A mechanistic review. *Pharmacol. Res.* **2020**, *159*, 104966. [CrossRef]
21. Sztretye, M.; Dienes, B.; Gönczi, M.; Czirják, T.; Csernoch, L.; Dux, L.; Szentesi, P.; Keller-Pintér, A. Astaxanthin: A potential mitochondrial-targeted antioxidant treatment in diseases and with aging. *Oxid. Med. Cell. Longev.* **2019**, *2019*, 3849692. [CrossRef]
22. Elvira-Torales, L.I.; García-Alonso, J.; Periago-Castón, M.J. Nutritional Importance of Carotenoids and Their Effect on Liver Health: A Review. *Antioxidants* **2019**, *8*, 229. [CrossRef] [PubMed]
23. Li, J.; Guo, C.; Wu, J. Astaxanthin in liver health and disease: A potential therapeutic agent. *Drug Des. Devel. Ther.* **2020**, *14*, 2275–2285. [CrossRef] [PubMed]
24. Tominaga, K.; Hongo, N.; Karato, M.; Yamashita, E. Cosmetic benefits of astaxanthin on human subjects. *Acta Biochim. Pol.* **2012**, *59*, 43–47. [CrossRef] [PubMed]
25. Davinelli, S.; Nielsen, M.; Scapagnini, G. Astaxanthin in skin health, repair, and disease: A comprehensive review. *Nutrients* **2018**, *10*, 522. [CrossRef]
26. Milito, A.; Castellano, I.; Damiani, E. From Sea to Skin: Is There a Future for Natural Photoprotectants? *Mar. Drugs* **2021**, *19*, 379. [CrossRef]
27. Roark, M.W.; Stringham, J.M. Visual performance in the "Real World": Contrast sensitivity, visual acuity, and effects of macular carotenoids. *Mol. Nutr. Food Res.* **2019**, *63*, e1801053. [CrossRef]
28. Comhaire, F.H.; Mahmoud, A. The role of food supplements in the treatment of the infertile man. *Reprod. BioMed. Online* **2003**, *7*, 385–391. [CrossRef]
29. Kurashova, N.A.; Dashiev, B.G.; Kolesnikov, S.I.; Dmitrenok, P.S.; Kozlovskaya, E.P.; Kasyanov, S.P.; Epur, N.V.; Usov, N.G.; Kolesnikova, L.I. Processes, and antioxidant protection in men with pathozoospermia after COVID-19. The effectiveness of correction with a promising antioxidant complex. *Bull. Exp. Biol. Med.* **2022**, *173*, 606–610. [CrossRef]
30. Gasmi, A.; Peana, M.; Pivina, L.; Srinath, S.; Gasmi Benahmed, A.; Semenova, Y.; Menzel, A.; Dadar, M.; Bjorklund, G. Interrelations between COVID-19 and other disorders. *Clin. Immunol.* **2021**, *224*, 108651. [CrossRef]
31. Talukdar, J.; Dasgupta, S.; Nagle, V.; Bhadra, B. COVID-19: Potential of Microalgae Derived Natural Astaxanthin as Adjunctive Supplement in Alleviating Cytokine Storm. 2020. Available online: https://dx.doi.org/10.2139/ssrn.3579738 (accessed on 20 August 2023).
32. Fakhri, S.; Nouri, Z.; Moradi, S.Z.; Farzaei, M.H. Astaxanthin, COVID -19 and immune response: Focus on oxidative stress, apoptosis and autophagy. *Phytother. Res.* **2020**, *34*, 2790–2792. [CrossRef] [PubMed]
33. Rodrigo-Baños, M.; Garbayo, I.; Vílchez, C.; Bonete, M.J.; Martínez-Espinosa, R.M. Carotenoids from Haloarchaea and Their Potential in Biotechnology. *Mar. Drugs* **2015**, *13*, 5508–5532. [CrossRef] [PubMed]
34. Novoveská, L.; Ross, M.E.; Stanley, M.S.; Pradelles, R.; Wasiolek, V.; Sassi, J.-F. Microalgal carotenoids: A review of production, current markets, regulations, and future direction. *Mar. Drugs* **2019**, *17*, 640. [CrossRef] [PubMed]
35. Langi, P.; Kiokias, S.; Varzakas, T.; Proestos, C. Carotenoids: From plants to food and feed industries. In *Microbial Carotenoids. Methods in Molecular Biology*; Barreiro, C., Barredo, J.L., Eds.; Humana Press: New York, NY, USA, 2018; Volume 1852. [CrossRef]
36. Hashimoto, H.; Uragami, C.; Cogdell, R.J. Chapter Carotenoids and photosynthesis. In *Carotenoids in Nature. Subcellular Biochemistry 79*; Stange, C., Ed.; Springer International Publishing: New York, NY, USA, 2016; pp. 111–139. [CrossRef]
37. Maoka, T. Carotenoids in marine animals. *Mar. Drugs* **2011**, *9*, 278–293. [CrossRef]
38. Eggersdorfer, M.; Wyss, A. Carotenoids in human nutrition and health. *Arch. Biochem. Biophys.* **2018**, *652*, 18–26. [CrossRef]
39. Maoka, T. Carotenoids as natural functional pigments. *J. Nat. Med.* **2020**, *74*, 1–16. [CrossRef] [PubMed]
40. D'Orazio, N.; Gemello, E.; Gammone, M.A.; De Girolamo, M.; Ficoneri, C.; Riccioni, G. Fucoxantin: A Treasure from the Sea. *Mar. Drugs* **2012**, *10*, 604–616. [CrossRef]
41. Arunkumar, R.; Gorusupudi, A.; Bernstein, P.S. The macular carotenoids: A biochemical overview. *Biochim. Biophys. Acta Mol. Cell Biol. Lipids* **2020**, *1865*, 158617. [CrossRef]
42. Genç, Y.; Bardakci, H.; Yücel, Ç.; Karatoprak, G.Ş.; Akkol, E.K.; Barak, T.H.; Sobarzo-Sánchez, E. Oxidative Stress and Marine Carotenoids: Application by Using Nanoformulations. *Mar. Drugs* **2020**, *18*, 423. [CrossRef]
43. Barros, M.P.; Rodrigo, M.J.; Zacarias, L. Dietary carotenoid roles in redox homeostasis and human health. *J. Agric. Food Chem.* **2018**, *66*, 5733–5740. [CrossRef]
44. Davinelli, S.; Saso, L.; D'Angeli, F.; Calabrese, V.; Intrieri, M.; Scapagnini, G. Astaxanthin as a Modulator of Nrf2, NF-κB, and Their Crosstalk: Molecular Mechanisms and Possible Clinical Applications. *Molecules* **2022**, *27*, 502. [CrossRef] [PubMed]

45. Luo, L.; Huang, F.; Zhong, S.; Ding, R.; Su, J.; Li, X. Astaxanthin attenuates ferroptosis via Keap1-Nrf2/HO-1 signaling pathways in LPS-induced acute lung injury. *Life Sci.* **2022**, *311 Pt A*, 121091. [CrossRef] [PubMed]
46. Drozdov, A.L.; Artyukov, A.A.; Drozdov, K.A. Mass Species of the Sea Star *Asterina Pectinifera* as a Potential Object of Mariculture. *Int. J. Oceanogr. Aquac.* **2020**, *4*, 000195. [CrossRef]
47. Dai, Y.; Prithiviraj, N.; Gan, J.; Zhang, X.A.; Yan, J. Tissue Extract Fractions from Starfish Undergoing Regeneration Promote Wound Healing and Lower Jaw Blastema Regeneration of Zebrafish. *Sci. Rep.* **2016**, *6*, 38693. [CrossRef] [PubMed]
48. World Register of Marine Species. Available online: https://www.marinespecies.org/aphia.php?p=taxdetails&id=459559 (accessed on 20 August 2023).
49. Artyukov, A.A.; Rutskova, T.A.; Kupera, E.V.; Makhankov, V.V.; Glazunov, V.P.; Kozlovskaya, E.P. Method for Obtaining a Carotenoid Complex from Starfish. Russian Patent 2469732 C1, 20 December 2012.
50. Artyukov, A.A.; Kupera, E.V.; Rutskova, T.A.; Makhankov, V.V.; Glazunov, V.P.; Klimovich, A.A.; Popov, A.M.; Kozlovskaya, E.P. An Agent Based on Biologically Active Compounds of Marine Hydrobionts, Which Has a Cancer-Preventive Effect and Increases the Therapeutic Activity of Antitumor Antibiotics. Russian Patent 2659682 C1, 3 July 2018.
51. Maoka, T.; Tsushima, M.; Matsuno, T. New acetylenic carotenoids starfishes from the *Asterina pectinifera* and *Asterias amurensis*. *Comp. Biochem. Physiol.* **1989**, *93*, 829–834. [CrossRef]
52. Popov, A.M.; Krivoshapko, O.N.; Artyukov, A.A. Prospects for the clinical use of astaxanthin and another oxygenated carotenoides. *Biopharm. J.* **2013**, *5*, 13–30.
53. Wang, T.; Gavin, H.M.; Arlt, V.M.; Lawrence, B.P.; Fenton, S.E.; Medina, D.; Vorderstrasse, B.A. Aryl hydrocarbon receptor activation during pregnancy, and in adult nulliparous mice, delays the subsequent development of DMBA-induced mammary tumors. *Int. J. Cancer* **2010**, *128*, 1509–1523. [CrossRef]
54. Vorontsova, J.E.; Cherezov, R.O.; Kuzin, B.A.; Simonova, O.B. Aryl-hydrocarbon receptor as a potential target for anticancer therapy. *Biochem. (Mosc.) Suppl. Ser. B Biomed. Chem.* **2019**, *13*, 36–54. [CrossRef]
55. Aleksandrova, A.V. The role of matrix metalloproteinases in the healing of burn wounds. *Eur. Stud. Sci. J.* **2013**, *2*, 40.
56. Meephansan, J.; Rungjang, A.; Yingmema, W.; Deenonpoe, R.; Ponnikorn, S. Effect of astaxanthin on cutaneous wound healing. *Clin. Cosmet. Investig. Dermatol.* **2017**, *10*, 259–265. [CrossRef]
57. Chou, H.Y.; Lee, C.; Pan, J.L.; Wen, Z.H.; Huang, S.H.; Lan, C.W.; Liu, W.T.; Hour, T.C.; Hseu, Y.C.; Hwang, B.H.; et al. Enriched Astaxanthin Extract from *Haematococcus pluvialis* Augments Growth Factor Secretions to Increase Cell Proliferation and Induces MMP1 Degradation to Enhance Collagen Production in Human Dermal Fibroblasts. *Int. J. Mol. Sci.* **2016**, *17*, 955. [CrossRef] [PubMed]
58. Berezovskaya, I.V. Classification of chemicals according to the parameters of acute toxicity in parenteral routes of administration. *Rus. Khim. Pharm. J.* **2003**, *37*, 32–34.
59. Park, J.S.; Chyun, J.H.; Kim, Y.K.; Line, L.L.; Chew, B.P. Astaxanthin decreased oxidative stress and inflammation and enhanced immune response in humans. *Nutr. Metab.* **2010**, *7*, 18. [CrossRef] [PubMed]
60. Mironov, A.N.; Bunyatyan, N.D.; Vasiliev, A.N.; Verstakova, O.L.; Zhuravleva, M.V.; Lepakhin, V.K.; Korobov, N.V.; Merkulov, V.A.; Orekhov, S.N.; Sakaeva, I.V.; et al. *Guidelines for Conducting Preclinical Studies of Drugs*; Grif and K: Tula, Russia, 2012; 944p, ISBN 9785812514663.
61. Dorozhkin, V.I.; Biryukova, N.P.; Bakhmutova, T.V. Modern Requirements to the Study of General Toxic Effect of Pharmacological Substances. *Probl. Vet. Sanit. Hyg. Ecol.* **2019**, *2*, 205–215. [CrossRef]

marine drugs

Article

Development of Novel Pharmaceutical Forms of the Marine Bioactive Pigment Echinochrome A Enabling Alternative Routes of Administration

Stefanos Kikionis [1], Paraskevi Papakyriakopoulou [2], Panagiotis Mavrogiorgis [1], Elena A. Vasileva [3], Natalia P. Mishchenko [3], Sergey A. Fedoreyev [3], Georgia Valsami [2,*], Efstathia Ioannou [1] and Vassilios Roussis [1,*]

1 Section of Pharmacognosy and Chemistry of Natural Products, Department of Pharmacy, National and Kapodistrian University of Athens, Panepistimiopolis Zografou, 15771 Athens, Greece; skikionis@pharm.uoa.gr (S.K.); panagiotismavro1991@gmail.com (P.M.); eioannou@pharm.uoa.gr (E.I.)
2 Section of Pharmaceutical Technology, Department of Pharmacy, National and Kapodistrian University of Athens, Panepistimiopolis Zografou, 15784 Athens, Greece; ppapakyr@pharm.uoa.gr
3 Laboratory of the Chemistry of Natural Quinonoid Compounds of the G. B. Elyakov Pacific Institute of Bioorganic Chemistry, Far-Eastern Branch of the Russian Academy of Science, Prospect 100 let Vladivostoku, 159, 690022 Vladivostok, Russia; vasilieva_el_an@mail.ru (E.A.V.); mischenkonp@mail.ru (N.P.M.); fedoreev-s@mail.ru (S.A.F.)
* Correspondence: valsami@pharm.uoa.gr (G.V.); roussis@pharm.uoa.gr (V.R.)

Abstract: Echinochrome A (EchA), a marine bioactive pigment isolated from various sea urchin species, is the active agent of the clinically approved drug Histochrome®. EchA is currently only available in the form of an isotonic solution of its di- and tri-sodium salts due to its poor water solubility and sensitivity to oxidation. Electrospun polymeric nanofibers have lately emerged as promising drug carriers capable of improving the dissolution and bioavailability of drugs with limited water solubility. In the current study, EchA isolated from sea urchins of the genus *Diadema* collected at the island of Kastellorizo was incorporated in electrospun micro-/nanofibrous matrices composed of polycaprolactone and polyvinylpyrrolidone in various combinations. The physicochemical properties of the micro-/nanofibers were characterized using SEM, FT-IR, TGA and DSC analyses. The fabricated matrices exhibited variable dissolution/release profiles of EchA, as evidenced in in vitro experiments using gastrointestinal-like fluids (pH 1.2, 4.5 and 6.8). Ex vivo permeability studies using the EchA-loaded micro-/nanofibrous matrices showed an increased permeation of EchA across the duodenum barrier. The results of our study clearly show that electrospun polymeric micro-/nanofibers represent promising carriers for the development of new pharmaceutical formulations with controlled release, as well as increased stability and solubility of EchA, suitable for oral administration, while offering the potential for targeted delivery.

Keywords: Echinochrome A; micro-/nanofibers; electrospinning; controlled release; dissolution; permeability

Citation: Kikionis, S.; Papakyriakopoulou, P.; Mavrogiorgis, P.; Vasileva, E.A.; Mishchenko, N.P.; Fedoreyev, S.A.; Valsami, G.; Ioannou, E.; Roussis, V. Development of Novel Pharmaceutical Forms of the Marine Bioactive Pigment Echinochrome A Enabling Alternative Routes of Administration. *Mar. Drugs* **2023**, *21*, 250. https://doi.org/10.3390/md21040250

Academic Editors: Vladimir I. Kalinin, Pavel S. Dmitrenok and Natalia V. Ivanchina

Received: 11 March 2023
Revised: 13 April 2023
Accepted: 14 April 2023
Published: 18 April 2023

1. Introduction

Echinochrome A (7-ethyl-2,3,5,6,8-pentahydroxy-1,4-naphthoquinone, EchA) is a marine pentahydroxyethyl naphthoquinone and is the most abundant pigment found in various sea urchin species [1]. It is isolated as a red crystalline powder (Russian Federation registration number P N002362/01-2003), and, in the form of its sodium salts, it is used as the active ingredient of the commercially available drug Histochrome® [2]. Histochrome® (Russian Federation registration number P N002363/01-2003) is used as a cardioprotective and antioxidant drug for the treatment of various cardiovascular diseases, such as coronary heart disease and reduction of the necrotic zone in myocardial infarction [3]. Histochrome® (Russian Federation registration number P N002363/02-2003) is also used in ophthalmology for the treatment of ocular diseases, such as macular degeneration, cornea and retina degenerative diseases, primary open-angle glaucoma, post-traumatic hemorrhages, diabetic retinopathy and dyscirculatory disorders in the central artery and vein of the retina [4,5].

EchA can simultaneously block a number of free radical reactions by the neutralization of reactive oxygen species, nitric oxide and peroxide radicals, the chelation of metal ions, the inhibition of lipid peroxidation and the regulation of antioxidant enzymes' levels [6,7]. Over the last years, there has been an increasing interest in the expansion of the commercial applications of EchA since, besides its antioxidant and cardioprotective activities [8–12], EchA has been reported to exhibit anti-inflammatory [13–16], antiviral [17] and antibacterial [18] activities, among others [19–25]. Recently, the multifaceted clinical effects and the mechanisms of action of EchA in the treatment of various cardiovascular, ophthalmic, cerebrovascular, metabolic and inflammatory diseases have been described, highlighting its remarkable pharmacological activities and therapeutic potential [26].

Currently, EchA is used only in the form of an isotonic solution of its di- and tri-sodium salts. The poor water solubility of EchA, in combination with its low stability in solution due to its sensitivity to oxidation, restrict its use in the pharmaceutical sector, prohibiting, up to now, its oral, buccal, nasal or transdermal administration.

Water solubility is a crucial parameter in drug formulation, influencing drug pharmacodynamics and pharmacokinetics. Most of the failures in the development of new drugs are usually associated with the poor water solubility of the active ingredient and the low stability of its solutions. Poor solubility can lead to low bioavailability and, as a result, to suboptimal drug delivery and low efficacy. About 40% of the drugs in the market and almost 90% of drug candidate molecules are poorly soluble in water [27]. Taking into account the range of activities displayed by EchA, the development of new formulations that could increase its solubility and enhance its stability while maintaining its efficacy is of significance. Recently, complexes of β-cyclodextrin–histochrome [28] and carrageenan–EchA [29–31] exhibiting increased stability/solubility in aqueous media have been reported as promising pharmaceutical forms of EchA.

Polymeric micro-/nanofibers have attracted significant interest in drug delivery applications as suitable drug carriers of both water-soluble and poorly water-soluble drugs due to their high encapsulation efficiency and high loading capacity [32–34]. Fibrous scaffolds can be easily produced from electrically charged polymeric solutions or melts by electrospinning, the most efficient, scalable and versatile technique for the preparation of polymeric fibers with diameters ranging from the submicron down to the nanometer scale [35–37]. Various active pharmaceutical ingredients can be incorporated into electrospun micro-/nanofibers for the preparation of alternative solid forms with modified and controlled release characteristics [38–40]. The rapid evaporation of solvents during electrospinning facilitates the incorporation of active ingredients into fibrous networks in the amorphous physical state. Due to homogeneous drug distribution and restricted molecular motion within the polymer matrix, electrospun micro-/nanofibers can maintain the embedded poorly water-soluble drugs in an amorphous state for prolonged periods, improving drug dissolution and bioavailability [41,42].

In the framework of our research interests towards the development of electrospun micro-/nanofibrous matrices for drug delivery and other biomedical applications [43–46], we invested in the development of new pharmaceutical formulations of EchA through its incorporation in polymeric micro-/nanofibers. Electrospun micro-/nanofibrous matrices composed of polycaprolactone (PCL) or/and polyvinylpyrrolidone (PVP) in various ratios loaded with EchA were fabricated and characterized by SEM, FT-IR, TGA and DSC analyses. The dissolution/release profile of EchA from the electrospun matrices was evaluated in vitro in gastrointestinal-like fluids, while the permeation of EchA across the duodenum barrier was assessed in ex vivo permeability studies on the small intestine of young rabbits.

2. Results and Discussion

Micro-/nanofibers represent alternative drug delivery systems that allow for the encapsulation and controlled release of various active ingredients. In the present study, EchA was isolated from sea urchins of the genus *Diadema* collected at the island of Kastellorizo and successfully incorporated into electrospun micro-/nanofibers composed of PCL or/and PVP at different ratios that were evaluated for their efficacy to serve as controlled release systems of EchA suitable for oral administration.

2.1. Physicochemical Characterization of Electrospun Micro-/Nanofibrous Patches

PCL and PVP were selected to serve as EchA-polymer carriers since they are non-toxic polymers that can be easily electrospun, offering, due to their high hydrophobic and hydrophilic character, respectively, different release profiles. In total, six different micro-/nanofibrous matrices incorporating in all cases EchA in a 10% *w/w* (weight to matrix weight) final concentration were fabricated, including two scaffolds based solely on PCL or PVP (PCL-EchA and PVP-EchA, respectively), three blended fiber mats resulting from the co-electrospinning of PCL and PVP in different ratios in an antiparallel setup [PCL-EchA/PVP-EchA (1:3), PCL-EchA/PVP-EchA (1:1) and PCL-EchA/PVP-EchA (3:1)] and one composite fiber mat resulting from the electrospinning of a single spinning solution of PCL and PVP [[PCL-PVP(1:3)]-EchA].

The chemical integrity of EchA following electrospinning was verified by ^{1}H NMR and UV-Vis analyses of the recovered compound after extraction of the fabricated fiber mats.

The electrospinning parameters were optimized to allow for the production of uniform fibrous matrices with bead-free fibers in all cases. The morphological characteristics of the fabricated micro-/nanofibrous matrices were evaluated through analyses of the obtained SEM images (Figure 1). The examination of the PCL-EchA fiber mat revealed a network of cylindrical-shaped fibers with diameters ranging from 185 nm to 1.15 μm and an average diameter size of 521 ± 96 nm. In the case of the PVP-EchA fiber mat, smooth fibers of a cylindrical morphology and a higher diameter size were observed, with diameters ranging from 24 nm to 2.28 μm and an average diameter size of 1.05 ± 0.16 μm. The PCL-EchA/PVP-EchA (1:3), PCL-EchA/PVP-EchA (1:1) and PCL-EchA/PVP-EchA (3:1) blended fiber mats exhibited similar homogeneous fibrous networks with the fiber size differences attributed to the different feeding rates of the PCL-EchA and PVP-EchA spinning solutions during the electrospinning process and the different ratio of the corresponding blended fibers. The PCL-EchA/PVP-EchA (1:3) fiber mat consisted of fibers with diameters ranging from 77 nm to 1.92 μm with an average diameter size of 964 ± 183 nm, whereas in the case of the PCL-EchA/PVP-EchA (1:1) fiber mat, the fiber diameters measured from 105 nm to 1.92 μm with an average diameter size of 857 ± 150 nm. In the case of PCL-EchA/PVP-EchA (3:1), the mat fibers' diameters ranged from 34 nm to 1.16 μm with an average diameter size of 592 ± 115 nm. A uniform micro-/nanofibrous network with the diameters of the fibers ranging from 78 nm to 1.87 μm and an average diameter size of 1.03 ± 0.16 μm was also observed in the case of the composite [PCL-PVP(1:3)]-EchA fiber mat, indicating that the combination of the polymers in the same spinning solution did not affect their spinability.

The FT-IR spectrum of EchA included a broad absorption band centered approximately at 3363 cm^{-1} assigned to –OH stretching vibrations, a –C=C stretching at 1675 cm^{-1}, a characteristic carbonyl –C=O absorption band at 1560 cm^{-1} and in-plane bending vibrations of the hydroxyl groups at 1420 cm^{-1} (Figure 2a) [29]. In the FT-IR spectrum of PCL, the absorption bands observed at 2944 and 2867 cm^{-1} were attributed to asymmetric and symmetric –CH_2 stretching vibrations, whereas the carbonyl –C=O stretching was recorded at 1725 cm^{-1} [47]. The FT-IR spectrum of PVP exhibited a characteristic broad absorption band at 3434 cm^{-1} assigned to –OH stretching vibrations and bands at 2948 and 1651 cm^{-1} attributed to –CH_2 and –C=O stretching vibrations, respectively. –C–H and –C–N bending vibrations were recorded at 1422 and 1284 cm^{-1}, respectively, whereas the –N–C=O bending was observed at 570 cm^{-1} [48].

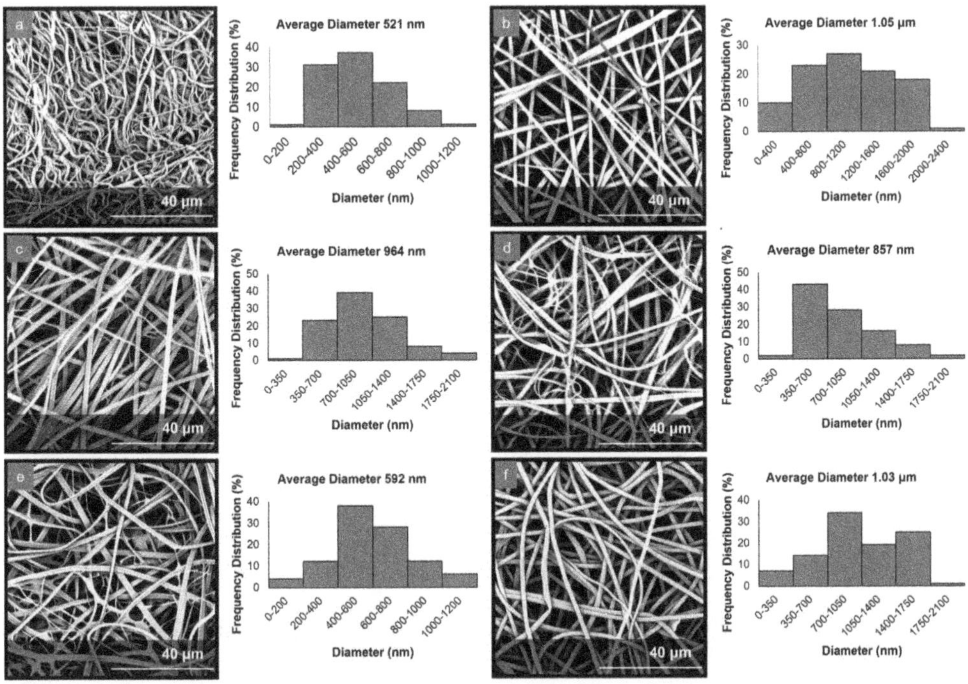

Figure 1. SEM images and diameter distribution histograms of (**a**) PCL-EchA, (**b**) PVP-EchA, (**c**) PCL-EchA/PVP-EchA (1:3), (**d**) PCL-EchA/PVP-EchA (1:1), (**e**) PCL-EchA/PVP-EchA (3:1) and (**f**) [PCL-PVP(1:3)]-EchA fibers.

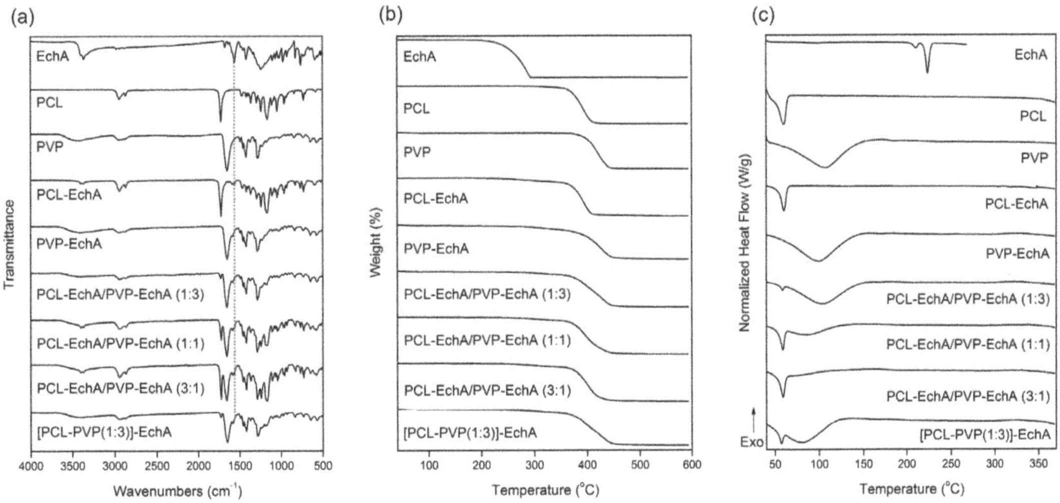

Figure 2. (**a**) FT-IR spectra, (**b**) TGA and (**c**) DSC thermograms of EchA, PCL, PVP, PCL-EchA, PVP-EchA, PCL-EchA/PVP-EchA (1:3), PCL-EchA/PVP-EchA (1:1), PCL-EchA/PVP-EchA (3:1) and [PCL-PVP(1:3)]-EchA fibrous matrices.

The FT-IR spectra of the fabricated matrices revealed the characteristic signals of their ingredients (Figure 2a). Due to the large amount of PCL and PVP dominating the fibrous matrices, all scaffolds exhibited mainly the characteristic absorption bands of the polymeric components. The incorporation of EchA into the polymeric fibers was evident from the absorption band at 1560 cm^{-1} attributed to carbonyl –C=O stretching vibrations since the other signals overlapped with those of PCL and/or PVP. The PCL-EchA/PVP-EchA (1:3), PCL-EchA/PVP-EchA (1:1), PCL-EchA/PVP-EchA (3:1) and [PCL-PVP(1:3)]-EchA fiber mats exhibited similar absorption bands in their FT-IR spectra, with the intensity of the –C=O band of PCL at 1725 cm^{-1} being analogous to the proportion of PCL in the fiber mats.

The physicochemical properties of the fabricated micro-/nanofibrous patches, as well as those of the utilized raw materials, were characterized by TGA and DSC analyses. As shown in the TGA thermograms (Figure 2b), EchA started to decompose at 220 °C, and its degradation was completed at approx. 294 °C. The thermal decomposition of PCL was observed between 355 and 411 °C, whereas PVP was recorded to decompose between 373 and 445 °C. The thermogravimetric curves of the designed matrices revealed the synergistic degradation phenomena of the combined ingredients. The PCL-EchA fibers started to decompose at 248 °C due to the degradation of EchA, with their main mass loss occurring from 357 °C until complete degradation at around 411 °C. In the case of PVP-EchA, the fibers started to decompose at 282 °C, whereas the main decomposition of the fiber mat occurred from 371 to 449 °C. The different ratios of the combined polymers in the blended and composite fibers resulted in different thermal degradation patterns for the corresponding matrices. In the case of the blended fiber mats, the PCL-EchA/PVP-EchA (1:3) scaffold started to decompose at 289 °C due to the presence of EchA, and its main degradation occurred between 359 and 448 °C. In the case of the PCL-EchA/PVP-EchA (1:1) fiber mat, the presence of EchA initiated the decomposition of the fibers at 275 °C, with the main mass loss occurring from 360 to 444 °C. The PCL-EchA/PVP-EchA (3:1) fiber mat started to decompose at 261 °C due to the degradation of EchA, and its major decomposition was recorded between 360 and 441 °C. The decomposition of the composite [PCL-PVP(1:3)]-EchA fiber mat was initiated at 285 °C due to the decomposition of EchA, and its main degradation occurred from 361 to 448 °C.

The DSC thermogram of EchA revealed a weak endotherm at 210.5 °C and a sharp endothermic peak at 223.9 °C attributed to the melting of the compound (Figure 2c). PCL exhibited a sharp melting endotherm at 59.9 °C and PVP showed a broad dehydration endotherm at 105.9 °C. In the thermograms of the various micro-/nanofibrous matrices, only the sharp melting peak of PCL and the broad dehydration curve of PVP were clearly evident at slightly shifted temperatures and different curve patterns due to different synergistic thermal events were attributed to their different compositions, fabrication methods and structural fiber characteristics. None of the characteristic thermal events of EchA were evident in the thermograms of the micro-/nanofiber matrices, indicating the absence of crystalline EchA within the fabricated matrices due to its conversion to an amorphous state during electrospinning [49].

2.2. Dissolution Studies

In vitro dissolution testing is widely used to assess the drug release of developed pharmaceutical products. According to the European Medicine Agency (EMA) [50] and the US Food and Drug Administration (FDA) [51] guidelines for solid oral dosage forms, dissolution testing should be performed in aqueous media with a pH range from 1.2 to 6.8. More specifically, the effect of the dissolution medium on the in vitro drug release should be evaluated at three different pH values (1.2, 4.5 and 6.8), corresponding to the different pH conditions that the drug undergoes while traversing the gastrointestinal tract. Therefore, all developed micro-/nanofibrous patches, as well as neat EchA, were tested at pH 1.2, 4.5 and 6.8 using a USP dissolution apparatus-II with the paddle method.

In all tested dissolution media, the PVP-EchA fiber mats showed a burst release of EchA corresponding to 91.1 ± 4.7, 95.1 ± 4.0 and 93.3 ± 0.7% of the loading dose at pH 1.2, 4.5 and 6.8, respectively (Figure 3). These release values are reported for the 30 min time point, indicating a pH-independent release of EchA that rapidly occurred in all three dissolution media. The conversion of EchA to an amorphous state during the electrospinning process, as suggested by the DSC thermograms, enabled a higher solubility of EchA released from the fibrous scaffolds in comparison to that of neat EchA at all tested pH conditions, as demonstrated by the performance of the PVP-EchA fiber mats. Specifically, after the immediate release of EchA during the deconstruction of the PVP fiber mats at the time point of 30 min, no precipitation of the substance was observed in any of the pH conditions. The high hydrophilicity of PVP enabled the effective wetting of the fiber mat by the water molecules [52] and allowed the immediate release of the incorporated EchA, which remained dissolved in the dissolution media. On the contrary, the lipophilic PCL employed in the case of the PCL-EchA fiber mats impeded the hydration of the scaffold, retarding the release of EchA, as depicted in the dissolution profiles at pH 1.2 and 4.5 (Figure 3).

At pH 1.2 and 4.5, all patches containing PCL showed sustained release profiles. The percentage of EchA released from the blended PCL-EchA/PVP-EchA fiber mats at both pH values increased gradually as the ratio of PVP/PCL in the patch increased. At pH 1.2, this trend was obvious when comparing PCL-EchA to PCL-EchA/PVP-EchA (3:1), PCL-EchA/PVP-EchA (1:1) and PCL-EchA/PVP-EchA (1:3), even though the level of variance in some cases resulted in a non-significant statistical difference (Figure 3a). It is worth noting that at pH 1.2, the blended PCL-EchA/PVP-EchA (1:3) performed markedly better than the composite [PCL-PVP(1:3)]-EchA mats, displaying almost double the release rate of the latter, even though the two fibrous scaffolds differed only in the method of fabrication. On the contrary, at pH 4.5, both fibrous scaffolds displayed similar dissolution profiles (Figure 3b).

EchA is known to be more soluble in higher pH values, as was also observed in the current study (Figure 3c). In particular, the dissolution of neat EchA at pH 6.8 approached the fast-release profile of the PVP-EchA mats, while the PCL-containing patches maintained the sustained release that was also observed at pH 1.2 and 4.5. The lower stability of EchA at less-acidic pH conditions led to a gradual degradation of EchA from the 60 min time point of the dissolution experiment and onwards. When PCL was present in the nanofibrous formulations at an equal or greater ratio to PVP (PCL-EchA, PCL-EchA/PVP-EchA (3:1) and PCL-EchA/PVP-EchA (1:1) fiber mats), the degradation of EchA was slightly delayed due to its slower release rate from the hydrophobic PCL-dominated scaffolds. At pH 6.8, the dissolution profiles of the PCL-containing mats were not significantly different at all time points of the study. The similar trend observed for their dissolution profiles at pH 6.8 may point to an effect of the pH on the polymer's integrity, as also reported in the literature [53]. Indeed, the degradation process of PCL has been found to be more favored as the medium environment becomes more alkaline due to OH^- nucleophilic substitution on the carbonyl group [54,55]. Accordingly, the faster release of EchA from the PCL-containing mats at pH 6.8 can probably be attributed to a possible alteration of the polymer's structure at this pH value.

2.3. Permeability Studies

The dissolution studies revealed that the combination of two polymers, one with a high hydrophilicity and one highly lipophilic, led to the development of fibrous scaffolds that offered a controlled release of EchA. The release rate was determined by the ratio of the two polymers, being higher when the amount of PVP prevailed. The burst release of EchA from the PVP-EchA mats was not considered desirable since the active compound is mostly absorbed in the intestine. Therefore, the fast release in the acidic conditions of the stomach would probably favor the degradation of EchA, without significantly contributing to the absorption process. Accordingly, the PCL-EchA/PVP-EchA (1:3) and the [PCL-PVP(1:3)]-EchA fiber mats, fabricated as either blended or composite, respectively, were selected to be tested in an ex vivo permeability study using rabbit intestine tissues as the biological barrier.

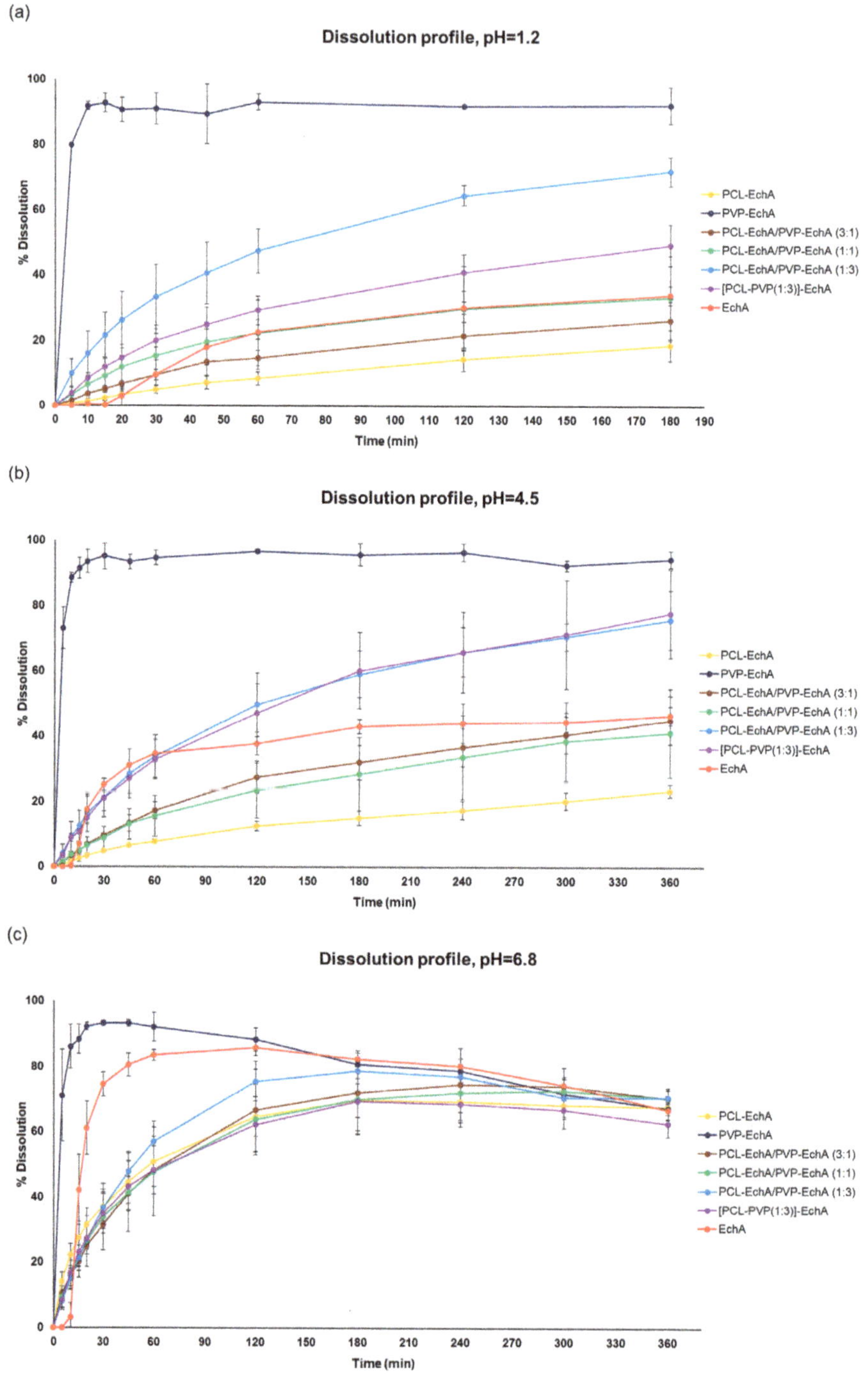

Figure 3. Dissolution/release profiles of neat EchA and PCL-EchA, PVP-EchA, PCL-EchA/PVP-EchA (3:1), PCL-EchA/PVP-EchA (1:1), PCL-EchA/PVP-EchA (1:3) and [PCL-PVP(1:3)]-EchA fibrous matrices at pH (**a**) 1.2, (**b**) 4.5 and (**c**) 6.8 (mean ± SD, $n = 3$).

The ex vivo permeation experiments with PCL-EchA/PVP-EchA (1:3) and [PCL-PVP(1:3)]-EchA patches were performed in the duodenum, the most proximal portion of the small intestine. The permeability of neat EchA was also assessed as a control in order to be compared with the values obtained from the fibrous mats. During the first four hours of the study, no significant differences were noted among the tested formulations. However, at the final time point of the 24 h, the amount of EchA transported across the intestine barrier was higher in the case of the composite [PCL-PVP(1:3)]-EchA patch compared to that of the blended PCL-EchA/PVP-EchA (1:3) mat and neat EchA. Specifically, the permeation from the [PCL-PVP(1:3)]-EchA patch, expressed as a % amount of the loading dose of EchA, was found to be 50% higher than that of both the PCL-EchA/PVP-EchA (1:3) mat and neat EchA (Figure 4).

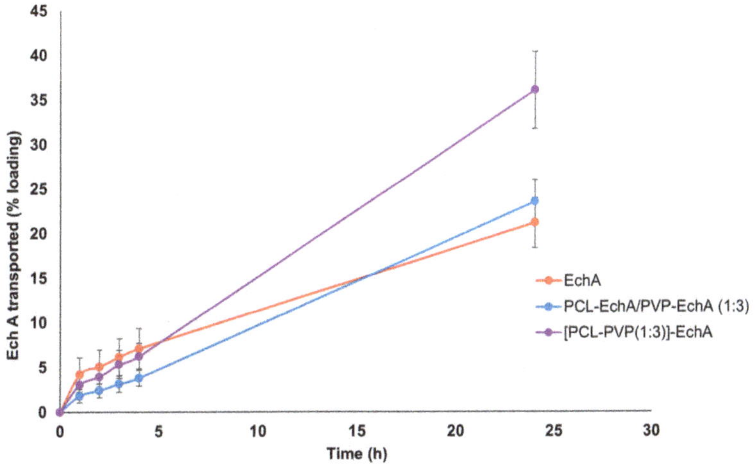

Figure 4. Permeation profiles of EchA from the PCL-EchA/PVP-EchA (1:3) and [PCL-PVP(1:3)]-EchA fibrous matrices, as well as that of neat EchA, through the intestine barrier of rabbits, expressed as quantity permeated per unit area (mean ± SD, $n = 7$).

The mixing of polymers in a single spinning solution electrospun towards the fabrication of the composite [PCL-PVP(1:3)]-EchA fibrous mats led to surfaces of uniform fibers composed of both hydrophilic and lipophilic parts. In contrast, the antiparallel electrospinning employed for the fabrication of the blended PCL-EchA/PVP-EchA (1:3) scaffold was expected to result in a network of PCL and PVP fibers. This structural difference between the two formulations could also be observed in the SEM images (Figure 1), which depicted different architectures for the two nanofibrous mats that may explain their different ex vivo performances. As previously reported [56], PCL presents adequate mucoadhesion at the beginning of the intestine mediated by hydrophobic interactions. Thus, the presence of PCL in all fibers of the [PCL-PVP(1:3)]-EchA mat would probably allow a more efficient interaction of the composite fibers with the intestinal mucosa, as was also revealed by the higher % of EchA transported across the intestine barrier. On the contrary, the hydrophilicity of the PVP fibers that dominated the structure of the PCL-EchA/PVP-EchA (1:3) patch and favored its dissolution, as established by the in vitro experiments, may render the meshing of PCL with the mucus layer more difficult. It is important to note that the structure of the mucosa affects the diffusivity and partition of EchA, evidencing the differences between the formulations that could not be observed in the in vitro dissolution experiments.

In terms of EchA stability, the higher mass balance of the PCL-EchA/PVP-EchA (1:3) and the [PCL-PVP(1:3)]-EchA patches indicated a protective effect of PCL against the pH-induced degradation. Specifically, the mass balance, calculated as the sum of EchA in the receptor, donor and mucosa, was equal to 42.57 ± 8.44% in the case of neat EchA, which was markedly increased when EchA was formulated in the fiber mats, as revealed by the 62.54 ± 9.43 and 70.11 ± 17.6% recovery in the cases of the PCL-EchA/PVP-EchA (1:3) and [PCL-PVP(1:3)]-EchA patches, respectively.

According to our results, it is evident that the selection of the polymers used as carriers of EchA can affect the release profile, allowing for the preparation of micro-/nanofibrous patches according to the desired properties for different routes of administration, while holding the potential for targeted delivery.

3. Materials and Methods

3.1. Materials

Polycaprolactone (PCL) (Mw 80,000), polyvinylpyrrolidone (PVP) (Mw 1,300,000), dichloromethane (DCM), dimethylformamide (DMF) and ethanol (EtOH) were purchased from Sigma-Aldrich (Darmstadt, Germany). All chemicals were of reagent grade and used directly without further purification. Disodium hydrogen phosphate and monosodium phosphate were acquired from Merck KGaA (Darmstadt, Germany). Sodium citrate dihydrate and citric acid were purchased from Fischer (Waltham, MA, USA). Triple-deionized water was used for all preparations. Clear hydroxypropyl methylcellulose, capsules size 0, were purchased from Capsule Fillers (Bologna, Italy). Normal-phase and reversed-phase C_{18} silica gel for column chromatography and TLC plates with Kieselgel 60 F_{254} and RP-18 F_{254s} on aluminum support were purchased from Merck (Darmstadt, Germany).

3.2. Biological Material

Sea urchins of the genus *Diadema* were collected by SCUBA diving at a depth of 5–7 m, at Agios Georgios bay, Kastellorizo Island, Greece, in September 2021. The urchins were immediately frozen, transferred to the Section of Pharmacognosy and Chemistry of Natural Products, Department of Pharmacy, National and Kapodistrian University of Athens, and stored at −20 °C until analyzed.

3.3. Extraction and Isolation of Echinochrome A

After washing with cold fresh water, the sea urchins were dissected, and the entrails were removed. The shells and spines (7.5 kg) were crushed into small pieces and exhaustively extracted with an ethanol solution (70% *v/v*) containing sulfuric acid (8–10% *v/v*) at room temperature for 24 h three times. Subsequently, the acidified extract was filtered and concentrated under reduced pressure to afford a viscous residue which was diluted with distilled water and sequentially extracted with CH_2Cl_2 and EtOAc. After evaporation of the organic solvents in vacuo, the CH_2Cl_2 and EtOAc extracts were pooled together and subjected to vacuum column chromatography over silica gel acidified with oxalic acid in ethanol, using mixtures of cHex with increasing amounts of CH_2Cl_2 and subsequently CH_2Cl_2 with increasing amounts of MeOH as the mobile phase, to yield 8 fractions (A–H). Fraction F (12 g) was subjected to vacuum column chromatography over C_{18} silica gel, using H_2O with increasing amounts of MeOH and finally EtOAc as eluent, to yield 12 fractions (F1–F12). Fraction F8 (H_2O/MeOH 20:80, 4 g) was further purified using vacuum column chromatography over C_{18} silica gel, using H_2O with increasing amounts of MeOH as the mobile phase, to yield 8 fractions (F8a–F8h). Fraction F8d (H_2O/MeOH 20:80, 2.95 g) was identified as EchA in pure form by comparison of its spectroscopic and physical data with those reported in the literature [57].

3.4. Preparation of Electrospun Micro-/Nanofibrous Patches

Electrospinning was conducted using a γ-High Voltage Research DC power supply generator (Gamma High Voltage Research, Ormond Beach, FL, USA) with the spinning solutions being loaded into 10 mL disposable syringes fitted with stainless steel blunt needles (23G). The syringes were mounted on a horizontally positioned programmable syringe pump (Harvard PHD 2000, Harvard Apparatus, Holliston, MA, USA), and the produced micro-/nanofibers were deposited on aluminum foil wrapped on an RC-6000

(NaBond Technologies, Hong Kong) rotating drum collector rotating at 500 rpm. For the fabrication of the blended fiber mats, electrospinning was performed using two horizontally opposed programmable syringe pumps so that the corresponding spinning solutions were co-electrospun in an antiparallel setup to ensure a homogeneous blending of the polymer fibers. Temperature and relative humidity were set at 21 ± 2 °C and 60 ± 5%, respectively. Electrospinning was performed with the solution feeding rate, applied voltage and tip-to-collector distance fixed appropriately in order to obtain the polymeric micro-/nanofibers.

All spinning solutions were prepared by dissolving the corresponding polymers and EchA at appropriate organic solvent systems at room temperature under stirring for 24 h to ensure their homogeneity. Appropriate quantities of EchA were added to each polymer solution so as to result in a 10% *w/w* (weight to matrix weight) final concentration of EchA. Specifically, the spinning solution of PCL-EchA was prepared by dissolving PCL at a concentration of 12% *w/v* in DCM:DMF 8:2 *v/v* followed by the addition of EchA at a concentration of 1.333% *w/v*. The spinning solution of PVP-EchA was prepared by dissolving PVP at a concentration of 12% *w/v* in ethanol followed by the addition of EchA at a concentration of 1.333% *w/v*. The [PCL-PVP(1:3)]-EchA spinning solution was prepared by dissolving PCL at a concentration of 3% *w/v* and PVP at a concentration of 9% *w/v* in DCM:EtOH 7:3 *v/v*, followed by the addition of EchA at a concentration of 1.333% *w/v*.

For the fabrication of the PCL-EchA and PVP-EchA fiber mats, the PCL-EchA and PVP-EchA spinning solutions were separately electrospun with the solution feeding rate, applied voltage and tip-to-collector distance fixed at 3 mL/h, 25 kV and 15 cm, respectively. For the fabrication of the blended PCL-EchA/PVP-EchA (1:3), PCL-EchA/PVP-EchA (1:1) and PCL-EchA/PVP-EchA (3:1) fiber mats, the PCL-EchA and PVP-EchA spinning solutions were co-electrospun in an antiparallel setup to ensure a homogeneous blending of the PCL and PVP polymer fibers at different ratios. Electrospinning was performed with the applied voltage and tip-to-collector distance fixed at 25 kV and 15 cm, respectively, with the syringes mounted on two horizontally opposed programmable syringe pumps. For the fabrication of the PCL-EchA/PVP-EchA (1:3) fiber mat, the feeding rate of the PCL-EchA spinning solution was adjusted to 1.5 mL/h, whereas the feeding rate of the PVP-EchA spinning solution was fixed at 4.5 mL/h. For the fabrication of the PCL-EchA/PVP-EchA (1:1) fiber mat, the feeding rates of the PCL-EchA and PVP-EchA spinning solutions were both adjusted to 3 mL/h. For the fabrication of the PCL-EchA/PVP-EchA (3:1) fiber mat, the feeding rate of the PCL-EchA spinning solution was adjusted to 4.5 mL/h, whereas the feeding rate of the PVP-EchA spinning solution was fixed at 1.5 mL/h. For the fabrication of the composite [PCL-PVP(1:3)]-EchA fiber mat, the [PCL-PVP(1:3)]-EchA spinning solution was electrospun with the solution feeding rate, applied voltage and tip-to-collector distance fixed at 3 mL/h, 25 kV and 15 cm, respectively.

3.5. Scanning Electron Microscopy (SEM)

For the morphological characterization of the micro-/nanofibers, the samples were precoated with a conductive layer of sputtered gold and examined using a PhenomWorld (Thermo Fischer Scientific, Waltham, MA, USA) desktop scanning electron microscope with a tungsten filament (10 kV) and a charge-reduction sample holder. The average fiber diameter was determined using the embedded image analysis software (Phenom Pro Suite/Fibermetric) by evaluating at least 100 measurements per sample.

3.6. Fourier Transform Infrared Spectroscopy (FT-IR)

FT-IR spectra were recorded using the attenuated total reflection (ATR) method on a Bruker Alpha II (Billerica, MA, USA) FT-IR spectrometer.

3.7. Ultraviolet-Visible Spectroscopy (UV-Vis)

UV-Vis spectra were recorded on an Infinite M200 PRO TECAN plate reader (Männedorf, Zürich, Switzerland). The wavelength range was 230-500 nm, and EchA was detected in the samples via short-lived signals (flash = 2) with steps equal to 2. Calibration curves of EchA were prepared for each buffer medium (pH 1.2, 4.5, 6.8 and 7.4). The calibration curve concentrations ranged from 0.5 to 50 μg/mL of EchA and were prepared using appropriate volumes of a starting stock solution of EchA in methanol diluted with the respective buffer medium.

3.8. Thermogravimetric Analysis (TGA)

Thermogravimetric analyses (TGA) were performed using a TGA 55 Thermogravimetric Analyzer (TA Instruments, New Castle, DE, USA) from 40 to 600 °C at a heating rate of 10 °C/min under a 25 mL/min nitrogen flow with the sample weight, sample temperature and heat flow recorded continuously.

3.9. Differential Scanning Calorimetry Analysis (DSC)

Differential scanning calorimetry (DSC) analyses were conducted using a Discovery DSC 25 Thermal Analyzer (TA instruments, New Castle, DE, USA). Sealed samples of 6–7 mg in aluminum pans were heated from 40 to 370 °C for the fibrous scaffolds and from 40 to 270 °C for neat EchA at a constant rate of 10 °C/min under a 50 mL/min nitrogen flow.

3.10. Dissolution Studies

Dissolution tests for the micro-/nanofibrous patches and neat EchA were performed in three different media using a Vankel 750D dissolution apparatus with the paddle method. The experiments were carried out in a 500 mL final volume of buffer at 37 °C and 50 rpm. The patches were introduced in capsule sinkers, while cellulose capsules were employed for neat EchA to avoid floating of the material during the experiment. Specifically, 20 mg (approx. 4 cm^2) of patch (containing 2 mg of EchA) or 2 mg of neat EchA were dispersed in 500 mL of HCl 0.1 M (pH 1.2), citric buffer 0.1 M (pH 4.5) or phosphate buffer 0.1 M (pH 6.8). At defined time intervals (5, 10, 15, 20, 30, 45, 60, 120 and 180 min for the dissolution study at pH 1.2 and 5, 10, 15, 20, 30, 45, 60, 120, 180, 240, 300 and 360 min for the dissolution study at pH 4.5 and 6.8), a 3 mL sample was withdrawn from the dissolution medium of each container and replaced with equal volume of fresh dissolution medium. The withdrawn samples were filtered via regenerated cellulose filters (Whatman, Spartan syringe filters, 0.45 μm) to remove undissolved EchA before the absorbance measurements, using 1 mL for each filter's saturation. The filtered volume was transferred in a UV transparent Corning 96 flat plate, and the absorbance of EchA was measured at 470 nm using an Infinite M200 PRO TECAN (Männedorf, Zürich, Switzerland) plate reader. The dissolution experiments for each patch and neat EchA were run in triplicate at the three tested pH values.

3.11. Permeability Studies

The small intestine of young rabbits (*Oryctolagus cuniculus*) was selected as the most appropriate tissue for the ex vivo diffusion experiments. The duodenum of small intestine was extracted on the day of the experiment from rabbits collected from a local slaughterhouse (Athens, Greece). For the isolation of the intestine tissue, a surgical scissor was used to remove the mesentery and cut the intestine on both sides of the duodenum area. Then, a longitudinal section of the isolated part of the intestinal tract allowed for the thorough cleansing of the tissue with water. The serosa and the muscle layers were gently removed using surgical forceps. During the isolation process, the intestinal mucosa was kept hydrated with saline solution. The extracted intestine was cut into squares of a 1 cm^2 surface to cover completely the Franz cells' diffusion area (0.636 cm^2). Each intestine section was mounted between the donor and receptor compartments of the Franz diffusion cell (Crown Glass, Somerville, MA, USA), with the mucosal interior side facing the donor. To assure the proper cell assembly and the integrity of the mucosa, the receptor compartment was filled with phosphate buffer saline (PBS) solution to check that no liquid could pass to the empty donor due to inappropriate mounting or lack of tissue integrity. The Franz cells were assembled by filling the receptor compartment with 5 mL of PBS (pH 7.4) and the membrane was mounted between the receptor and donor compartments. A magnetic stirrer was added into the receptor, and the two parts were kept together with a metal clamp. The assembled system was allowed to equilibrate at 37 °C for 15 min. Then, 2 mg of each patch (containing 0.2 mg of EchA) or 0.2 mg of neat EchA were placed in the donor compartment and were wetted with 1 mL of the pH 6.8 buffer to resemble the intestinal conditions. Both the donor and the receptor compartments were covered with Parafilm® to prevent evaporation. All experiments lasted for 24 h. At specific time intervals (1, 2, 3, 4 and 24 h), 0.5 mL were sampled from the receptor compartment and replaced by an equal volume of fresh PBS. The absorbance of the samples was measured on an Infinite M200 PRO TECAN (Männedorf, Zürich, Switzerland) plate reader. At the end of the experiment, the residual formulation in the donor compartment was quantitatively collected and diluted to determine the remaining EchA and calculate the mass balance. The amount accumulated in the tissue was recovered by comminuting the mucosa with a surgical blade and homogenizing it with a small pestle three times, using 300 µL of water for 30 sec each time. Then, it was further homogenized with 300 µL of methanol for 30 sec. After homogenization, the extract was diluted and centrifuged before the absorbance was measured. The amounts of EchA recovered from the mucosa, receptor and donor compartments allowed for the calculation of the mass balance.

4. Conclusions

In the present study, we investigated the development of new pharmaceutical forms of EchA through its incorporation into polymeric micro-/nanofibers. EchA was successfully loaded into electrospun micro-/nanofibrous matrices composed of PCL or/and PVP in various combinations. The fabricated matrices were characterized using SEM, FT-IR, TGA and DSC analyses, and the release profile of EchA from the matrices was evaluated in vitro at three different pH values (1.2, 4.5 and 6.8), simulating the different conditions of the gastrointestinal tract. The dissolution studies revealed variable release profiles of EchA from each polymeric matrix in the different pH environments. Subsequently, ex vivo permeation experiments in the duodenum of the small intestine of young rabitts showed that the incoproration of EchA into the electrospun fibers enhanced its permeation across the duodenum barrier.

According to the obtained results, electrospun polymeric micro-/nanofibers represent a promising carrier for EchA that can control its release either in the stomach or the intestine, while at the same time increasing the solubility and stability of the bioactive compound. The selection of the polymers used as carriers of EchA can affect the release profile, thus allowing for the preparation of micro-/nanofibrous patches tailored according to the desired specifications/properties for different routes of administration and different therapeutic targets, offering the potential for targeted delivery. For example, when EchA should be released in the highly acidic pH environment of the stomach, polymers soluble in acidic pH could be blended with polymers of variable solubility in order to achieve the desired release profiles. In contrast, when EchA should be released in the more alkaline environment of the duodenum and the other parts of the small intestine or in the colon, polymers insoluble in acidic but soluble in neutral to slightly basic media should be selected and blended with polymers of variable solubility in order to achieve prolonged release profiles.

Author Contributions: Conceptualization, E.A.V., E.I. and V.R.; methodology, S.K., P.P., G.V., E.I. and V.R.; investigation, S.K., P.P., P.M. and E.A.V.; resources, N.P.M., S.A.F., G.V., E.I. and V.R.; writing—original draft preparation, S.K. and P.P.; writing—review and editing, E.A.V., G.V., E.I. and V.R.; visualization, S.K. and P.P.; supervision, G.V., E.I. and V.R.; project administration, V.R.; funding acquisition, E.A.V., E.I. and V.R. All authors have read and agreed to the published version of the manuscript.

Funding: This research was partially funded by the research projects MARINOVA (grant number 70/3/14684) and BioNP (grant number 70/3/14685), as well as the Russian Science Foundation (grant number 21-75-00069, https://rscf.ru/en/project/21-75-00069/).

Institutional Review Board Statement: Not applicable.

Data Availability Statement: The data presented in this study are available in the present article.

Conflicts of Interest: The authors declare no conflict of interest. The funders had no role in the design of the study; in the collection, analyses, or interpretation of data; in the writing of the manuscript; or in the decision to publish the results.

References

1. Vasileva, E.A.; Mishchenko, N.P.; Tran, V.T.T.; Vo, H.M.N.; Fedoreyev, S.A. Spinochrome identification and quantification in Pacific Sea urchin shells, coelomic fluid and eggs using HPLC-DAD-MS. *Mar. Drugs* **2021**, *19*, 21. [CrossRef]
2. Mishchenko, N.P.; Fedoreev, S.A.; Bagirova, V.L. Histochrome: A new original domestic drug. *Pharm. Chem. J.* **2003**, *37*, 48–52. [CrossRef]
3. Elyakov, G.B.; Maximov, O.B.; Mischenko, N.P.; Koltsova, E.A.; Fedoreev, S.A.; Glebko, L.I.; Krasovskaya, N.P.; Artjukov, A.A. Drug Preparation "Histochrome" for Treating Acute Myocardial Infarction and Ischemic Heart Diseases. European Patent EP1121930B1, 14 November 2007.
4. Elyakov, G.B.; Maximov, O.B.; Mischenko, N.P.; Koltsova, E.A.; Fedoreev, S.A.; Glebko, L.I.; Krasovskaya, N.P.; Artjukov, A.A. Histochrome and Its Therapeutic Use in Acute Myocardial Infarction and Ischemic Heart Disease. U.S. Patent US6410601B2, 25 June 2002.
5. Elyakov, G.B.; Maximov, O.B.; Mischenko, N.P.; Koltsova, E.A.; Fedoreev, S.A.; Glebko, L.I.; Krasovskaya, N.P.; Artjukov, A.A. Composition Comprising Di-and Trisodium Salts of Echinochrome for Treating Ocular Conditions. European Patent EP1121929B1, 3 November 2004.
6. Vasileva, E.A.; Mishchenko, N.P.; Zadorozhny, P.A.; Fedoreyev, S.A. New aminonaphthoquinone from the sea urchins *Strongylocentrotus pallidus* and *Mesocentrotus nudus*. *Nat. Prod. Commun.* **2016**, *11*, 821–824. [CrossRef]
7. Lebedev, A.V.; Ivanova, M.V.; Levitsky, D.O. Echinochrome, a naturally occurring iron chelator and free radical scavenger in artificial and natural membrane systems. *Life Sci.* **2005**, *76*, 863–875. [CrossRef]
8. Jeong, S.H.; Kim, H.K.; Song, I.-S.; Lee, S.J.; Ko, K.S.; Rhee, B.D.; Kim, N.; Mishchenko, N.P.; Fedoryev, S.A.; Stonik, V.A.; et al. Echinochrome A protects mitochondrial function in cardiomyocytes against cardiotoxic drugs. *Mar. Drugs* **2014**, *12*, 2922–2936. [CrossRef]
9. Kim, H.K.; Youm, J.B.; Jeong, S.H.; Lee, S.R.; Song, I.S.; Ko, T.H.; Pronto, J.R.; Ko, K.S.; Rhee, B.D.; Kim, N.; et al. Echinochrome A regulates phosphorylation of phospholamban Ser16 and Thr17 suppressing cardiac SERCA2A Ca^{2+} reuptake. *Pflugers Arch.* **2015**, *467*, 2151–2163. [CrossRef]
10. Park, J.H.; Lee, N.-K.; Lim, H.J.; Mazumder, S.; Rethineswaran, V.K.; Kim, Y.-J.; Jang, W.B.; Ji, S.T.; Kang, S.; Kim, D.Y.; et al. Therapeutic cell protective role of histochrome under oxidative stress in human cardiac progenitor cells. *Mar. Drugs* **2019**, *17*, 368. [CrossRef] [PubMed]
11. Kim, R.; Hur, D.; Kim, H.K.; Han, J.; Mishchenko, N.P.; Fedoreyev, S.A.; Stonik, V.A.; Chang, W. Echinochrome A attenuates cerebral ischemic injury through regulation of cell survival after middle cerebral artery occlusion in rat. *Mar. Drugs* **2019**, *17*, 501. [CrossRef] [PubMed]
12. Hwang, J.-W.; Park, J.-H.; Park, B.-W.; Kim, H.; Kim, J.-J.; Sim, W.-S.; Mishchenko, N.P.; Fedoreyev, S.A.; Vasileva, E.A.; Ban, K.; et al. Histochrome attenuates myocardial ischemia-reperfusion injury by inhibiting ferroptosis-induced cardiomyocyte death. *Antioxidants* **2021**, *10*, 1624. [CrossRef]
13. Artyukov, A.A.; Zelepuga, E.A.; Bogdanovich, L.N.; Lupach, N.M.; Novikov, V.L.; Rutckova, T.A.; Kozlovskaya, E.P. Marine polyhydroxynaphthoquinone, echinochrome A: Prevention of atherosclerotic inflammation and probable molecular targets. *J. Clin. Med.* **2020**, *9*, 1494. [CrossRef] [PubMed]
14. Lennikov, A.; Kitaichi, N.; Noda, K.; Mizuuchi, K.; Ando, R.; Dong, Z.; Fukuhara, J.; Kinoshita, S.; Namba, K.; Ohno, S.; et al. Amelioration of endotoxin-induced uveitis treated with the sea urchin pigment echinochrome in rats. *Mol. Vis.* **2014**, *20*, 171–177. [PubMed]
15. Seol, J.E.; Ahn, S.W.; Seol, B.; Yun, H.R.; Park, N.; Kim, H.K.; Vasileva, E.A.; Mishchenko, N.P.; Fedoreyev, S.A.; Stonik, V.A.; et al. Echinochrome A protects against ultraviolet b-induced photoaging by lowering collagen degradation and inflammatory cell infiltration in hairless mice. *Mar. Drugs* **2021**, *19*, 550. [CrossRef]
16. Oh, S.-J.; Seo, Y.; Ahn, J.-S.; Shin, Y.Y.; Yang, J.W.; Kim, H.K.; Han, J.; Mishchenko, N.P.; Fedoreyev, S.A.; Stonik, V.A.; et al. Echinochrome A reduces colitis in mice and induces in vitro generation of regulatory immune cells. *Mar. Drugs* **2019**, *17*, 622. [CrossRef]
17. Fedoreyev, S.A.; Krylova, N.V.; Mishchenko, N.P.; Vasileva, E.A.; Pislyagin, E.A.; Iunikhina, O.V.; Lavrov, V.F.; Svitich, O.A.; Ebralidze, L.K.; Leonova, G.N. Antiviral and antioxidant properties of echinochrome A. *Mar. Drugs* **2018**, *16*, 509. [CrossRef]
18. Sadek, S.A.; Hassanein, S.S.; Mohamed, A.S.; Soliman, A.M.; Fahmy, S.R. Echinochrome pigment extracted from sea urchin suppress the bacterial activity, inflammation, nociception, and oxidative stress resulted in the inhibition of renal injury in septic rats. *J. Food Biochem.* **2022**, *46*, e13729. [CrossRef]

19. Park, G.-T.; Yoon, J.-W.; Yoo, S.-B.; Song, Y.-C.; Song, P.; Kim, H.-K.; Han, J.; Bae, S.-J.; Ha, K.-T.; Mishchenko, N.P.; et al. Echinochrome A treatment alleviates fibrosis and inflammation in bleomycin-induced scleroderma. *Mar. Drugs* **2021**, *19*, 237. [CrossRef]
20. Yun, H.R.; Ahn, S.W.; Seol, B.; Vasileva, E.A.; Mishchenko, N.P.; Fedoreyev, S.A.; Stonik, V.A.; Han, J.; Ko, K.S.; Rhee, B.D.; et al. Echinochrome A treatment alleviates atopic dermatitis-like skin lesions in NC/Nga mice via IL-4 and IL-13 suppression. *Mar. Drugs* **2021**, *19*, 622. [CrossRef] [PubMed]
21. Lee, S.R.; Pronto, J.R.D.; Sarankhuu, B.-E.; Ko, K.S.; Rhee, B.D.; Kim, N.; Mishchenko, N.P.; Fedoreyev, S.A.; Stonik, V.A.; Han, J. Acetylcholinesterase inhibitory activity of pigment echinochrome A from sea urchin *Scaphechinus mirabilis*. *Mar. Drugs* **2014**, *12*, 3560–3573. [CrossRef] [PubMed]
22. Soliman, A.M.; Mohamed, A.S.; Marie, M.A.S. Comparative study between the hypoglycemic and antioxidant effects of echinochrome on type 1 and type 2 diabetes mellitus. *J. Biol. Res.* **2016**, *4*, 2322.
23. Park, G.-B.; Kim, M.-J.; Vasileva, E.A.; Mishchenko, N.P.; Fedoreyev, S.A.; Stonik, V.A.; Han, J.; Lee, H.S.; Kim, D.; Jeong, J.-Y. Echinochrome A promotes ex vivo expansion of peripheral blood-derived $CD34^+$ cells, potentially through downregulation of ROS production and activation of the Src-Lyn-p110δ pathway. *Mar. Drugs* **2019**, *17*, 526. [CrossRef]
24. Jeong, S.H.; Kim, H.K.; Song, I.-S.; Noh, S.J.; Marquez, J.; Ko, K.S.; Rhee, B.D.; Kim, N.; Mishchenko, N.P.; Fedoreyev, S.A.; et al. Echinochrome A increases mitochondrial mass and function by modulating mitochondrial biogenesis regulatory genes. *Mar. Drugs* **2014**, *12*, 4602–4615. [CrossRef]
25. Seo, D.Y.; McGregor, R.A.; Noh, S.J.; Choi, S.J.; Mishchenko, N.P.; Fedoreyev, S.A.; Stonik, V.A.; Han, J. Echinochrome A improves exercise capacity during short-term endurance training in rats. *Mar. Drugs* **2015**, *13*, 5722–5731. [CrossRef]
26. Kim, H.K.; Vasileva, E.A.; Mishchenko, N.P.; Fedoreyev, S.A.; Han, J. Multifaceted clinical effects of echinochrome. *Mar. Drugs* **2021**, *19*, 412. [CrossRef] [PubMed]
27. Kalepu, S.; Nekkanti, V. Insoluble drug delivery strategies: Review of recent advances and business prospects. *Acta Pharm. Sin. B* **2015**, *5*, 442–453. [CrossRef]
28. Bikbov, M.M.; Nikitin, N.A.; Surkova, V.K.; Farkhutdinov, R.R.; Khalilov, L.M.; Tulyabaev, A.R.; Nikitina, A.F.; Fedoreev, S.A.; Mishchenko, N.P. Experimental antioxidant activity of a β-Cyclodextrin—Histochrome complex. *Pharm. Chem. J.* **2018**, *51*, 980–984. [CrossRef]
29. Yermak, I.M.; Gorbach, V.I.; Karnakov, I.A.; Davydova, V.N.; Pimenova, E.A.; Chistyulin, D.A.; Isakov, V.V.; Glazunov, V.P. Carrageenan gel beads for echinochrome inclusion: Influence of structural features of carrageenan. *Carbohydr. Polym.* **2021**, *272*, 118479. [CrossRef] [PubMed]
30. Yermak, I.M.; Gorbach, V.I.; Glazunov, V.P.; Kravchenko, A.O.; Mishchenko, N.P.; Pimenova, E.A.; Davydova, V.N. Liposomal form of the echinochrome-carrageenan complex. *Mar. Drugs* **2018**, *16*, 324. [CrossRef]
31. Yermak, I.M.; Mischchenko, N.P.; Davydova, V.N.; Glazunov, V.P.; Tarbeeva, D.V.; Kravchenko, A.O.; Pimenova, E.A.; Sorokina, I.V. Carrageenans-sulfated polysaccharides from red seaweeds as matrices for the inclusion of echinochrome. *Mar. Drugs* **2017**, *15*, 337. [CrossRef]
32. Singh, A.; Rath, G.; Singh, R.; Goyal, A.K. Nanofibers: An effective tool for controlled and sustained drug delivery. *Curr. Drug Deliv.* **2018**, *15*, 155–166. [CrossRef]
33. Torres-Martinez, E.J.; Cornejo Bravo, J.M.; Serrano Medina, A.; Pérez González, G.L.; Villarreal Gómez, L.J. A Summary of electrospun nanofibers as drug delivery system: Drugs loaded and biopolymers used as matrices. *Curr. Drug Deliv.* **2018**, *15*, 1360–1374. [CrossRef]
34. Hu, X.; Liu, S.; Zhou, G.; Huang, Y.; Xie, Z.; Jing, X. Electrospinning of polymeric nanofibers for drug delivery applications. *J. Control. Release* **2014**, *185*, 12–21. [CrossRef]
35. Kenry; Lim, C.T. Nanofiber technology: Current status and emerging developments. *Prog. Polym. Sci.* **2017**, *70*, 1–17. [CrossRef]
36. Bhardwaj, N.; Kundu, S.C. Electrospinning: A fascinating fiber fabrication technique. *Biotechnol. Adv.* **2010**, *28*, 325–347. [CrossRef] [PubMed]
37. Greiner, A.; Wendorff, J.H. Electrospinning: A fascinating method for the preparation of ultrathin fibers. *Angew. Chem. Int. Ed.* **2007**, *46*, 5670–5703. [CrossRef] [PubMed]
38. Vlachou, M.; Kikionis, S.; Siamidi, A.; Tragou, K.; Ioannou, E.; Roussis, V.; Tsotinis, A. Modified in vitro release of melatonin loaded in nanofibrous electrospun mats incorporated into monolayered and three-layered tablets. *J. Pharm. Sci.* **2019**, *108*, 970–976. [CrossRef] [PubMed]
39. Al-Enizi, A.M.; Zagho, M.M.; Elzatahry, A.A. Polymer-based electrospun nanofibers for biomedical applications. *Nanomaterials* **2018**, *8*, 259. [CrossRef]
40. Cheng, H.; Yang, X.; Che, X.; Yang, M.; Zhai, G. Biomedical application and controlled drug release of electrospun fibrous materials. *Mater. Sci. Eng. C* **2018**, *90*, 750–763. [CrossRef]
41. Vlachou, M.; Kikionis, S.; Siamidi, A.; Kyriakou, S.; Tsotinis, A.; Ioannou, E.; Roussis, V. Development and characterization of Eudragit®-based electrospun nanofibrous mats and their formulation into nanofiber tablets for the modified release of furosemide. *Pharmaceutics* **2019**, *11*, 480. [CrossRef]
42. Yu, D.-G.; Li, J.-J.; Williams, G.R.; Zhao, M. Electrospun amorphous solid dispersions of poorly water-soluble drugs: A review. *J. Control. Release* **2018**, *292*, 91–110. [CrossRef]

43. Terezaki, A.; Kikionis, S.; Ioannou, E.; Sfiniadakis, I.; Tziveleka, L.-A.; Vitsos, A.; Roussis, V.; Rallis, M. Ulvan/gelatin-based nanofibrous patches as a promising treatment for burn wounds. *J. Drug Deliv. Sci. Technol.* **2022**, *74*, 103535. [CrossRef]
44. Kyritsi, A.; Kikionis, S.; Tagka, A.; Koliarakis, N.; Evangelatou, A.; Papagiannis, P.; Stratigos, A.; Karalis, V.; Dallas, P.; Vitsos, A.; et al. Management of acute radiodermatitis in non-melanoma skin cancer patients using electrospun nanofibrous patches loaded with *Pinus halepensis* bark extract. *Cancers* **2021**, *13*, 2596. [CrossRef]
45. Vlachou, M.; Kikionis, S.; Siamidi, A.; Tragou, K.; Kapoti, S.; Ioannou, E.; Roussis, V.; Tsotinis, A. Fabrication and characterization of electrospun nanofibers for the modified release of the chronobiotic hormone melatonin. *Curr. Drug Deliv.* **2019**, *16*, 79–85. [CrossRef]
46. Toskas, G.; Hund, R.-D.; Laourine, E.; Cherif, C.; Smyrniotopoulos, V.; Roussis, V. Nanofibers based on polysaccharides from the green seaweed *Ulva rigida*. *Carbohydr. Polym.* **2011**, *84*, 1093–1102. [CrossRef]
47. Elzein, T.; Nasser-Eddine, M.; Delaite, C.; Bistac, S.; Dumas, P. FTIR study of polycaprolactone chain organization at interfaces. *J. Colloid Interface Sci.* **2004**, *273*, 381–387. [CrossRef]
48. Rahma, A.; Munir, M.M.; Khairurrijal; Prasetyo, A.; Suendo, V.; Rachmawati, H. Intermolecular interactions and the release pattern of electrospun curcumin-polyvinyl(pyrrolidone) fiber. *Biol. Pharm. Bull.* **2016**, *39*, 163–173. [CrossRef] [PubMed]
49. Li, J.; Li, C.; Zhang, H.; Gao, X.; Wang, T.; Wang, Z.; Zheng, A. Preparation of azithromycin amorphous solid dispersion by hot-melt extrusion: An advantageous technology with taste masking and solubilization effects. *Polymers* **2022**, *14*, 495. [CrossRef]
50. European Medicines Agency. Dissolution Specification for Generic Oral Immediate Release Products—Scientific Guideline. 2017. Available online: https://www.ema.europa.eu/en/dissolution-specification-generic-oral-immediate-release-products-scientific-guideline (accessed on 5 March 2023).
51. Food and Drug Administration. Dissolution Testing of Immediate Release Solid Oral Dosage Forms. 1997. Available online: https://www.fda.gov/regulatory-information/search-fda-guidance-documents/dissolution-testing-immediate-release-solid-oral-dosage-forms (accessed on 5 March 2023).
52. Rahmani, F.; Ziyadi, H.; Baghali, M.; Luo, H.; Ramakrishna, S. Electrospun PVP/PVA nanofiber mat as a novel potential transdermal drug-delivery system for buprenorphine: A solution needed for pain management. *Appl. Sci.* **2021**, *11*, 2779. [CrossRef]
53. White, J.M.; Jurczyk, J.; Van Horn, R.M. Physical structure contributions in pH degradation of PEO-b-PCL films. *Polym. Degrad. Stab.* **2021**, *183*, 109468. [CrossRef]
54. Hernández, A.R.; Contreras, O.C.; Acevedo, J.C.; Navarro Moreno, L.G. Poly(ε-caprolactone) degradation under acidic and alkaline conditions. *Am. J. Polym. Sci.* **2013**, *3*, 70–75. [CrossRef]
55. Łysik, D.; Mystkowska, J.; Markiewicz, G.; Deptuła, P.; Bucki, R. The influence of mucin-based artificial saliva on properties of polycaprolactone and polylactide. *Polymers* **2019**, *11*, 1880. [CrossRef]
56. Abid, Z.; Dalskov Mosgaard, M.; Manfroni, G.; Singh Petersen, R.; Hagner Nielsen, L.; Müllertz, A.; Boisen, A.; Sylvest Keller, S. Investigation of mucoadhesion and degradation of PCL and PLGA microcontainers for oral drug delivery. *Polymers* **2019**, *11*, 1828. [CrossRef] [PubMed]
57. Vasileva, E.A.; Mishchenko, N.P.; Tran, V.T.T.; Vo, H.M.N.; Bui, L.M.; Denisenko, V.A.; Fedoreyev, S.A. Quinoid pigments from the sea urchin *Astropyga radiata*. *Chem. Nat. Compd.* **2017**, *53*, 356–358. [CrossRef]

marine drugs

MDPI

Communication

1-*O*-Alkylglycerol Ethers from the Marine Sponge *Guitarra abbotti* and Their Cytotoxic Activity

Sergey A. Dyshlovoy [1,2,3,*,†], Sergey N. Fedorov [4,*,†], Vasily I. Svetashev [5], Tatiana N. Makarieva [4], Anatoliy I. Kalinovsky [4], Olga P. Moiseenko [4], Vladimir B. Krasokhin [4,‡], Larisa K. Shubina [4], Alla G. Guzii [4], Gunhild von Amsberg [1,2] and Valentin A. Stonik [4]

1 Department of Oncology, Hematology and Bone Marrow Transplantation with Section Pneumology, Hubertus Wald Tumorzentrum—University Cancer Center Hamburg (UCCH), University Medical Center Hamburg-Eppendorf, 20251 Hamburg, Germany; g.von-amsberg@uke.de
2 Martini-Klinik, Prostate Cancer Center, University Hospital Hamburg-Eppendorf, 20251 Hamburg, Germany
3 Institute of Science-Intensive Technologies and Advanced Materials, Far Eastern Federal University, 690091 Vladivostok, Russia
4 G.B. Elyakov Pacific Institute of Bioorganic Chemistry, Far-East Branch of the Russian Academy of Sciences, 690022 Vladivostok, Russia; makarieva@piboc.dvo.ru (T.N.M.); kaaniw@piboc.dvo.ru (A.I.K.); olga-moiseenko@yandex.ru (O.P.M.); shubina@piboc.dvo.ru (L.K.S.); gagry@rambler.ru (A.G.G.); stonik@piboc.dvo.ru (V.A.S.)
5 National Scientific Center of Marine Biology, Far-East Branch of the Russian Academy of Sciences, 690041 Vladivostok, Russia; vsvetashev@mail.ru
* Correspondence: s.dyshlovoy@uke.de (S.A.D.); fedorov@piboc.dvo.ru (S.N.F.)
† These authors contributed equally to this work.
‡ Deceased.

Citation: Dyshlovoy, S.A.; Fedorov, S.N.; Svetashev, V.I.; Makarieva, T.N.; Kalinovsky, A.I.; Moiseenko, O.P.; Krasokhin, V.B.; Shubina, L.K.; Guzii, A.G.; von Amsberg, G.; et al. 1-*O*-Alkylglycerol Ethers from the Marine Sponge *Guitarra abbotti* and Their Cytotoxic Activity. *Mar. Drugs* **2022**, *20*, 409. https://doi.org/10.3390/md20070409

Academic Editors: Vladimir I. Kalinin, Pavel S. Dmitrenok and Natalia V. Ivanchina

Received: 3 June 2022
Accepted: 19 June 2022
Published: 22 June 2022

Abstract: The cytotoxicity-bioassay-guided fractionation of the ethanol extract from the marine sponge *Guitarra abbotti*, whose 1-*O*-alkyl-*sn*-glycerol ethers (AGEs) have not been investigated so far, led to the isolation of a complex lipid fraction containing, along with previously known compounds, six new lipids of the AGE type. The composition of the AGE fraction as well as the structures of 6 new and 22 previously known compounds were established using ^{1}H and ^{13}C NMR, GC/MS, and chemical conversion methods. The new AGEs were identified as: 1-*O*-(Z-docos-15-enyl)-*sn*-glycerol (**1**), 1-*O*-(Z-docos-17-enyl)-*sn*-glycerol (**2**), 1-*O*-(Z-tricos-15-enyl)-*sn*-glycerol (**3**), 1-*O*-(Z-tricos-16-enyl)-*sn*-glycerol (**4**), 1-*O*-(Z-tricos-17-enyl)-*sn*-glycerol (**5**), and 1-*O*-(Z-tetracos-15-enyl)-*sn*-glycerol (**6**). The isolated AGEs show weak cytotoxic activity in THP-1, HL-60, HeLa, DLD-1, SNU C4, SK-MEL-28, and MDA-MB-231 human cancer cells. A further cytotoxicity analysis in JB6 P^{+} Cl41 cells bearing mutated MAP kinase genes revealed that ERK2 and JNK1 play a cytoprotective role in the cellular response to the AGE-induced cytotoxic effects.

Keywords: marine sponge; 1-*O*-alkylglycerol ethers; saturated and unsaturated; NMR; GC/MS; diacetates; bismethylsulfides; trimethylsilyl ethers; cytotoxic activity; MAP kinases

1. Introduction

It is well known that many marine invertebrates, including sponges [1–13], corals [14–19], mollusks [14,20–25], starfish [26], holothurians [27,28], crabs [21], and ascidians [29] as well as some marine algae [30] are well-established sources of a variety of natural 1-*O*-alkyl-*sn*-glycerol ethers (AGEs); for a review, see [31,32]. AGE molecules consist of a long-chain alkyl moiety linked to the glycerol by an ether bond at the *sn*-1 position. Previously, it was established that all the natural AGEs are enantiomerically pure, with an *S* configuration of the asymmetric carbon in the glycerol moiety [33,34]. The AGEs metabolism is controlled by the activity of alkylglycerol monooxygenase (AGMO), which is capable of cleavage of the ether bond of AGEs [35,36]. In marine invertebrates, AGEs mostly present as complex inseparable mixtures of ethers containing different alkyl radicals of various lengths and levels of unsaturation [1–32]. It is highly likely that AGEs are used by invertebrates as a part of their chemical defense against predators. Some reports showed the toxicity of AGEs against fish [10] as well as an antifeedant effect against starfish [14,20]. Both natural and synthetic AGEs possess various biological activities; for a review, see [37]. Among others, there are such useful properties as anticancer [2,3,17,18,21,29,38–45], anti-influenza [46], antibacterial [16,30,47,48], antifungal [49], and antifouling [50] activities found in these metabolites. AGEs have been described to reduce cardiovascular and rheumatoid arthritis risk factors [51], the side effects of radiotherapy [52–54], obesity [55–58], microglial activation, and a neuropathic pain [22,59]. Additionally, AGEs are effective adjuvants [60,61], which modulate endothelial cell permeability [62], the immune response in vitro and in vivo [63–66], open the blood–brain barrier [67,68], are able to penetrate the skin [69], and improve sperm motility [70]. The search for new structural variants of these lipids and the study of their diversity and biological activities represent an interesting aspect of the marine natural products research field.

In a continuation of the studies on cytotoxic marine natural products and their synthetic analogues [71–74], we have examined the ethanol extract of the cold-water marine sponge *Guitarra abbotti*, which exhibited a cytotoxic activity against human leukemia THP-1 cells in a screening assay. The AGEs of this sponge have never been studied before. A bioassay-guided fractionation of the crude extract led to the isolation of an AGE mixture containing 6 new (**1–6**, Figure 1) and 22 known (**7–28**, Tables 1 and 2) compounds of this class. The structures of the AGEs and the composition of the mixture as well as its cytotoxic properties and the partial molecular mechanism of its cytotoxic action are reported.

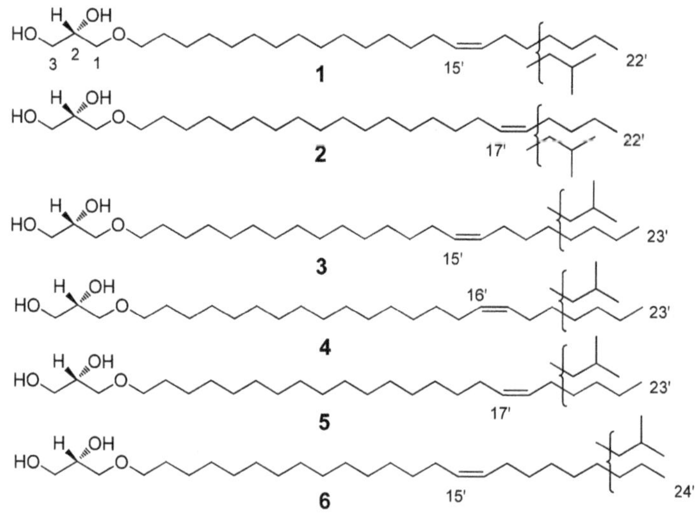

Figure 1. Structures of the new compounds, **1–6**: 1-*O*-(Z-docos-15-enyl)-sn-glycerol (**1**), 1-*O*-(Z-docos-17-enyl)-sn-glycerol (**2**), 1-*O*-(Z-tricos-15-enyl)-sn-glycerol (**3**), 1-*O*-(Z-tricos-16-enyl)-sn-glycerol (**4**), 1-*O*-(Z-tricos-17-enyl)-sn-glycerol (**5**), and 1-*O*-(Z-tetracos-15-enyl)-sn-glycerol (**6**).

Table 1. GC data and characteristic MS fragmentation of TMS derivatives of AGEs.

# of Compound	Retention Time	Fatty Alcohol Residue	Peak Area, % of Total	Characteristic Ion Fragment				
				M^+–CH_3	M^+–$HOSi(CH_3)_3$–$CH_2OSi(CH_3)_3$	M^+–$HOSi(CH_3)_3$	M^+–$HOSi(CH_3)_3$–C_4H_9	M^+
	13.06	N/I *	0.647					
	13.78	N/I *	0.890					
7	16.07	14:0	0.436	417	-	-	285	-
8 (*iso*)	16.58	15:0	0.404	431	-	356	299	-
9 (*n*)	16.87	15:0	1.113	431	-	356	299	-
10 (*iso*)	17.35	16:0	2.514	445	-	370	313	-
17 or **18** or **19** or **20**	17.51	16n:1	1.973	443	265	-	311	458
17 or **18** or **19** or **20**	17.59	16n:1	1.168	-	265	-	311	458
11 (*n*)	17.64	16:0	21.116	445		370	313	-
12 (*iso*)	18.10	17:0	13.648	459		384	327	-
13 (*anteiso*)	18.18	17:0	2.181	459		384	327	-
14 (*n*)	18.38	17:0	1.414	459		384	327	-
15 (*iso*)	18.80	18:0	0.700	473		398	341	-
21 or **22** or **23** or **24**	18.93	18n:1	1.243	471	293	396	339	486
21 or **22** or **23** or **24**	18.99	18n:1	2.118	471	293	396	339	486
16 (*n*)	19.10	18:0	5.403	473	-	398	341	-
25 or **26** or **27**	20.38	20n:1	0.400	-	321	-	-	514
1	21.74	22n:1	16.036	527	349	-	395	542
2	21.81	22n:1	2.860	527	349	-	395	542
3 or **4** or **5**	22.56	23n:1	0.731	-	363	-	-	-
	23.13	N/I *	1.496					
6	23.35	24n:1	1.722	-	377	480	-	-
28	23.43	24n:1	18.241	-	377	480	423	-
	25.35	N/I *	1.548					

*—not identified.

Table 2. GC data and characteristic MS fragmentation of DMDS derivatives of acetylated AGEs.

# of Compound	Retention Time	Fatty Alcohol Residue	Ion Fragment		
			Ion A	Ion B	M^+
17	44.69	16:1n-7	173	259	492
18	45.00	16:1n-9	145	287	492
19	45.69	16:1n-11	117	315	492
20	47.69	16:1n-13	89	343	492
21	51.10	18:1n-9	173	287	520
22	51.40	18:1n-11	145	315	520
23	51.62	18:1n-12	131	329	520
24	52.13	18:1n-13	117	343	520
25	58.48	20:1n-11	173	315	548
26	59.39	20:1n-13	145	343	548
27	60.29	20:1n-15	117	371	548

Table 2. *Cont.*

# of Compound	Retention Time	Fatty Alcohol Residue	Ion Fragment		
			Ion A	Ion B	M^+
1	70.67	22:1n-15	145	371	576
2	72.07	22:1n-17	117	399	576
3	77.50	23:1n-15	159	371	590
4	77.90	23:1n-16	145	385	590
5	78.67	23:1n-17	131	399	590
6	86.27	24:1n-15	173	371	604
28	87.30	24:1n-17	145	399	604

2. Results

2.1. Composition of the Isolated AGE Fraction

The isolated AGE mixture exhibited the characteristic signals in the ^{1}H and 13C NMR spectra, similar to those previously published [5,6,9,10,12] (see Materials and Methods).

To establish the composition and the structures of the isolated AGEs, including the percentages of the compounds, the AGEs were converted to their corresponding TMS ethers and further analyzed with GC/MS. The GC data and characteristic MS fragmentation of the TMS derivatives of AGEs are shown in Table 1. The data analysis revealed that 22 of the identified AGEs (compounds **7–28**) exhibit very similar fragmentations to compounds isolated earlier from other biological sources [1–32]. This fragmentation, in the majority of cases, was related to the loss of a methyl group, trimethylsilanol, (trimethylsilanoxy)methylene, and tetramethylene groups. Nevertheless, six previously unknown compounds (Figure 1) required a more detailed analysis of their structures.

Next, to determine the exact double bond position in the unsaturated AGEs, correspondent acetates and then their dimethyldisulfide derivatives were synthesized and further investigated with GC/MS. The fragmentation scheme is represented in Figure 2, and the generated GC data and characteristic MS fragmentations are shown in Table 2. The S configuration of the stereogenic center at C-2 of the glycerol moiety of AGEs is the same as was determined for the structural analogues previously isolated from marine invertebrates [1–32]. Thus, 18 unsaturated substances were identified, 6 of which, with carbon chain lengths of C_{22}-C_{24}, were identified as new compounds: 1-*O*-(Z-docos-15′-enyl)-*sn*-glycerol (**1**); 1-*O*-(Z-docos-17′-enyl)-*sn*-glycerol (**2**); 1-*O*-(Z-tricos-15′-enyl)-*sn*-glycerol (**3**); 1-*O*-(Z-tricos-16′-enyl)-*sn*-glycerol (**4**); 1-*O*-(Z-tricos-17′-enyl)-*sn*-glycerol (**5**); and 1-*O*-(Z-tetracos-15′-enyl)-*sn*-glycerol (**6**) (Figure 1, Table 2).

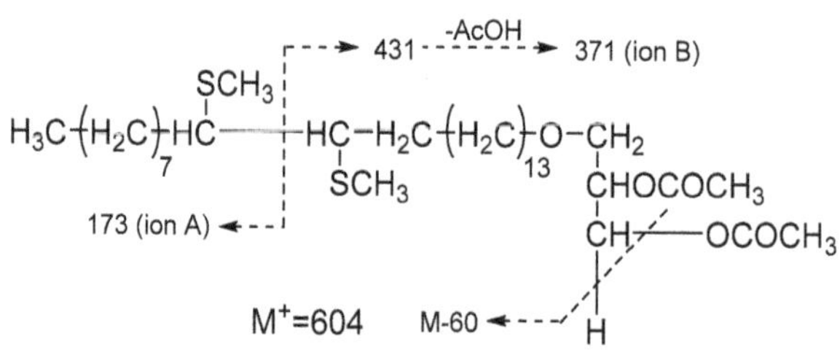

Figure 2. Characteristic MS fragmentation of DMDS derivative of acetylated 1-*O*-(Z-tetracos-15′-enyl)-*sn*-glycerol (**6**).

Apart from the 6 new compounds, **1–6**, the 22 previously known AGEs (10 saturated and 12 unsaturated) were identified in the isolated mixture: 1-*O*-(tetradecanyl)-*sn*-glycerol (**7**); 1-*O*-(pentadecanyl)-*sn*-glycerol (T_R 16.58 of TMS derivarive (TMS-d), Table 1) (**8**); 1-*O*-

(pentadecanyl)-*sn*-glycerol (T_R 16.87 of TMS-d, Table 1) (**9**); 1-*O*-(hexadecanyl)-*sn*-glycerol (T_R 17.35 of TMS-d, Table 1) (**10**); 1-*O*-(hexadecanyl)-*sn*-glycerol (T_R 17.64 of TMS-d, Table 1) (**11**); 1-*O*-(heptadecanyl)-*sn*-glycerol (T_R 18.10 of TMS-d, Table 1) (**12**); 1-*O*-(heptadecanyl)-*sn*-glycerol (T_R 18.18 of TMS-d, Table 1) (**13**); 1-*O*-(heptadecanyl)-*sn*-glycerol (T_R 18.38 of TMS-d, Table 1) (**14**); 1-*O*-(octadecanyl)-*sn*-glycerol (T_R 18.80 of TMS-d, Table 1) (**15**); 1-*O*-(octadecanyl)-*sn*-glycerol (T_R 19.10 of TMS-d, Table 1) (**16**); 1-*O*-(Z-hexadec-7′-enyl)-*sn*-glycerol (T_R 44.69 of DMDS-d, Table 2) (**17**); 1-*O*-(Z-hexadec-9′-enyl)-*sn*-glycerol (T_R 45.00 of DMDS-d, Table 2) (**18**); 1-*O*-(Z-hexadec-11′-enyl)-*sn*-glycerol (T_R 45.69 of DMDS-d, Table 2) (**19**); 1-*O*-(Z-hexadec-13′-enyl)-*sn*-glycerol (T_R 47.69 of DMDS-d, Table 2) (**20**); 1-*O*-(Z-octadec-9-enyl)-*sn*-glycerol (T_R 51.10 of DMDS-d, Table 2) (**21**); 1-*O*-(Z-octadec-11-enyl)-*sn*-glycerol (T_R 51.40 of DMDS-d, Table 2) (**22**); 1-*O*-(Z-octadec-12-enyl)-*sn*-glycerol (T_R 51.62 of DMDS-d, Table 2) (**23**); 1-*O*-(Z-octadec-13-enyl)-*sn*-glycerol (T_R 52.13 of DMDS-d, Table 2) (**24**); 1-*O*-(Z-cos-11′-enyl)-*sn*-glycerol (T_R 58.48 of DMDS-d, Table 2) (**25**); 1-*O*-(Z-cos-13′-enyl)-*sn*-glycerol (T_R 59.39 of DMDS-d, Table 2) (**26**); 1-*O*-(Z-cos-15′-enyl)-*sn*-glycerol (T_R 60.29 of DMDS-d, Table 2) (**27**); and 1-*O*-(Z-tetracos-17′-enyl)-*sn*-glycerol (T_R 87.30 of DMDS-d, Table 2) (**28**) [1–32].

2.2. *Anticancer Effects of the Isolated AGE Fraction*

The cytotoxicity of the isolated AGE mixture was evaluated using the MTS viability assay in seven human cancer cell lines [75]. The calculated IC_{50} values are shown in Table 3.

Table 3. Cytotoxic activity of the isolated AGE mixture against human cancer cell lines. Cisplatin was used as a positive control.

Cell Line	Cancer Type	IC_{50} (AGE), µg/mL	IC_{50} (Cisplatin), µg/mL
HL-60	promyelocytic leukemia	87.4 ± 23.9	0.7 ± 0.09
THP-1	monocytic leukemia	35.9 ± 4.4	3.31 ± 0.74
HeLa	cervix carcinoma	85.9 ± 17	1.55 ± 0.21
DLD-1	colon cancer	103.3 ± 21.9	9.24 ± 1.43
SNU C4	colon cancer	117.4 ± 33.1	4.01 ± 1.21
SK-MEL-28	melanoma	85.8 ± 4.7	0.89 ± 0.04
MDA-MB-231	breast cancer	137 ± 23.8	60.6 ± 26.4

We used dominant negative mutant (DNM) JB6 Cl41 cells to elucidate the roles of the three major MAP kinases (MAPKs) in the cytotoxic effects of the isolated AGE mixture. The cytotoxic activity of the isolated AGE fraction against the normal mouse epidermal cell line JB6 Cl41 and its stable transfectants JB6 Cl41 DN-JNK1, JB6 Cl41 DN-p38, and JB6 Cl41 DN-ERK2 cells is shown in Figure 3. The DN cell lines contain a mutation in the kinase coding gene, which leads to the inactivation of the corresponding kinase in the cells. An examination of the drug effect in the cells bearing inactivated kinase may help to reveal a role of the kinase in the therapeutic effect of the drug. Our experiments indicated that an inactivation of ERK2 and JNK1 results in higher cytotoxicity of the tested AGE fraction. These data suggest that ERK2 and JNK1 (but not p38) play a cytoprotective role in the cellular response to the AGE treatment.

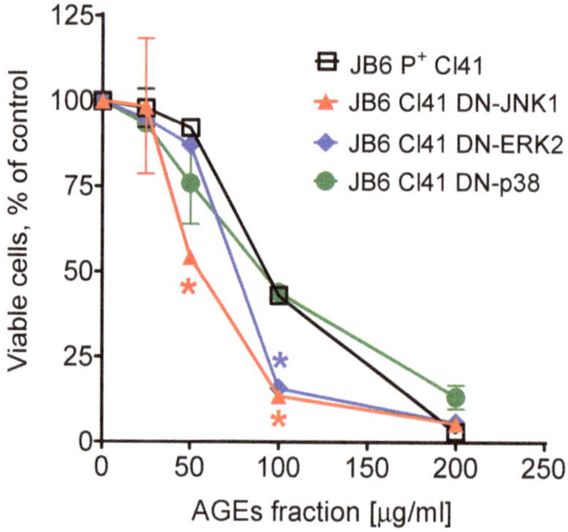

Figure 3. Cytotoxic effect of the isolated AGE mixture in JB6 P+ Cl41 cells and its stable transfectants, JB6 Cl41 DN-JNK1, JB6 Cl41 DN-p38, and JB6 Cl41 DN-ERK2 cells. * Significant different ($p < 0.05$, Student's t-test) in the viability of the cells bearing the mutant kinase compared to the viability measured in JB6 P+ Cl41 cells (wild-type) exposed to the same concentration of the AGE mixture.

3. Discussion

The mixture of the native AGEs was isolated from the extracts of the sponge *G. abbotti* without any prior derivatization or hydrolysis. A comparison with the literature data [5,6,9,10,12] for the ^{1}H NMR spectrum of the AGE mixture (see Materials and Methods) made it possible to correlate the signals at δ 3.55–3.86 ppm with the protons of the glyceride group, those at δ 3.46 ppm with the protons at C1′ of the alkyl chains, those at δ 1.57 ppm with the protons at C2′ of the alkyl chains, those at δ 1.22–1.38 ppm with the protons of other CH_2 groups in the alkyl chains, those at δ 5.34 ppm with the protons at double bonds, those at δ 2.01 ppm with the allylic protons, and those at 0.85–0.88 ppm with protons of the terminal methyl groups of the alkyl chain. Thus, the analysis of the NMR spectra showed that the mixture consists of saturated and unsaturated AGEs, the side chains of which may contain *n*- and *iso*- terminal methyl groups (Figures S1–S3, Supplementary Materials). The Z-geometry of the double bonds in the unsaturated lipids was determined by the small coupling constant, J = 4.6 Hz, of the olefinic proton signal in the ^{1}H spectrum (according to 4.5–6.3 Hz for the Z-configuration in [6,9]) as well as by the shielded chemical shifts of the allylic (δ 26.1 ppm) and olefinic (δ 129.9 ppm) carbon atoms in the ^{13}C NMR spectrum of the AGE mixture [6,76]. The *S* configuration at the asymmetric carbon of the glycerol moiety was previously established for all the natural AGEs [1–32]. Therefore, we assume the same configuration for our isolates. To establish the chemical structures of the lipids of the isolated AGE fraction, first their TMS ethers were obtained and analyzed by the GC/MS method (Figures S4 and S5). The GC data and characteristic MS fragmentation of the TMS derivatives of AGEs (see Table 1) revealed the structures of the 10 previously known saturated AGEs that are part of the isolated alkylglycerol mixture [1–32]. The GC/MS data not only confirmed the presence of both saturated and unsaturated AGEs in the mixture but also indicate their approximately equal ratio. Thus, the two main saturated AGEs, containing side chains with 16 and 17 carbon atoms, accounted for 21% and 13% of the total mixture, while the two main unsaturated AGEs with 22 and 24 carbon atoms in the side chains accounted for 16% and 18%, respectively (Table 1). The positions of the double bonds in the unsaturated AGEs were determined based on the analysis of the characteristic MS fragmentation of the DMDS derivatives of the acetylated AGEs (Table 2, Figure 2, Figures S6 and S7). As a result, after analyzing the mass spectra of the TMS and DMDS derivatives of the unsaturated lipids that are part of the AGE mixture, 18 unsaturated substances were structurally identified, 6 of which, with carbon chain lengths of C_{22}-C_{24}, are new compounds. Their chemical structures were established as follows: 1-*O*-(Z-docos-15′-enyl)-*sn*-glycerol (**1**), 1-*O*-(Z-docos-17′-enyl)-*sn*-glycerol (**2**), 1-*O*-(Z-tricos-15′-enyl)-*sn*-glycerol (**3**), 1-*O*-(Z-tricos-16′-enyl)-*sn*-glycerol (**4**), 1-*O*-(Z-tricos-17′-enyl)-*sn*-glycerol (**5**), and 1-*O*-(Z-tetracos-15′-enyl)-*sn*-glycerol (**6**) (Figure 1).

Different studies suggest that in a living organism AGEs serve as precursors and can be enzymatically converted into various types of biologically active ether lipid compounds [77]. At the same time, enzymatically unmodified AGEs also exhibit biological activities. Ether lipids execute various biological functions, including the modulation of several important signaling pathways [78,79]. Thus, AGEs can be used to generate plasmalogens or platelet-activating factor (PAF) in various biological systems in vitro and in vivo [80,81]. Plasmalogens are important membrane constituents as well as regulators of cholesterol biosynthesis and transport; they play a role in intercellular communication, cell migration, and signal transduction [77,80]. PAF, in turn, is involved in a wide range of membrane-dependent processes and has potent biological activities towards various cell types and systems of an organism, including inflammation, circulation, reproduction, and development; for a review, see [82,83]. A deficiency of plasmalogens can impair some membrane-associated signaling, such as AKT/PKB, leading to a myelination defect [84], while supplementation with plasmalogens modulates important signaling pathways associated with ERK, AKT, p38, and JNK [85–87]. On the other hand, there are some studies which highlight the bioactivity of non-modified AGEs. For example, AGEs were found to accumulate in adipocytes upon differentiation and regulate adipogenesis [58], induce calcium influx in human lymphocytes [88], and inhibit PKC in vitro and in vivo [89].

MAP kinases play roles in various biological processes [85–87]. Due to the broad spectrum of biological targets and processes affected by AGEs, we have evaluated the importance of MAPK-dependent pathways for the cytotoxic effect of the isolated AGE fraction. In our study, using mouse epidermal JB6 Cl41 cells we demonstrated that ERK2- and JNK1- (but not p38) related pathways may play a cytoprotective role in the cytotoxic response to AGE exposure (Figure 3). Thus, the cells bearing knocked-out ERK2 and JNK1 genes are more sensitive to the cytotoxic effect of the isolated AGE fraction compared to the original JB6 Cl41 cells expressing nonmutant MAPK genes. Further experiments, including those examining an effect on MAPK activation, would be necessary to validate this speculation. Within the current study, the further examination of biological activity could not be performed due to the limited amount of the isolated AGE fraction.

The inhibitory effect of AGEs on protein kinase C [89] suggests their potent action against proliferative diseases. Indeed, it was shown that the treatment with AGEs prevents tumor growth in vivo due to the inhibition of angiogenesis [38,39,41,44]. Interestingly, unsaturated AGEs in some assays are significantly more active when compared to saturated molecules [38,39].

We demonstrated here that the AGE mixture inhibits the viability of seven human cancer cell lines representing different cancer entities (Table 3). In all the cell lines, apart from HL-60 and SK-MEL-28, the cytotoxic effects of AGEs from marine invertebrates were reported for the very first time. The highest cytotoxic activity, with IC_{50} = 35.9 μg/mL (Table 3), was shown for THP-1 cells, highlighting the potential of these and related compounds for the treatment of human leukemia. However, it should be noted that the cytotoxic activity of the isolated AGE fraction was overall rather weak.

4. Materials and Methods

4.1. General Procedures

The ^{1}H and ^{13}C NMR spectra were recorded on a Bruker DRX-300 spectrometer (Bruker GmbH, Bremen, Germany) at 300 and 75 MHz, respectively, in $CDCl_3$ with tetramethylsilane as an internal standard.

The GC/MS data for the TMS derivatives were obtained using a Hewlett Packard GC HP6890 instrument (Agilent Technologies Inc., Santa Clara, CA, USA) with an HP5973 mass-selective detector. The injector and transfer line temperatures were 270 °C. A Hewlett Packard HP-5MS capillary column (Agilent Technologies Inc., Santa Clara, CA, USA), 30 m × 0.25 mm, phase layer 0.25 µm, was used at 100 °C with a 2 °C/min ramp to 270 °C, which was held for 30 min. The column contained 5% phenylmethylsiloxane and it was used as the mobile phases at a flow rate of 1 mL/min. The sample was dissolved in chloroform at a concentration of 10 mg/mL. The injection volume was 0.2 µL, and the split ratio was 15:1. The mass spectra were recorded at 70 eV.

The GC/MS data for the DMDS-derived AGEs were obtained using a Shimadzu GCMS-QP5050 instrument (Shimadzu, Kyoto, Japan). An MDN-5S capillary column (Shimadzu, Kyoto, Japan), 30 m × 0.25 mm, phase layer 0.25 µm, was used at 200 °C with a 2 °C/min ramp to 300 °C, which was held for 45 min. The split ratio was 15:1, and the flow rate was 1 mL/min. The injector temperature was 270 °C. The mass spectra were recorded at 70 eV.

Low-pressure column liquid chromatography was performed using KSK silica gel (50–100 µm, Sorbpolymer, Krasnodar, Russia). Sorbfil silica gel plates (4.5 × 6.0 cm, 5–17 µm, Sorbpolymer, Krasnodar, Russia) were used for TLC. HPLC was performed using an Agilent 1100 instrument (Agilent Technologies Inc., Santa Clara, CA, USA) equipped with a differential refractometer on a Diasorb-60-Silicagel (250 × 4.6 mm) column (BioChemMak, Moscow, Russia). Cells were counted using an Olympus inverted research microscope (Olympus, Tokyo, Japan). The absorption of MTS/farmazan was measured spectrophotometrically using the µQuant microplate reader (Bio-Tek Instruments Inc., Winooski, VT, USA).

4.2. Reagents

Minimum essential medium (MEM), DMEM, and RPMI medium were purchased from BioloT (Sankt-Peterburg, Russian Federation); fetal bovine serum (FBS) was purchased from Thermo Fisher Scientific (Cramlington, Northumberland, UK); penicillin/streptomycin was purchased from Bio-Whittaker (Walkersville, MD, USA); and L-glutamine was purchased from Mediatech Inc. (Herndon, VA, USA). The Cell Titer 96 Aqueous One Solution Reagent (MTS) kit for the cell viability assay was purchased from Promega (Madison, WI, USA).

4.3. Animal Material

The marine sponge *Guitarra abbotti* (family Guitarridae, order Poecilosclerida) was collected by dredging at the depth of 109 m at 48°00′08″ N, 153°20′07″ E, Kuril Islands, the Sea of Okhotsk, Pacific Ocean. A voucher specimen is kept in the collection of the G.B. Elyakov Pacific Institute of Bioorganic Chemistry. The taxonomic identification was performed by V.B. Krasokhin.

4.4. Extraction and Isolation

Animal materials (1050 g, wet weight) were extracted with EtOH (2 L) immediately after collection. After evaporation in vacuo, the ethanol extract was re-dissolved in 200 mL of EtOH/H_2O (5:1, *v*/*v*) and extracted with 3 × 200 mL of *n*-hexane. The aqueous-ethanolic and combined *n*-hexane fractions after evaporation were subjected to an evaluation of cytotoxic activity against human leukemia THP-1 cells by the MTS method and showed IC_{50} = 0.6 and 0.5 mg/mL, respectively. The *n*-hexane fraction was selected for the further isolation of anticancer compounds. This fraction (3.525 g) was evaporated and subjected to column chromatography on a silica gel column (diameter/length = 6:14 cm) using an *n*-hexane/AcOEt gradient as an eluent with *n*-hexane to AcOEt ratios of 19:1, 9:1, 4:1, 3:2, 1:1, 2:3, 1:4, 1:9, and 1:19 (*v*/*v*), then with 100% EtOAc and then with 100% EtOH. A 200 mL volume of eluent was used for the elution of each fraction. A subsequent evaluation of the cytotoxic activity revealed that the fraction eluted with *n*-hexane/AcOEt (2:3, *v*/*v*) possessed the highest cytotoxicity towards THP-1 cells, IC_{50} = 125 µg/mL. This fraction

(299 mg) was further subjected to the same silica gel column chromatography using 330 mL of *n*-hexane/AcOEt (3:2, *v*/*v*) and 400 mL of *n*-hexane/AcOEt (1:1, *v*/*v*) as eluents. The collected fractions (10 mL each) were examined using TLC (SiO_2 and *n*-hexane/AcOEt (1:1, *v*/*v*)) as a chromatographic system, with the further detection of spots on the TLC chromatograms using H_2SO_4/EtOH (1:9, *v*/*v*). Following the TLC analysis, three fractions were obtained, i.e., (i) fraction #1 (one grey spot on the TLC, R_f = 0.5); (ii) fraction #2 (one violet spot on the TLC with R_f = 0.2); and (iii) fraction #3 (two spots on the TLC, R_f = 0.5 and R_f = 0.2). The examination of the cytotoxic activity against human THP-1 cells revealed fraction #2 to be the most active, with IC_{50} = 62.5 µg/mL. Further, fraction #2 (185 mg, dry weight) was additionally purified by HPLC using a Diasorb-60-Silica gel column and *n*-hexane/AcOEt (3:2, *v*/*v*) as the eluent. As a result, the more active mixture (IC_{50} = 35.9 µg/mL in THP-1 cells) of the purified 1-*O*-alkylglycerol ethers (AGEs) was obtained (160 mg, dry weight).

4.5. Characterization of the Purified 1-O-Alkylglycerol Ether (AGE) Mixture

The chemical nature of the isolated purified fraction was established using their NMR spectra. Generally, the next spectroscopic information was obtained.

^{1}H NMR ($CDCl_3$, 300 MHz) δ 0.85 d, *J* = 6.6 Hz, *iso*-$C\underline{H}_3$; 0.88 t, *J* = 6.7 Hz, *n*-$C\underline{H}_3$; 1.22-1.38, m, $(C\underline{H}_2)_n$, aliphatic chain; 1.57, quint, *J* = 6.9 Hz, 2H-2′; 2.01, q, *J* = 6.3 Hz, 2H allylic protons; 3.46, td, *J* = 6.6 Hz, *J* = 1.6 Hz, 2H-1′; 3.55, dd, *J* = 9.2 Hz, *J* = 3.5 Hz, 2H-1; 3.64 dd, *J* = 11.4 Hz, *J* = 5.2 Hz, 1H-3a; 3.72 dd, *J* = 11.4 Hz, *J* = 3.8 Hz, 1H-3b; 3.86, m, 1H-2; 5.34, t, *J* = 4.6 Hz, 2H, olefinic protons.

^{13}C NMR ($CDCl_3$, 75 MHz) δ 14.10 (CH_3); 22.64 (CH_2); 26.05(2CH_2, allylic carbons); 27.19 (CH_2); 28.97 (CH_2); 29.07 (CH_2); 29.30 (CH_2); 29.34 (CH_2); 29.44 (CH_2); 29.54 (CH_2); 29.66 (CH_2); 31.76 (CH_2); 31.90 (CH_2); 64.24 (CH_2-3); 70.40 (CH-2); 71.84 (CH_2-1); 72.48 (CH_2-1′); 129.88 (CH=CH).

These data indicate the AGE nature of the compounds in the isolated mixture. These compounds contain a glycerol moiety linked by an ether bond with a fatty alcohol residue. In the analyzed compounds, the fatty alcohol residues contain either normal or iso-ends, and some of them have an additional disubstituted double bond. The majority of the identified metabolite compounds have previously been isolated from different biological sources. However, six of them (**1**–**6**) were new natural products (see below).

4.6. Preparation of Trimethylsilyl Derivatives (TMS-d) and Their GC/MS Analysis

First, 0.1 mg of dried AGE mixture was treated with 0.1 mL of BSTFA (Supelco, Bellefonte, PA, USA) at 60 °C for 1 h to convert the AGEs to their trimethylsilyl derivatives (TMS-d). The obtained TMS-d were analyzed by the GC/MS method (Figures S4 and S5, Supplementary Materials). The mass spectra of TMS-d of the new and previously known AGEs, although the exact position of the double bond could not be precisely determined in some cases, provide valuable information about their structures.

The mass spectra data of the TMS-d for the previously known compounds (**7**–**28**) are given below.

TMS-d of 1-*O*-(tetradecanyl)-*sn*-glycerol (**7**), T_R 16.07, m/z 417, 285, 205, 147, 117, 103, 73; 57; TMS-d of 1-*O*-(pentadecanyl)-*sn*-glycerol (**8**), T_R 16.58, m/z 431, 356, 299, 205, 147, 117, 103, 73; 57; TMS-d of 1-*O*-(pentadecanyl)-*sn*-glycerol (**9**), T_R 16.87, m/z 431, 356, 299, 205, 147, 117, 103, 73; 57; TMS-d of 1-*O*-(hexadecanyl)-*sn*-glycerol (**10**), T_R 17.35, m/z 445, 370, 313, 205, 147, 117, 103, 73, 57; TMS-d of 1-*O*-(Z-hexadec-7′ or 9′ or 11′ or 13′-enyl)-*sn*-glycerol (**17** or **18** or **19** or **20**), T_R 17.51, m/z 458, 443, 355, 311, 265, 205, 147, 117, 103, 73, 55; TMS-d of 1-*O*-(Z-hexadec-7′ or 9′ or 11′ or 13′-enyl)-*sn*-glycerol (**17** or **18** or **19** or **20**), T_R 17.59, m/z 458, 355, 311, 265, 205, 147, 117, 103, 73, 55; TMS-d of 1-*O*-(hexadecanyl)-*sn*-glycerol (**11**), T_R 17.64, m/z 445, 370, 313, 205, 147, 117, 103, 73, 57; TMS-d of 1-*O*-(heptadecanyl)-*sn*-glycerol (**12**), T_R 18.10, m/z 459, 384, 327, 205, 147, 117, 103, 73, 57; TMS-d of 1-*O*-(heptadecanyl)-*sn*-glycerol (**13**), T_R 18.18, m/z 459, 384, 327, 205, 147, 117, 103, 73, 57; TMS-d of 1-*O*-(heptadecanyl)-*sn*-glycerol (**14**), T_R 18.38, m/z 459, 384, 327,

205, 147, 117, 103, 73, 57; TMS-d of 1-*O*-(octadecanyl)-*sn*-glycerol (**15**), T_R 18.80, m/z 473, 398, 341, 205, 147, 117, 103, 73, 57; TMS-d of 1-*O*-(Z-octadec-9′ or 11′ or 12′ or 13′-enyl)-*sn*-glycerol (**21** or **22 or 23** or **24)**, T_R 18.93, m/z 486, 471, 293, 205, 147, 117, 103, 73, 55; TMS-d of 1-*O*-(Z-octadec-9′ or 11′ or 12′ or 13′-enyl)-*sn*-glycerol (**21** or **22** or **23** or **24)**, T_R 18.99, m/z 486, 471, 293, 205, 147, 117, 103, 73, 55.; TMS-d of 1-*O*-(octadecanyl)-*sn*-glycerol (**16**), T_R 19.10, m/z 473, 398, 341, 205, 147, 117, 103, 73, 57; TMS-d of 1-*O*-(Z-cos-11′ or 13′ or 15′-enyl)-*sn*-glycerol (**25** or **26** or **27)**, T_R 20.38, m/z 514, 321, 205, 147, 117, 103, 73, 55; TMS-d of 1-*O*-(Z-tetracos-17′-enyl)-*sn*-glycerol (**28**), T_R 23.43, m/z 480, 467, 390, 377, 205, 147, 117, 103, 73, 55.

For the six new identified compounds (**1–6**) (Figure 1), the mass spectra of their TMS-d are given below.

TMS-d of 1-*O*-(Z-docos-15′-enyl)-*sn*-glycerol (**1**), retention time (T_R) 21.74, m/z 542, 527, 395, 349, 205, 147, 117, 103, 73, 55; TMS-d of 1-*O*-(Z-docos-17′-enyl)-*sn*-glycerol (**2**), T_R 21.81, m/z 542, 527, 395, 349, 205, 147, 117, 103, 73, 55; TMS-d of 1-*O*-(Z-tricos-15′-enyl)-*sn*-glycerol (**3**) or 1-*O*-(Z-tricos-16′-enyl)-*sn*-glycerol (**4**), or 1-*O*-(Z-tricos-17′-enyl)-*sn*-glycerol (**5**), T_R 22.56, m/z 363, 205, 147, 117, 103, 73, 55; TMS-d of 1-*O*-(Z-tetracos-15′-enyl)-*sn*-glycerol (**6**), T_R 23.35, m/z 480, 467, 390, 377, 205, 147, 117, 103, 73, 55.

4.7. Preparation of Acetate Derivatives

First, 1.3 mg of the dried AGE mixture was treated with 0.4 mL of Ac_2O/Py (1:1, *v*/*v*) at RT overnight to convert the AGEs to their diacetate derivatives. Then, 2 mL of EtOH was added, and the sample was dried in vacuo.

4.8. Preparation of Dimethyldisulfide Acetate Derivatives (DMDS-d) of AGEs and Their GC/MS Analysis

First, 1.2 mg of dried AGE acetate mixture (see above) was mixed with 0.2 mL of dimethyldisulfide (DMDS) and 0.05 mL of iodine solution in Et_2O (60 mg/mL) and incubated at RT overnight. Then, 5 mL of *n*-hexane was added, and the mixture was washed with 5 mL of an aqueous 5% solution of $Na_2S_2O_3 \times 5H_2O$ until the color of iodine disappeared. The *n*-hexane fraction was separated, and the reaction products were extracted one more time from the polar fraction using *n*-hexane (1 mL). The combined *n*-hexane fractions were dried over sodium sulfate, evaporated in vacuo, and re-dissolved in hexane for further analysis.

Mass spectra of the DMDS-d of the new AGEs, **1–6**:

DMDS-d of 1-*O*-(Z-docos-15′-enyl)-*sn*-glycerol (**1**), T_R 70.67, m/z 576, 529, 431, 371, 311, 285, 263, 255, 206, 159, 145, 97, 83, 69, 55; DMDS-d of 1-*O*-(Z-docos-17′-enyl)-*sn*-glycerol (**2**), T_R 72.07, m/z 576, 529, 459, 399, 339, 313, 291, 283, 234, 159, 117, 97, 83, 69, 55; DMDS-d of 1-*O*-(Z-tricos-15′-enyl)-*sn*-glycerol (**3**), T_R 77.50, m/z 590, 436, 371, 341, 311, 173, 159, 145, 117, 97, 83, 69, 55; DMDS-d of 1-*O*-(Z-tricos-16′-enyl)-*sn*-glycerol (**4**), T_R 77.90, m/z 590, 436, 385, 341, 297, 159, 145, 117, 97, 83, 69, 55; DMDS-d of 1-*O*-(Z-tricos-17′-enyl)-*sn*-glycerol (**5**), T_R 78.67, m/z 590, 436, 399, 341, 283, 159, 145, 131, 117, 97, 83, 69, 55; DMDS-d of 1-*O*-(Z-tetracos-15′-enyl)-*sn*-glycerol (**6**), T_R 86.27, m/z 604, 450, 371, 311, 255, 207, 173, 159, 109, 97, 83, 69, 55.

Mass spectra of the DMDS-d of the previously known AGEs:

DMDS-d of 1-*O*-(Z-hexadec-7′-enyl)-*sn*-glycerol (**17**), T_R 44.69, m/z 492, 319, 259, 173, 159, 143, 95, 89, 81, 69, 61, 55; DMDS-d of 1-*O*-(Z-hexadec-9′-enyl)-*sn*-glycerol (**18**), T_R 45.00, m/z 492, 347, 287, 171, 159, 145, 95, 89, 81, 69, 61, 55; DMDS-d of 1-*O*-(Z-hexadec-11′-enyl)-*sn*-glycerol (**19**), T_R 45.69, m/z 492, 375, 315, 199, 159, 117, 95, 89, 81, 69, 61, 55; DMDS-d of 1-*O*-(Z-hexadec-13′-enyl)-*sn*-glycerol (**20**), T_R 47.69, m/z 492, 403, 343, 227, 159, 95, 89, 81, 69, 61, 55; DMDS-d of 1-*O*-(Z-octadec-9′-enyl)-*sn*-glycerol (**21**), T_R 51.10, m/z 520, 347, 287, 227, 173, 159, 95, 81, 69, 55; DMDS-d of 1-*O*-(Z-octadec-11′-enyl)-*sn*-glycerol (**22**), T_R 51.40, m/z 520, 375, 315, 255, 145, 159, 95, 81, 69, 55; DMDS-d of 1-*O*-(Z-octadec-12′-enyl)-*sn*-glycerol (**23**), T_R 51.62, m/z 520, 389, 329, 269, 131, 159, 95, 81, 69, 55; DMDS-d of 1-*O*-(Z-octadec-13′-enyl)-*sn*-glycerol (**24**), T_R 52.13, m/z 520, 403, 343, 283, 117, 159, 95,

81, 69, 55; DMDS-d of 1-O-(Z-cos-11′-enyl)-sn-glycerol (**25**), T_R 58.48, m/z 548, 375, 315, 255, 199, 159, 117, 109, 95, 83, 69, 55; DMDS-d of 1-O-(Z-cos-13′-enyl)-sn-glycerol (**26**), T_R 59.39, m/z 548, 403, 343, 227, 159, 117, 109, 95, 83, 69, 55; DMDS-d of 1-O-(Z-cos-15′-enyl)-sn-glycerol (**27**), T_R 60.29, m/z 548, 431, 371, 255, 199, 159, 117, 109, 95, 83, 69, 55; DMDS-d of 1-O-(Z-tetracos-17′-enyl)-sn-glycerol (**28**), T_R 87.30, m/z 604, 459, 399, 339, 283, 159, 145, 97, 83, 69, 55.

4.9. Cell Culture

The JB6 P^+ Cl41 mouse epidermal cell line and its stable transfectants JB6 Cl 41 DN-JNK1, JB6 Cl 41 DN-p38, and JB6 Cl 41 DN-ERK2, which have the knockout JNK1 p38 and ERK2 genes, respectively, were cultured as monolayers at 37 °C and 5% CO_2 in MEM containing 5% fetal bovine serum (FBS), 2 mM L-glutamine, 100 units/mL penicillin, and 100 mg/mL streptomycin. The human cancer cell lines HL-60 (promyelocytic leukemia), THP-1 (monocytic leukemia), HeLa (cervix carcinoma), SNU C4 (colon cancer), DLD-1 (colon cancer), MDA-MB-231 (breast adenocarcinoma), and SK-MEL-28 (melanoma) were obtained from the American Type Culture Collection (Rockville, MD, USA). The HL-60, THP-1, HeLa, DLD-1, SNU C4, and SK-MEL-28 cancer cell lines were cultured at 37 °C and 5% CO_2 in RPMI medium containing 10% FBS, 2 mM L-glutamine, 100 units/mL penicillin, and 100 mg/mL streptomycin. The MDA-MB-231 cancer cell line was cultured at 37 °C and 5% CO_2 in DMEM containing 10% FBS, 2mM L-glutamine, 100 units/mL penicillin, and 100 mg/mL streptomycin. HL-60 and THP-1 cells were cultured as suspensions, and the other cell lines were cultured as monolayers. Information regarding the genetic background of these cell lines is available online at the ATCC website.

4.10. Cell Viability Test

The effect of the obtained AGE mixture on the viability of the THP-1, HL-60, HeLa, DLD-1, SNU C4, SK-MEL-28, MDA-MB-231, JB6 P^+ Cl41, JB6 Cl 41 DN-JNK1, JB6 Cl 41 DN-p38, and JB6 Cl 41 DN-ERK2 cell lines was evaluated using the MTS assay [75]. Briefly, corresponding cells were seeded in 96-well plates (6000 cells per well) and incubated overnight in 100 μL of medium per well for adherent cells, or 50 μL/well for non-adherent cells (THP-1, HL-60). For adherent cells, the media were then replaced with fresh media containing the AGE mixture at various concentrations in a total volume of 0.1 mL per well, and the cells were incubated for 22 h. For suspension cells, 50 μL /well of fresh medium containing the AGE mixture was added, and the cells were incubated for 22 h. Then, 10 mL of the MTS reagent was added into each well, and the MTS reduction was measured spectrophotometrically 2 h later at 492 nm and 690 nm (background) using the μQuant microplate reader.

4.11. Statistical Analysis

The statistical analyses were performed using Statistica 6.0 (StatSoft, Inc., Tulsa, OK, USA) or GraphPad Prism v.9.1.1 software (GraphPad Software, San Diego, CA, USA). The results of two independent experiments, each performed in triplicate, were used for the analyses. Significant differences from control were calculated using Student's t-test. The method of regressions was used to calculate the IC_{50} values.

5. Conclusions

In conclusion, 6 new and 22 previously known 1-O-alkylglycerol ethers were identified in the AGE fraction isolated from the marine sponge *Guitarra abbotti*. Their structures were established using ^{1}H and ^{13}C NMR spectroscopy and a GC/MS analysis of their TMS and DMDS derivatives as well as a comparison with the literature data. The isolated AGE fraction consisted of both saturated and unsaturated AGEs, which presented at a nearly equimolar ratio. The isolated AGEs exhibited a rather weak cytotoxic activity towards seven human cancer cell lines. Moreover, the active MAP kinases ERK2 and JNK1 were shown to play a cytoprotective role in the cellular response to the AGE-induced cytotoxic effects.

Supplementary Materials: The following supporting information can be downloaded at: https://www.mdpi.com/article/10.3390/md20070409/s1, Figure S1: 1H NMR spectrum of AGE mixture in CDCl3 (300 MHz); Figure S2: 13C NMR spectrum of AGE mixture in CDCl3 (75 MHz); Figure S3: DEPT spectrum of AGE mixture in CDCl3; Figure S4: GC data for the TMS-derivatives of AGEs; Figure S5: GC/MS data for the TMS-derivatives of AGEs; Figure S6: GC data for the DMDS-derivatized AGEs; Figure S7: GC/MS data for the DMDS-derivatized AGEs.

Author Contributions: Conception and design, S.A.D., S.N.F., T.N.M., G.v.A. and V.A.S.; Development of methodology, T.N.M., S.A.D., S.N.F., and V.A.S.; Acquisition of data, V.I.S., L.K.S., A.G.G., A.I.K., O.P.M., S.N.F. and S.A.D.; Data analysis, all authors; Data interpretation, all authors; Taxonomic identification of the animal material, V.B.K.; Anticancer activity examination, S.A.D. and S.N.F.; Writing—original draft preparation, S.A.D. and S.N.F.; Writing—review and editing, all authors; Review and/or revision of the final version of the manuscript, all authors; Artwork, S.A.D. and S.N.F.; Fundraising, V.A.S. and G.v.A.; Study supervision, S.N.F. and S.A.D. All authors have read and agreed to the published version of the manuscript.

Funding: The chemical part of this study was supported by the Ministry of Science and Higher Education of the Russian Federation (grant 13.1902.21.0012; contract No. 075-15-2020-796).

Institutional Review Board Statement: Not applicable.

Data Availability Statement: The original data are available from the correspondent authors on request.

Acknowledgments: We devote this article to the 80th anniversary of the birth of an outstanding Russian scientist, Academician of the Russian Academy of Sciences, Professor Valentin A. Stonik, who made a great contribution to the study of various natural products. For many years Professor Stonik has been leading a scientific school known worldwide working on marine and terrestrial natural compounds from diverse chemical classes such as steroids, terpenoids, alkaloids, glycosides, peptides, and lipids as well as products of mixed biosynthesis. The authors are grateful to Z. Dong (Hormel Institute, Austin, Minnesota, USA), who kindly provided JB6 cells used in this work. The study was carried out using the equipment of the Collective Facilities Center "The Far Eastern Center for Structural Molecular Research (NMR/MS) (CSMR PIBOC FEB RAS)".

Conflicts of Interest: The authors declare no conflict of interest.

References

1. Liu, Y.; Jung, J.H.; Ji, H.; Zhang, S. Glicerolipids from a *Sarcotragus* species sponge. *Molecules* **2006**, *11*, 714–719. [CrossRef] [PubMed]
2. Lee, S.-Y.; Zhao, Q.; Choi, K.; Hong, J.; Lee, D.S.; Lee, C.-O.; Jung, J.H. A new glycerol ether from a marine sponge *Stelleta* species. *Nat. Prod. Sci.* **2003**, *9*, 232–234.
3. Seo, Y.; Cho, K.W.; Lee, H.-S.; Rho, J.-R.; Shin, J. New acetylenic enol ethers of glycerol from the sponge *Petrosia* sp. *J. Nat. Prod.* **1999**, *62*, 122–126. [CrossRef] [PubMed]
4. Mishra, P.D.; Wahidullah, S.; D'Souza, L.D.; Kamat, S.Y. Steroids from marine sponges *Suberites vestigium* and *Chrotella australiensis*. *Indian J. Chem.* **1997**, *36B*, 719–721.
5. Quijano, L.; Cruz, F.; Navarrete, I.; Gomez, P.; Rios, T. Alkyl glycerol monoethers in the marine sponge *Desmapsamma anchorata*. *Lipids* **1994**, *29*, 731–734. [CrossRef]
6. Costantino, V.; Fattorusso, E.; Mangoni, A.; Aknin, M.; Fall, A.; Samb, A.; Miralles, J. An unusual ether glycolipid from the senegalese sponge *Trikentrion loeve* Carter. *Tetrahedron* **1993**, *49*, 2711–2716. [CrossRef]
7. Prinsep, M.R.; Blunt, J.W.; Munro, M.H.G. A new sterol sulfate from the marine sponge *Stylopus australis*. *J. Nat. Prod.* **1989**, *52*, 657–659. [CrossRef]
8. Smith, G.M.; Djerassi, C. Phospholipid studies of marine organisms: 14. Ether lipids of the sponge *Tethya aurantia*. *Lipids* **1987**, *22*, 236–240. [CrossRef]
9. Guella, G.; Mancini, I.; Pietra, F. (+)-Raspailyne-A, a novel, acid-sensitive acetylenic enol ether glyceride from the marine sponge *Raspailia pumila*. *J. Chem. Soc. Chem. Commun.* **1986**, *1*, 77–78. [CrossRef]
10. Myers, B.L.; Crews, P. Chiral ether glycerides from a marine sponge. *J. Org. Chem.* **1983**, *48*, 3583–3585. [CrossRef]
11. Schmitz, F.J.; Vanderah, D.J.; Hollenbeak, K.H.; Enwall, C.E.L.; Gopichand, Y.; SenGupta, P.K.; Hossain, M.B.; van der Helm, D. Metabolites from the marine sponge *Tedania ignis*. A new atisanediol and several known diketopiperazines. *J. Org. Chem.* **1983**, *48*, 3941–3945. [CrossRef]
12. Cardellina, J.H.; Graden, C.J.; Greer, B.J. 17Z-Tetracosenyl 1-glycerol ether from the sponges *Cinachyra alloclada* and *Ulosa ruetzleri*. *Lipids* **1983**, *18*, 107–110. [CrossRef]

13. Do, M.N.; Erickson, K.L. Branched chain mono-glycerol ethers from a Taiwanese marine sponge of the genus *Aaptos*. *Tetrahedron Lett.* **1983**, *24*, 5699–5702. [CrossRef]
14. McClintock, J.B.; Baker, B.J.; Slattery, M.; Neine, J.N.; Bryan, P.J.; Yoshida, W.; Davies-Coleman, M.T.; Faulkner, D.J. Chemical defence of common Antarctic shallow-water nudibranch *Tritoniella belli* Eliot (Mollusca: Tritonidae) and its prey, *Clavularia frankliniana* Rouel (Cnidaria: Octocorallia). *J. Chem. Ecol.* **1994**, *20*, 3361–3372. [CrossRef]
15. Imbs, A.B.; Demidkova, D.A.; Dautova, T.N. Lipids and fatty acids of cold-water soft corals and hydrocorals: A comparison with tropical species and implications for coral nutrition. *Mar. Biol.* **2016**, *163*, 202. [CrossRef]
16. Diaz, Y.M.; Laverde, G.V.; Gamba, L.R.; Wandurraga, H.M.; Arevalo-Ferro, C.; Rodriguez, F.R.; Bertran, C.D.; Hernandez, L.C. Biofilm inhibition activity of compounds isolated from two *Eunicea* species collected at the Caribbean Sea. *Rev. Bras. Farm.* **2015**, *25*, 605–611. [CrossRef]
17. Han, A.-R.; Song, J.-I.; Jang, D.S.; Min, H.-Y.; Lee, S.K.; Seo, E.-K. Cytotoxic constituents of the octocoral *Dendronephthya gidantea*. *Arch. Pharm. Res.* **2005**, *28*, 290–293. [CrossRef]
18. Pettit, G.R.; Fujii, Y. Antineoplastic agents. 81. The glycerol ethers of *Palythoa liscia*. *J. Nat. Prod.* **1982**, *45*, 640–643. [CrossRef]
19. Kind, C.A.; Bergmann, W. Contributions to the study of marine products. XI. The occurence of octadecyl alcohol, batyl alcohol, and cetyl palmitate in gorgonias. *J. Org. Chem.* **1942**, *7*, 424–427. [CrossRef]
20. Gustafson, K.; Andersen, R.J. Chemical studies of British Columbia nudibranchs. *Tetrahedron* **1985**, *41*, 1101–1108. [CrossRef]
21. Latyshev, N.A.; Ermakova, S.P.; Ermolenko, E.V.; Imbs, A.B.; Kasyanov, S.P.; Sultanov, R.M. 1-*O*-alkylglycerols from the hepatopancreas of the crab *Paralithodes camtschaticus*, liver of the squid *Berryteuthis magister*, and liver of the skate *Bathyraja parmifera*, and their anticancer activity on human melanoma cells. *J. Food Biochem.* **2019**, *43*, e12828. [CrossRef]
22. Tyrtyshnaia, A.A.; Manzhulo, I.V.; Sultanov, R.M.; Ermolenko, E.V. Adult hippocampal neurogenesis in neuropathic pain and alkyl glycerol ethers treatment. *Acta Histochem.* **2017**, *119*, 812–821. [CrossRef]
23. Hayashi, K.; Okawa, Y.; Kawasaki, K. Liver lipids of gonatid squid *Berryteuthis magister*: A rich source of alkyi glyceryl ethers. *Bull. Jap. Soc. Sci. Fish.* **1985**, *51*, 1523–1526. [CrossRef]
24. Phleger, C.F.; Nichols, P.D.; Virtue, P. Lipids and buoyancy in Southern ocean pteropods. *Lipids* **1997**, *32*, 1093–1100. [CrossRef]
25. Thompson, G.A.; Lee, P. Studies of the α-glyceryl ether lipids occurring in molluscan tissues. *Biochim. Biophys. Acta* **1965**, *98*, 151–159. [CrossRef]
26. Hayashi, K. Content and composition of glyceryl ethers in the pyloric ceca and ovaries of the starfish *Distolasterias nippon*, *Asterina pectinifera*, and *Lysastrosoma anthosticta*. *Fish. Sci.* **1998**, *64*, 852–853. [CrossRef]
27. Rybin, V.; Pavel, K.; Mitrofanov, D. 1-*O*-Alkylglycerol ether lipids in two holothurian species: *Cucumaria japonica* and *C. okhotensis*. *Nat. Prod. Commun.* **2007**, *2*, 933–936. [CrossRef]
28. Santos, V.L.C.; Billett, D.S.M.; Wolff, G.A. 1-*O*-Alkylglyceryl ether lipids of the gut walls and contents of an Abyssal holothurian (*Oneirophanta mutabilis*). *J. Braz. Chem. Soc.* **2002**, *13*, 653–657. [CrossRef]
29. Takeara, R.; Jimenez, P.C.; Wilke, D.V.; de Moraes, M.O.; Pessoa, C.; Lopes, N.P.; Lopes, J.L.C.; da Cruz Lotufo, T.M.; Costa-Lotufo, L.V. Antileukemic effects of *Didemnum psammatodes* (Tunicata: Ascidiacea) constituents. *Comp. Biochem. Physiol. Part A* **2008**, *151*, 363–369. [CrossRef]
30. Othmani, A.; Bunet, R.; Bonnefont, J.-L.; Briand, J.-F.; Culioli, G. Settlement inhibition of marine biofilm bacteria and barnacle larvae by compounds isolated from the Mediterranean brown alga *Taonia atomaria*. *J. Appl. Phycol.* **2016**, *28*, 1975–1986. [CrossRef]
31. Magnusson, C.D.; Haraldsson, G.G. Ether lipids. *Chem. Phys. Lipids* **2011**, *164*, 315–340. [CrossRef] [PubMed]
32. Urate, K.; Takaishi, N. Ether lipids based on the glyceryl ether skeleton: Present state, future potential. *JAOCS* **1996**, *73*, 819–830. [CrossRef]
33. Baer, E.; Fisher, H.O.L. Studies on acetone-glyceraldehyde, and optically active glycerides IX. Configuration of the natural batyl, chimil, and selachyl alcohols. *J. Biol. Chem.* **1941**, *140*, 397–410. [CrossRef]
34. IUPAC-IUB Commission on Biochemical Nomenclature. The nomenclature of lipids (Recommendations 1976). *J. Lipid Res.* **1978**, *19*, 114–128. [CrossRef]
35. Taguchi, H.; Armarego, W.L.F. Glyceryl-ether monooxygenase [EC 1.14.16.5]. A microsomal enzyme of ether lipid metabolism. *Med. Res. Rev.* **1998**, *18*, 43–89. [CrossRef]
36. Watschinger, K.; Keller, M.A.; Golderer, G.; Hermann, M.; Maglione, M.; Sarg, B.; Lindner, H.H.; Hermetter, A.; Werner-Felmayer, G.; Konrat, R.; et al. Identification of the gene encoding alkylglycerol monooxygenase defines a third class of tetrahydrobiopterin-dependent enzymes. *Proc. Natl. Acad. Sci. USA* **2010**, *107*, 13672–13677. [CrossRef] [PubMed]
37. Iannitti, T.; Palmieri, B. An update on the therapeutic role of alkylglycerols. *Mar. Drugs* **2010**, *8*, 2267–2300. [CrossRef] [PubMed]
38. Deniau, A.-L.; Mosset, P.; Le Bot, D.; Legrand, A.B. Which alkylglycerols from shark liver oil have anti-tumour activities? *Biochimie* **2011**, *93*, 1–3. [CrossRef]
39. Deniau, A.-L.; Le Bot, D.; Mosset, P.; Legrand, A.B. Activités antitumorale et antimétastasique des alkylglycérols naturels: Relation structure-activité. *OCL Ol. Corps Gras Lipides* **2010**, *17*, 236–237. (In French) [CrossRef]
40. Pedrono, F.; Saıag, B.; Moulinoux, J.-P.; Legrand, A.B. 1-*O*-Alkylglycerols reduce the stimulating effects of bFGF on endothelial cell proliferation *in vitro*. *Cancer Lett.* **2007**, *251*, 317–322. [CrossRef]
41. Pedrono, F.; Martin, B.; Leduc, C.; Le Lan, J.; Saiag, B.; Legrand, P.; Moulinoux, J.-P.; Legrand, A.B. Natural alkylglycerols restrain growth and metastasis of grafted tumors in mice. *Nutr. Cancer* **2004**, *48*, 64–69. [CrossRef]

42. Krotkiewski, M.; Przybyszewska, M.; Janik, P. Cytostatic and cytotoxic effects of alkylglycerols (Ecomer). *Med. Sci. Monit.* **2003**, *9*, 131–135.
43. Hernández-Colina, M.; Cermeño, A.M.; García, A.D. Selective cytotoxic effect of 1-*O*-undecylglycerol in human melanoma cells. *J. Pharm. Pharmacog. Res.* **2016**, *4*, 84–94.
44. Brohult, A.; Brohult, J.; Brohult, S. Regression of tumor growth after administration of alkoxyglycerols. *Acta Obstet. Gynecol. Scand.* **1978**, *57*, 79–83. [CrossRef]
45. Brohult, A.; Brohult, J.; Brohult, S. Biochemical effects of alkylglycerols and their use in cancer therapy. *Acta Chem. Scand.* **1970**, *24*, 730–732. [CrossRef]
46. Di Iannitti, T.; Capone, S.; Palmieri, B. A telephone interview to assess alkylglycerols effectiveness in preventing influenza-like symptoms in Modena, Emilia Romagna, Italy, in the season 2009–2010. *Clin. Ter.* **2011**, *162*, e115–e118.
47. Daza, K.E.C.; Gómez, E.S.; Murillo, B.D.M.; Wandurraga, H.M. Natural and enantiopure alkylglycerols as antibiofilms against clinical bacterial isolates and quorum sensing inhibitors of *Chromobacterium violaceum* ATCC 12472. *Antibiotics* **2021**, *10*, 430.
48. Montoya, D.J.F.; Jordan, L.A.C.; Murillo, B.M.; Gomez, E.S.; Wandurraga, H.M. Enantiomeric synthesis of natural alkylglycerols and their antibacterial and antibiofilm activities. *Nat. Prod. Res.* **2021**, *35*, 2544–2550. [CrossRef]
49. Hayens, M.P.; Buckley, H.R.; Higgins, M.L.; Pieringer, R.A. Synergism between the antifungal agents amphotericin B and alkyl glycerol ethers. *Antimicrob. Agent. Chemother.* **1994**, *38*, 1523–1529. [CrossRef]
50. Nascimento, T.S.; Monteiro, L.G.; Braga, E.F.; Batista, W.R.; Albert, A.L.M.; Chantre, L.G.F.; de Machado, S.P.; Lopes, R.S.C.; Lopes, C.C. Synthesis of natural ether lipids and 1-*O*-hexadecylglycero-arylboronates via an epoxide-ring opening approach: Potential antifouling additives to marine paint coatings. *Int. J. Adv. Res. Sci. Eng. Technol.* **2018**, *5*, 326–332. [CrossRef]
51. Parri, A.; Fito, M.; Torres, C.F.; Munoz-Aguayo, D.; Schroder, H.; Cano, J.F.; Vazquez, L.; Reglero, G.; Covas, M.-I. Alkylglycerols reduce serum complement and plasma vascular endothelial growth factor in obese individuals. *Inflammopharmacology* **2016**, *24*, 127–131. [CrossRef] [PubMed]
52. Brohult, A.; Brohult, J.; Brohult, S.; Joelsson, I. Effect of alkoxyglycerols on the frequency of fistulas following radiation therapy for carcinoma of the uterine cervix. *Acta Obstet. Gynecol. Scand.* **1979**, *58*, 203–207. [CrossRef] [PubMed]
53. Alexander, P.; Connell, D.I.; Brohult, A.; Brohult, S. Reduction of radiation induced shortening of life span by a diet augmented with alkoxy glycerol esters and essential fatty acids. *Gerontologia* **1959**, *3*, 147–152. [CrossRef] [PubMed]
54. Bronhult, A.; Holmberg, J. Alkoxyglycerols in the treatment of leukopenia caused by irradiation. *Nature* **1954**, *174*, 1102–1103. [CrossRef]
55. Yu, H.; Dilbaz, S.; Cobmann, J.; Hoang, A.C.; Diedrich, V.; Herwig, A.; Harauma, A.; Hoshi, Y.; Moriguchi, T.; Landgraf, K.; et al. Breast milk alkylglycerols sustain beige adipocytes through adipose tissue macrophages. *J. Clin. Investig.* **2019**, *129*, 2485–2499. [CrossRef]
56. Hofer, D.C.; Pessentheiner, A.R.; Pelzmann, H.J.; Schlager, S.; Madreiter-Sokolowski, C.T.; Kolb, D.; Eichmann, T.O.; Rechberger, G.; Bilban, M.; Graier, W.F.; et al. Critical role of the peroxisomal protein PEX16 in white adipocyte development and lipid homeostasis. *Biochim. Biophys. Acta* **2017**, *1862*, 358–368. [CrossRef]
57. Zhang, M.; Sun, S.; Tang, N.; Cai, W.; Qian, L. Oral administration of alkylglycerols differentially modulates high-fat diet-induced obesity and insulin resistance in mice. *Evid. Based Complement. Alter. Med.* **2013**, *2013*, 834027. [CrossRef]
58. Homan, E.A.; Kim, Y.-G.; Cardia, J.P.; Saghatelian, A. Monoalkylglycerol ether lipids promote adipogenesis. *J. Am. Chem. Soc.* **2011**, *133*, 5178–5181. [CrossRef]
59. Tyrtyshnaia, A.A.; Manzhulo, I.V.; Kipryushina, Y.; Ermolenko, E.V. Neuroinflammation and adult hippocampal neurogenesis in neuropathic pain and alkyl glycerol ethers treatment in aged mice. *Int. J. Mol. Med.* **2019**, *43*, 2153–2163. [CrossRef]
60. Acevedo, R.; Gil, D.; del Campo, J.; Bracho, G.; Valdes, Y.; Perez, O. The adjuvant potential of synthetic alkylglycerols. *Vaccine* **2006**, *24*, S32–S33. [CrossRef]
61. Ngwenya, B.Z.; Foster, D.M. Enhancement of antibody production by lysophosphatidylcholine and alkylglycerol. *Proc. Soc. Exp. Biol. Med.* **1991**, *196*, 69–75. [CrossRef]
62. Marigny, K.; Pedrono, F.; Martin-Chouly, C.A.E.; Youmine, H.; Saiag, B.; Legrand, A.B. Modulation of endothelial permeability by 1-*O*-alkylglycerols. *Acta Physiol. Scand.* **2002**, *176*, 263–268. [CrossRef]
63. Qian, L.; Zhang, M.; Wu, S.; Zhong, Y.; Van Tol, E.; Cai, W. Alkylglycerols modulate the proliferation and differentiation of non-specific agonist and specific antigen-stimulated splenic lymphocytes. *PLoS ONE* **2014**, *9*, e96207. [CrossRef]
64. Homma, S.; Yamamoto, N. Activation process of macrophages after in vitro treatment of mouse lymphocytes with dodecylglycerol. *Clin. Exp. Immunol.* **1990**, *79*, 307–313. [CrossRef]
65. Yamamoto, N.; St Claire, D.A., Jr.; Homma, S.; Ngwenya, B.Z. Activation of mouse macrophages by alkylglycerols, inflammation products of cancerous tissues. *Cancer Res.* **1988**, *48*, 6044–6049.
66. Brohult, A.; Brohult, J.; Brohult, S. Effect of irradiation and alkoxyglycerol treatment on the formation of antibodies after *Salmonella* vaccination. *Experientia* **1972**, *8*, 954–955. [CrossRef]
67. Erdlenbruch, B.; Alipour, M.; Fricker, G.; Miller, D.S.; Kugler, W.; Eibl, H.; Lakomek, M. Alkylglycerol opening of the blood–brain barrier to small and large fluorescence markers in normal and C6 glioma-bearing rats and isolated rat brain capillaries. *Br. J. Pharm.* **2003**, *140*, 1201–1210. [CrossRef]
68. Erdlenbruch, B.; Jendrossek, V.; Eibl, H.; Lakomek, M. Transient and controllable opening of the blood-brain barrier to cytostatic and antibiotic agents by alkylglycerols in rats. *Exp. Brain Res.* **2000**, *135*, 417–422.

69. Bernal-Chávez, S.A.; Pérez-Carreto, L.Y.; Nava-Arzaluz, M.G.; Ganem-Rondero, A. Alkylglycerol derivatives, a new class of skin penetration modulators. *Molecules* **2017**, *22*, 185. [CrossRef]
70. Cheminade, C.; Gautier, V.; Hichami, A.; Allaume, P.; Le Lannou, D.; Legrand, A.B. 1-*O*-Alkylglycerols improve boar sperm motility and fertility. *Biol. Reprod.* **2002**, *66*, 421–428. [CrossRef]
71. Shubina, L.K.; Makarieva, T.N.; Denisenko, V.A.; Popov, R.S.; Dyshlovoy, S.A.; Grebnev, B.B.; Dmitrenok, P.S.; von Amsberg, G.; Stonik, V.A. Gracilosulfates A–G, monosulfated polyoxygenated steroids from the marine sponge *Haliclona gracilis*. *Mar. Drugs* **2020**, *18*, 454. [CrossRef] [PubMed]
72. Dyshlovoy, S.A.; Kaune, M.; Kriegs, M.; Hauschild, J.; Busenbender, T.; Shubina, L.K.; Makarieva, T.N.; Hoffer, K.; Bokemeyer, C.; Graefen, M.; et al. Marine alkaloid monanchoxymycalin C: A new specific activator of JNK1/2 kinase with anticancer properties. *Sci. Rep.* **2020**, *10*, 13178. [CrossRef] [PubMed]
73. Fedorov, S.N.; Kuzmich, A.S.; Sabutskii, Y.E.; Guzii, A.G.; Popov, R.S.; Ogurtsov, V.A.; Rakitin, O.A.; Polonik, S.G. Synthesis and studies of acetylthioglycoside conjugates of 4-chloro-1,2-dithiole-3-thione as potential antitumor agents. *Russ. Chem. Bull.* **2021**, *70*, 573–579. [CrossRef]
74. Fedorov, S.N.; Kuzmich, A.S.; Agafonova, I.G.; Sabutskii, Y.E.; Guzii, A.G.; Popov, R.S.; Ogurtsov, V.A.; Rakitin, O.A.; Polonik, S.G. Synthesis and study of thioglycoside conjugates of 4-chloro-1,2-dithiol-3-one as potential cancer-preventive substances in vitro and in vivo. *Russ. Chem. Bull.* **2022**, *71*, 489–495. [CrossRef]
75. Baltrop, J.A.; Owen, T.C.; Cory, A.H.; Cory, J.G. 5-(3-Carboxymethoxyphenyl)-2-(4,5-dimethylthiazolyl)-3-(4-sulfophenyl) tetrazolium, innersalt (MTS) and related analogs of 3-(4,5-dimethylthiazolyl)-2,5-diphenyltetrazolium bromide (MTT) reducing to purple water-soluble formazans as cell-viability indicators. *Bioorg. Med. Chem. Lett.* **1991**, *1*, 611–614.
76. Gunstone, F.D.; Pollard, M.R.; Scrimgeour, C.M.; Vedanayagam, H.S. Fatty acids. Part 50. ^{13}C Nuclear magnetic resonance studies of olefinic fatty acids and esters. *Chem. Phys. Lipids* **1977**, *18*, 115–129. [CrossRef]
77. Dorninger, F.; Forss-Petter, S.; Wimmer, I.; Berger, J. Plasmalogens, platelet-activating factor and beyond—ether lipids in signaling and neurodegeneration. *Neurobiol. Disease* **2020**, *145*, 105061. [CrossRef]
78. Dorninger, F.; Forss-Petter, S.; Berger, J. From peroxisomal disorders to common neurodegenerative diseases—the role of ether phospholipids in the nervous system. *FEBS Lett.* **2017**, *591*, 2761–2788. [CrossRef]
79. Dean, J.M.; Lodhi, I.J. Structural and functional roles of ether lipids. *Protein Cell* **2018**, *9*, 196–206. [CrossRef]
80. Paul, S.; Rasmiena, A.A.; Huynh, K.; Smith, A.A.T.; Mellett, N.A.; Jandeleit-Dahm, K.; Lancaster, G.I.; Meikle, P.J. Oral supplementation of an alkylglycerol mix comprising different alkyl chains effectively modulates multiple endogenous plasmalogen species in mice. *Metabolites* **2021**, *11*, 299. [CrossRef]
81. Hichami, A.; Duroudier, V.; Leblais, V.; Vernhet, L.; Le Goffic, F.; Ninio, E.; Legrand, A. Modulation of platelet-activating-factor production by incorporation of naturally occurring 1-*O*-alkylglycerols in phospholipids of human leukemic monocyte-like THP-1 cells. *Eur. J. Biochem.* **1997**, *250*, 242–248. [CrossRef]
82. Koltai, M.; Hosford, D.; Guinot, P.; Esanu, A.; Braquet, P. Platelet activating factor (PAF). A review of its effects, antagonists and possible future clinical implications (Part I). *Drugs* **1991**, *42*, 9–29. [CrossRef]
83. Koltai, M.; Hosford, D.; Guinot, P.; Esanu, A.; Braquet, P. Platelet activating factor (PAF). A review of its effects, antagonists and possible future clinical implications (Part II). *Drugs* **1991**, *42*, 174–204. [CrossRef]
84. da Silva, T.F.; Eira, J.; Lopes, A.T.; Malheiro, A.R.; Sousa, V.; Luoma, A.; Avila, R.L.; Wanders, R.J.A.; Just, W.W.; Kirschner, D.A.; et al. Peripheral nervous system plasmalogens regulate Schwann cell differentiation and myelination. *J. Clin. Investig.* **2014**, *124*, 2560–2570. [CrossRef]
85. Hossain, M.S.; Mineno, K.; Katafuchi, T. Neuronal orphan G-protein coupled receptor proteins mediate plasmalogens-induced activation of ERK and Akt signaling. *PLoS ONE* **2016**, *11*, e0150846. [CrossRef]
86. Ali, F.; Hossain, M.S.; Sejimo, S.; Akashi, K. Plasmalogens inhibit endocytosis of toll-like receptor 4 to attenuate the inflammatory signal in microglial cells. *Mol. Neurobiol.* **2019**, *56*, 3404–3419. [CrossRef]
87. Youssef, M.; Ibrahim, A.; Akashi, K.; Hossain, M.S. PUFA-plasmalogens attenuate the LPS-induced nitric oxide production by inhibiting the NF-kB, p38 MAPK and JNK pathways in microglial cells. *Neuroscience* **2019**, *397*, 18–30. [CrossRef]
88. Pedrono, F.; Khan, N.A.; Legrand, A.B. Regulation of calcium signaling by 1-*O*-alkylglycerols in human Jurkat T lymphocytes. *Life Sci.* **2004**, *74*, 2793–2801. [CrossRef]
89. McNeely, T.B.; Rosen, G.; Londner, M.V.; Turco, S.J. Inhibitory effects on protein kinase C activity by lipophosphoglycan fragments and glycosylphosphatidylinositol antigens of the protozoan parasite *Leishmania*. *Biochem. J.* **1989**, *259*, 601–604. [CrossRef]

marine drugs

MDPI

Article

Petrosamine Revisited. Experimental and Computational Investigation of Solvatochromism, Tautomerism and Free Energy Landscapes of a Pyridoacridinium Quaternary Salt

Christopher J. Gartshore [1], Xiao Wang [1], Yongxuan Su [1] and Tadeusz F. Molinski [1,2,*]

[1] Department of Chemistry and Biochemistry, University of California, 9500 Gilman Drive MC3568, La Jolla, San Diego, CA 92093, USA; gartcj@gmail.com (C.J.G.); xiao.wang1@merck.com (X.W.); yxsu@ucsd.edu (Y.S.)

[2] Skaggs School of Pharmacy and Pharmaceutical Sciences, University of California, 9500 Gilman Drive MC3568, La Jolla, San Diego, CA 92093, USA

* Correspondence: tmolinski@ucsd.edu

Abstract: Petrosamine (**1**)—a colored pyridoacridine alkaloid from the Belizean sponge, *Petrosia* sp., that is also a potent inhibitor of acetylcholine esterase (AChE)—was investigated by spectroscopic and computational methods. Analysis of the petrosamine-free energy landscapes, pK_a and tautomerism, revealed an accurate electronic depiction of the molecular structure of **1** as the di-keto form, with a net charge of $q = +1$, rather than a dication ($q = +2$) under ambient conditions of isolation-purification. The pronounced solvatochromism (UV-vis) reported for **1**, and related analogs were investigated in detail and is best explained by charge delocalization and stabilization of the ground state (HOMO) of **1** rather than an equilibrium of competing tautomers. Refinement of the molecular structure **1** by QM methods complements published computational docking studies to define the contact points in the enzyme active site that may improve the design of new AChE inhibitors based on the pyridoacridine alkaloid molecular skeleton.

Keywords: alkaloid; pyridoacridine; acetylcholine esterase; merocyanine; marine natural product; NMR; UV-vis spectroscopy

Citation: Gartshore, C.J.; Wang, X.; Su, Y.; Molinski, T.F. Petrosamine Revisited. Experimental and Computational Investigation of Solvatochromism, Tautomerism and Free Energy Landscapes of a Pyridoacridinium Quaternary Salt. *Mar. Drugs* **2023**, *21*, 446. https://doi.org/10.3390/md21080446

Academic Editor: Vladimir I. Kalinin

Received: 13 July 2023
Revised: 2 August 2023
Accepted: 3 August 2023
Published: 11 August 2023

1. Introduction

In 1988, Molinski and Faulkner reported the structure of petrosamine (**1**, Figure 1, a highly condensed pentacyclic alkaloid of intense color that was isolated from the sponge, *Petrosia* sp., collected at Carrie Bow Cay, Belize [1]. However, no bioactivity **1** was reported then, and a pronounced solvatochromism—the property of color changes (λ_{max}) in solvents of different polarities—was observed. At the time, **1** represented the most complex structure in a growing class of pyridoacridine alkaloids from marine invertebrates [2]. Structurally, **1** can be considered a methylated and oxidized analog of the pentacyclic amphimedine—the first member of this class of alkaloids, described by Schmitz and co-workers in 1983 [3]. Unlike the structures of most pyridoacridines where the 3 nitrogen atoms are sp^2 hybridized, **1** contains a quaternary ammonium salt: a sp^3 quaternized N. Pyridoacridines manifest a range of biological activities including cytotoxicity, antineoplastic properties, antibacterial activity, and enzyme inhibition; a subject that has been extensively reviewed [4].

Since its initial report, alkaloid **1** has been reisolated from a Thai sponge species, *Petrosia* n sp. [5], and two new analogues have been described; petrosamine B (**2**) [6], a regioisomer of **1** from the Australian sponge *Oceanapia* sp. with modest inhibitory activity against aspartate semialdehyde dehydrogenase, and debromopetrosamine (**3**) from *Xestospongia* cf *carbonaria* collected in Palau [7]. Numerous marine natural products have been reported with neuroprotective properties in experimental models for neurodegenerative diseases, including acetylcholine esterase (AChE) inhibition [8]. Suwanborirux and coworkers showed that **1** is a potent inhibitor of AChE from the Pacific electric ray, *Torpedo* *californica* (IC_{50} = 91 nM) [5]; approximately six times more potent than galanthamine (**4**), an alkaloid used in the past for treatment of patients with Alzheimer's disease (AD) to compensate neurotransmitter deficiency.

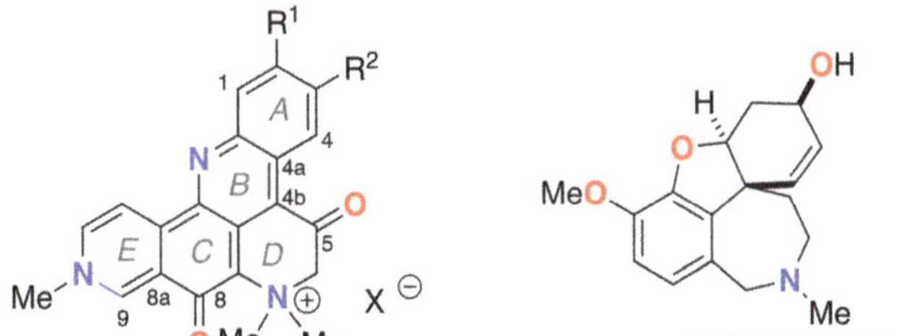

4 Galanthamine

R^1 = Br, R^2 = H, X = Cl^- **1** Petrosamine
R^1 = H, R^2 = Br, X = CF_3COO^- **2** Petrosamine B
R^1 = R^2 = H, X = CF_3COO^- **3** Debromopetrosamine

Figure 1. Structures of petrosamine (**1**), analogs **2**, **3** and galanthamine (**4**).

Preliminary docking studies of **1** with AChE revealed subtle electronic interactions with the putative receptor contacts [5]. Given the global rise of AD within aging populations and the therapeutic importance of AChE inhibitors for treatment, a detailed understanding of the electronic structure and molecular parameters for enzyme-inhibitor molecular contacts of **1** may advance design principles for new AD therapeutics. In this report, we present refined measurements of UV-vis, pK_a properties and quantum mechanical calculations of **1** that relate the observed solvatochromism through mapping of dipolar resonance forms to a simple model based on merocyanine dyes and stabilization of the ground state of the HOMO.

Petrosamine (**1**) exhibits several unusual physical and spectroscopic properties. The melting point of **1** is in excess of 300 °C suggesting high stabilization within the lattice energy of the crystalline form. Unlike most other pentacyclic pyridoacridines, the color of **1** is strikingly variable depending upon the physical state. Crystals of **1** are deep blue-purple but dilute solutions of **1** show highly complex UV-vis spectra due to long-wavelength absorption bands that manifest pronounced solvatochromism (changes in λ_{max} of **1** when measured in different solvents). For example, solutions of **1** in MeOH appear deep-blue (longest λ_{max} = 595 nm), but aqueous solutions of **1** appear purple λ_{max} = 574 nm) [1]. In dilute THF solutions (sparingly soluble) or DMSO, **1** appears green λ_{max} = 611 nm) [1]. In short, the range $\Delta\lambda_{max}$ of **1** is 36 nm in a range of solvents from H_2O to DMSO. Petrosamine also displays pH indicator properties: addition of excess alkali to purple aqueous or blue MeOH solutions of **1**, changes the color to green, suggestive of a hyposochromic shift ($\Delta\lambda_{max}$ < 0) mediated by the Brønsted acidity of the α-CH_2 next to the keto group and deprotonation to the corresponding enolate.

The report by Quinn and coworkers describing the isolation of **2** by "fractionation on C18" and elution with a stepped gradient of acidic (CF_3COOH) aqueous-MeOH [6] also reported solvatochromism in **2** very similar to that of **1**. Assignment of the structure of **2** was achieved through extensive analysis of ^1H, ^{13}C NMR and 2D NMR data (D_2O-TFA-1% DMSO-d_6). Compound **2** was depicted as a dication different from **1** in the resonance form of an O-protonated vinylogous amide embodying a fully aromatic quinolinium ring E [6]. Additional structural anomalies emerged. The assignment of C-8 in **2** is supported by an HMBC correlation from H-9 to C-8 ($^3J_{CH}$) based on the low ^{13}C NMR chemical shift (δ 155.8 ppm), but no inter-ring HMBC correlations were reported for C-5 [6]. In contrast, the X-ray crystal structure of **1** [1] reveals two keto groups in the solid state (*d* (C-8–O) = 1.256 Å; *d* (C-5–O) = 1.203 Å), *d* (C-5–C-6) = 1.495(21). The C-O and C-C bond lengths are only compatible with a di-keto structure [9]. As the structures of **1** and **2** are very similar, consequently, the solution properties (p*K*a, UV-vis) should also be closely matched. Faulkner and Molinski reported the ^{13}C NMR spectra (DMSO-d_6) of **1** but not the assignments of the signals [1]. Suwanborirux and coworkers, reporting the ^{13}C NMR spectrum of **1** in the latter solvent, assigned the signal at δ 161.3, s, the higher-field of the two most deshielded signals (δ 187.2, s; 161.3, s) to C-8 [5])

Comparisons of the NMR assignments of **1** and **2** are complicated further by measurements in different solvents and possible changes in tautomer or enolization states that accompany changes in pH and NMR solvent polarity, and H-bond donor-acceptor properties of the alkaloids. The latter factors affect the UV-vis properties of **1** and **2**. Surprisingly, neither the Quinn [6] nor Suwanborirux [5] groups claim to have observed the rapid enolization of **1** in D_2O reported by Faulkner and Molinski that resulted in complete deuteriation of C-9 and 'loss' of the C-9 signal [1,10].

To reconcile these apparent paradoxes, it is required that a re-examination of **1** be given separate considerations of resonance in the native structure and possible tautomerism of **1** at the α-CH_2 next to the C-5 C=O group. We undertook the task and completed additional solution spectroscopic measurements of **1** [NMR and electronic spectroscopy, UV-vis] and MS time-course studies and augmented by quantum mechanics calculations. The results presented here clarify several phenomena of **1**, including solvent dependence of HOMO-LUMO energies, kinetic and thermodynamic considerations of pK_a, and tautomerism.

2. Results and Discussion

The solvatochromism of **1** and **2** is similar to lower-order merocyanine dyes, the canonical substructures of which can be discerned within the molecular framework of both natural products (Figure 2). For example, the same conjugated sub-structure in Brooker's merocyanines [11,12] [generalized by Brooker as the vinylogous amide; neutral ***ia*** and ***ib*** dipolar (zwitterionic) resonance forms] is embedded in the molecular frameworks of **1** and **2**. Brooker merocyanines, for example, **5b** and **6b** (Figure 2), exhibit strongly positive solvatochromism (hypsochromism, or a blue shift in λ_{max} in polar solvents). It is not unreasonable to invoke the same spectroscopic and electronic properties of **5**—zwitterionic form with extended conjugation and stabilization of the ground state by solvation in polar solvents—as necessary and sufficient conditions to explain the solvatochromism of **1–3**.

While a complete description of the electronic properties of petrosamine (**1**) may be achieved in rigorous quantum mechanical calculations (see below), it is helpful for visualization purposes to consider Lewis's bond formalism and resonance structures [13]. For clarity, only two pairs of resonance forms are depicted for **1** to illustrate the zwitterionic contributions of merocyanine substructures: the non-charged resonance forms **1a**,**c**, and 'zwitterionic' resonance form **1b**,**d**. Forms **1a**,**b** depict the shorter bond path (n = 3, c.f. ***ia***, ***ib***, Figure 2) and dipolar resonance forms **1c**,**d** show a longer bond path in an 'aza-vinylogous amide' (n = 4). The shortest bond path, n = 1 (not shown), would involve only the atoms C-8–C8a–C9–N-10), while the longest path, n = 4, evoking 'particle-in-a-box' formalism [14], best explains the long-wavelength UV-vis bands of **1** giving rise to its colors.

Solvatochromism of Brooker merocyanines has been rationalized [**11b**], using semi-quantitative valence resonance and frontier orbital theories, as arising from more extensive stabilization of the dipolar form **4a** of the ground state relative to the excited state in polar solvents. Accordingly, this differential stabilization increases the electronic transition energy, ΔE, due to a larger gap between the HOMO and LUMO of the longest wavelength transitions. The HOMO-LUMO gap is predicted to increase (blue-shifted absorption) in those structures with higher contributions from the zwitterionic resonance form **4b**, leading to more pronounced solvatochromism. More recent semi-empirical calculations (COSMO, PM3) of a different set of substituted Brooker merocyanines (**5**) by Morley and coworkers [15] supported stabilization of the ground state as mostly responsible for solvatochromism and predicted a larger role for hydrogen bonding in the zwitterionic form **5b** over the neutral form **5a**.

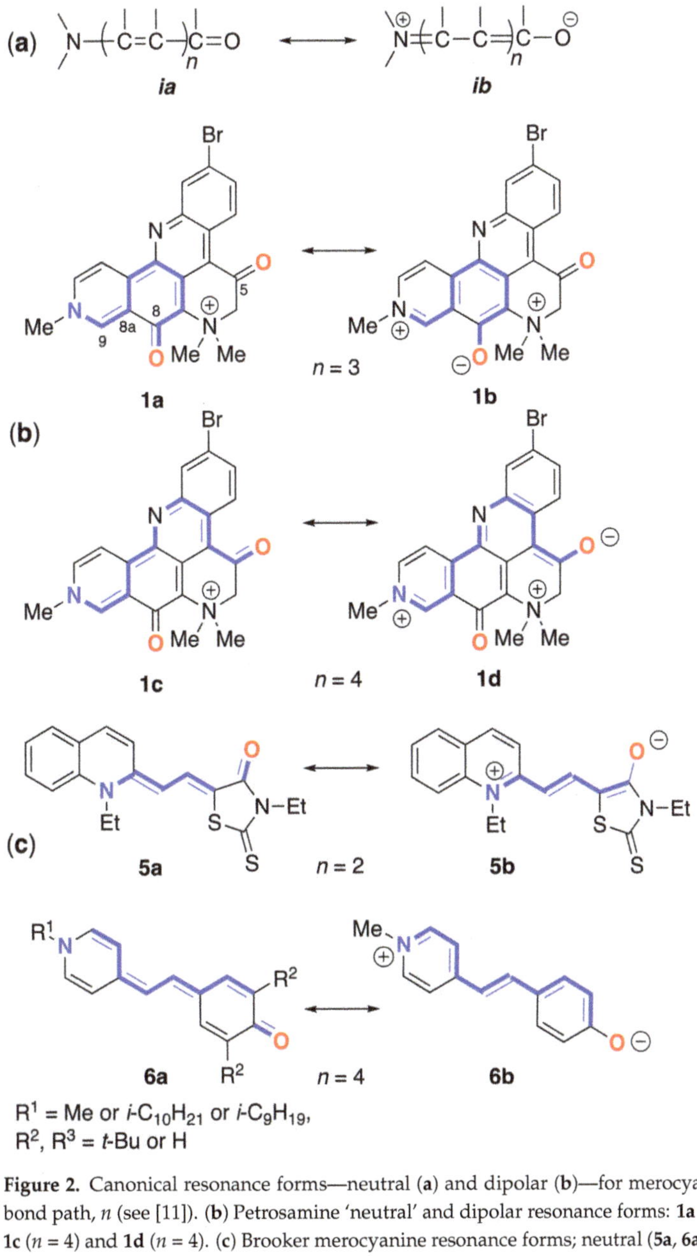

Figure 2. Canonical resonance forms—neutral (**a**) and dipolar (**b**)—for merocyanines defined by bond path, n (see [11]). (**b**) Petrosamine 'neutral' and dipolar resonance forms: **1a** (n = 3), **1b** (n = 2), **1c** (n = 4) and **1d** (n = 4). (**c**) Brooker merocyanine resonance forms; neutral (**5a**, **6a**) and dipolar (**5b**, **6b**) forms [11].

The structure of petrosamine (**1**) appears to fulfil the criteria for merocyanine-type solvatochromism. Measurements of the UV-vis spectrum of solutions of **1** prepared in DMSO–H_2O solvents of variable composition (Figure 3) exhibit changes in the λ_{max}. Most prominently, the longest wavelength absorption band with the largest hypsochromic shift between 100% DMSO and 100% H_2O ($\Delta\lambda_{max}$ − 78 nm) is assigned to the forbidden n–π* transition that lends color to **1**, analogous to that of merocyanines.

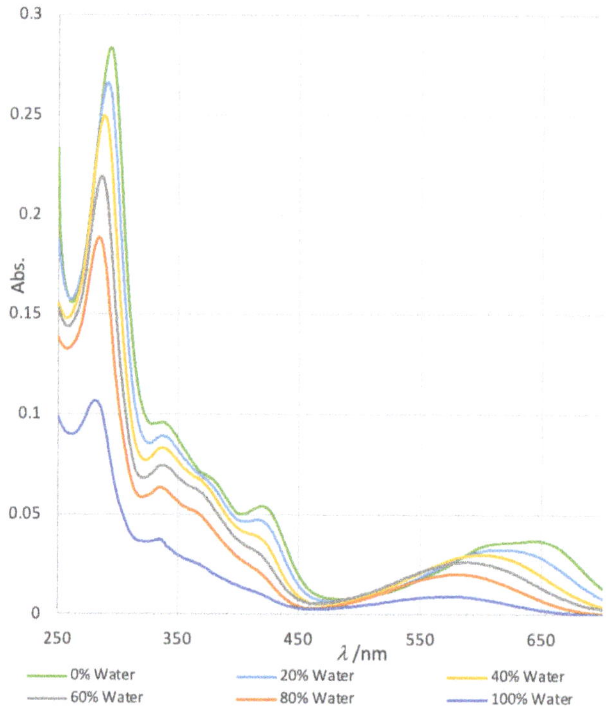

Figure 3. Solvatochromism in normalized electronic UV-vis spectra of petrosamine (**1**) in H_2O-DMSO solvents of variable composition (H_2O v/v = 0%, 20%, 40%, 60%, 80% and 100%).

A clear trend emerges band-1 (λ_{max1}, defined here for convenience, as the dominant π-π^* transition) shows a dramatic decrease in ε with increasing H_2O content of the solvent and a weak bathochromic shift between 100% H_2O to 100% DMSO ($\Delta\lambda_{max1}$ 15 nm, Table 1). In contrast, the corresponding changes in band-2 include a strong blue shift (hypsochromism, λ_{max2} − 78 nm). The band λ_{max2} is affected most by solvents with increasing H_2O content which parallels the reported behavior of Brooker merocyanines.

Table 1. UV-vis properties of λ_{max1} [1] and λ_{max2} [1] in petrosamine (**1**) in DMSO-H_2O [1].

% H_2O (v/v)	λ_{max1} (nm)	ε_1 [1]	λ_{max2} (nm)	ε_2 [1]
0	296	184,000	648	24,000
20	290	172,000	622	20,600
40	288	161,000	604	19,400
60	287	141,700	589	16,500
80	282	122,252	581	12,900
100	281	68,500	570	4700

[1] See text for definitions. $\varepsilon_{1,2}$ values normalized from the literature value of λ_{max2} (ε 4700) [1].

For comparison, we prepared the known merocyanine **6b** from 4-methylpyridine by the following sequence adapted from Minch and Sadiq Shah [16]: methylation (MeI, *i*PrOH, reflux), condensation of the resultant pyridinium methiodide with *p*-hydroxybenzaldehyde (piperidine, EtOH, reflux), and neutralization of the product **7** (nBu_4N^+ HO^-) to zwitterionic phenolate **6b**. Measurements of the UV-vis spectra of **6b** (nBu_4N^+ salt) in mixed solvents (acetone-H_2O, see Figures S3 and S4 in Supporting Information) showed a hypsochromic trend similar to **1** in DMSO-H_2O. The long-wavelength (band-2) varied from λ_{max} (acetone) 591 nm to λ_{max} (H_2O) 444 nm ($\Delta\lambda_{max}$ − 69 nm) while ε changed only slightly across the range of solvent mixtures [17].

2.1. QM Calculations

The energies of electronic states of **1** were calculated using QM methods (DFT). Starting with the X-ray coordinates of **1**, the structure's geometry was minimized using MMFF, then further refined by DFT (B3LYP 6-31G*, polarization continuum model = H_2O). The 3D model of **1**, overlaid with frontier MOs, is shown in Figure 4. As expected, the HOMO-LUMO gap is relatively small (ΔE = 2.4 eV). The orbital coefficients and calculated dipole moment of the excited state (μ = 19.2 D) are consistent with dominant contributions from the resonance form **1b** rather than **1a** and the strong donor properties of N-10. At first, this may seem anomalous, given the relatively low-field ^{13}C NMR signal measured for C-5 in **1** (δ 187.4 [1] or 187.2 in **2** [5]); however, the additional deshielding effect is likely attributed to the inductive effect of the quaternized N-7.

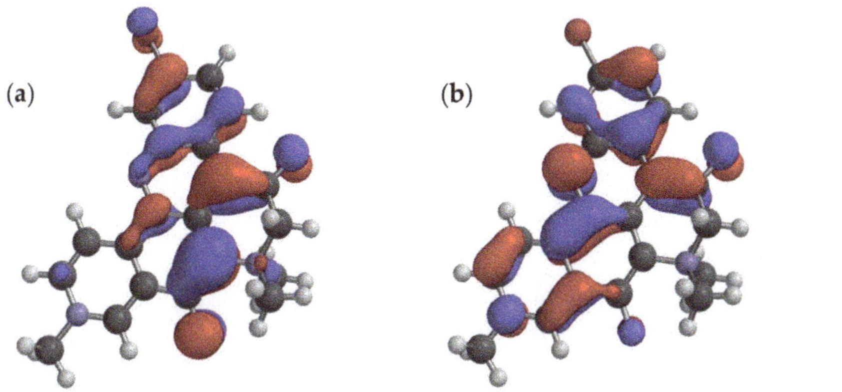

Figure 4. Calculated minimized structure and overlaid frontier π-orbitals of petrosamine (**1**) (DFT B3LYP 6-31G*, polarization continuum model = H_2O) (**a**) HOMO and (**b**) LUMO.

The electrostatic potential map of the minimized structure of **1** (Figure 5) clearly shows two loci of positive charge: one centered on the quaternized N-10, as expected, in ring D, and a second associated with the N atom in ring E. The latter supports charge-separated forms **1b,d** (Figure 2) in which N-10 participates as a donor group. Consequently, the formal bonding electron pair of N-10 is highly delocalized and can confer only weak basicity to **1**. Together with the UV-vis properties of **1**, the overall electron delocalization consolidates an electronic structure in which a zwitterionic partial structure strongly lends to charge separation, mostly in the ground state.

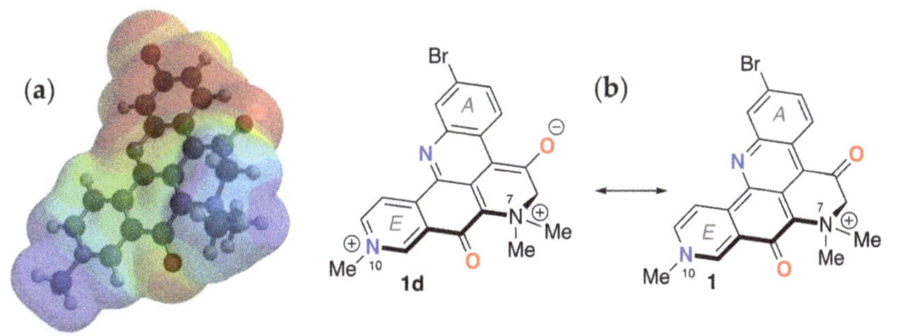

Figure 5. Minimized energy structure of petrosamine (**1**, DFT, Spartan'20, EDF2 6-31G, dipole moment μ = 19.2 D). (**a**) electrostatic potential surface and (**b**) the corresponding molecular framework of **1** (dipolar **1d** and 'charge-minimal' **1** resonance forms).

From an empirical viewpoint, the latter makes sense as the charge-separated forms **1b,d** preserve the aromaticity of rings A and E. A result of this delocalization is reduced basicity of N in ring E and, consequently, an expected lower pK_a for the conjugate Brønsted acid. Formally, adding H^+ **1** to give a dication should sooner favor the C-8 or C-5 oxygen as an acceptor rather than N-10 or N-13. We find no evidence (NMR) for protonation of **1** at pH ~ 2, which attests to the overall poor basicity of **1**; an unsurprising finding given the permanent formal charge of +1 in this quaternary ammonium salt in all but the most basic or the most acidic media.

2.2. pK_a of Petrosamine (1), Does the Structure of 1 Exhibit Substantial Enol Content?

To estimate the pK_a of petrosamine, the UV-vis spectra of **1** were measured in buffered D_2O at different pH. Across the pH 2–10 (Figure S1), the UV-vis spectrum of **1** is unchanged. From the Henderson-Hasselbach relationship (Equation (1)) [18] for Brønsted acid HA, the condition pH = pK_a is met when concentrations of conjugate species are equal ([HA] = [A^-]). We surmise the pK_a of **1** lies outside this pH range (pK_a > 10). Indeed, a bathochromic shift in the UV-vis spectrum of **1** is only observed when a methanolic solution is treated with strong alkali NaOH (2 M, pH > 13). Conversely, only when a sample of **1** is dissolved in a very strong Brønsted acid (neat CF_3COOH) are conditions met for a reasonable expectation of diprotonated petrosamine ([**1**•2H]$^{2+}$). In the event when **1** was dissolved in neat TFA, the blue-purple color changed to bright yellow. The latter observation contrasts with the supposition drawn by Quinn and coworkers of doubly-protonated petrosamine B ([**2**•2H]$^{2+}$) from their observation that **2** remains "*bright blue. . .when dissolved in methanol*", under the relatively benign conditions used in C_{18}-reversed phase isolation of the alkaloid (5% TFA-MeOH) [6]. In other words, N-10 in the monocationic molecules **1–3** is 'pyridinium-like' and non-basic (resembling **1d**), and the C-8 C=O is insufficiently Brønsted-basic to be substantially protonated under ordinary isolation conditions.

$$\mathrm{pH} = \mathrm{p}K_a + \log_{10} [A^-]/[HA] \quad (1)$$

It is evident from the ^{1}H and ^{13}C NMR (DMSO-d_6) that C-5 in petrosamine (**1**) is in the C8 keto form, a conclusion also reached by both the Suwanborirux and coworkers [5], and independently by Quinn and coworkers for petrosamine B (**2**) [6]. The single crystal X-ray crystallography of **1** is concordant with the C8 keto form. Solid state **1** is best represented by the di-keto tautomer, e.g., interatomic distances C8-O2, 1.256(17) Å; C5-O1, 1.203(15), and C5-C6, 1.495(21) [1].

2.3. Kinetic Measurements of Hydrogen-Deuterium Exchange of 1

None of the resonance forms of **1a–d** (Figure 2)—or more precisely, pathways of electron delocalization—explain the complete exchange of the H-6 protons by deuterium when **1** was dissolved in deuteric solvents (D_2O, CD_3OD) [1]. The latter can only be rationalized by consideration of the possibility of enolization (Figure 6), either through the positively charged **8a** in neutral to weakly acidic pH or the charge-neutral (zwitterionic) enolate **8b** under highly basic pH. The kinetic parameters for successive H replacement by D in **1**, defined by rate constants k_1 and k_2, are intrinsic properties as opposed to thermodynamic properties that relate to the equilibrium constants K_{eq1} and K_{eq2}; the latter are largely dependent upon the strength and concentration of added base, B^- (Figure 6).

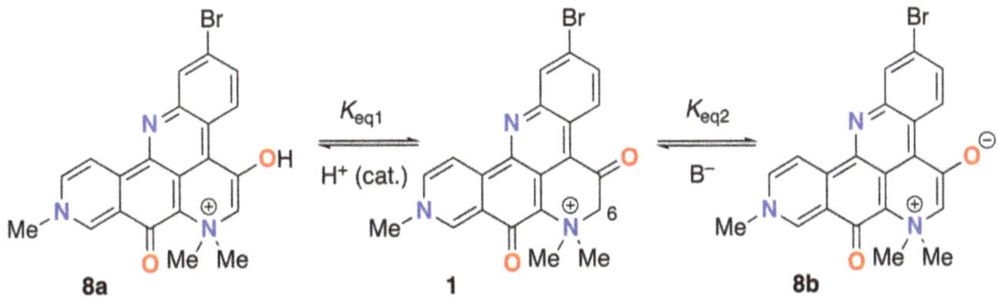

Figure 6. Keto-enol tautomerism of **1**. Enol **8a** (acid-catalyzed) and enolate **8b** (base-promoted).

In hydroxylic solvents, **1** also resides largely in the di-keto form. However, we found that in both D_2O and CD_3OD, the C-6 methylene protons undergo rapid exchange to give the C-6 CD_2-isotopomer (**1**-d_2) [1]; a rate so fast that we were unable to detect the C-6 CH_2 or the intermediate forms within the time frame between sample preparation and measurement of the ^1H NMR spectra. This observation was supported by ^1H NMR, HSQC and HMBC measurements of **1** in protic solvent (CD_3OH) where the C-6-CH_2 group is still observable (HSQC correlation: H-6 to C-6, δ_H 4.64, s; δ_C 70.4 ppm). The simplest explanation for both phenomena is catalytic H-D exchange α-to the C-5 C=O group favored by an intermediate, the extensively-conjugated enol tautomer **8a** (Figure 6).

To examine the kinetics of H-D exchange in **1** and place an upper bound on the rate of H-D exchange, the time-dependent appearance of the CD_2-isotopmer by ESI mass spectrometry upon rapid dissolution of **1** in CD_3OD (Figure 7). Under controlled conditions (23 °C, in situ measurement of the ESIMS of petrosamine with rapid sampling (sampled every t = 15 s) measurement—see Supplementary Materials (S6–S9 and Experimental (S14)), **1**-d_0 ($C_{21}H_{17}{}^{89}BrN_3O_2$, calculated *m/z* 422.0499 [M$^+$]) quickly disappeared and was replaced by an ephemeral isotopologue **1**-d_1 (*m*/*z* 423.0561), followed by convergence upon **1**-d_2 (*m*/*z* 424.0624). Complete exchange (>90%) was observed within 90 s of dissolution of **1**. From triplicate measurements and rapid sampling, we could fit the kinetic deuterium exchange data to a first-order rate law. We estimated the apparent first and second H-D exchange rate constants to be (Table 2), k_1 = 0.131(68) s^{-1} and k_2 = 0.0755(26) s^{-1}, respectively (best R^2; see S6–S10). As expected, rate constant k_2 is about half of k_1, consistent with the expected rate law (k_1/k_2 = 1.74(94)), negating the involvement of a substantial primary kinetic isotope effect.

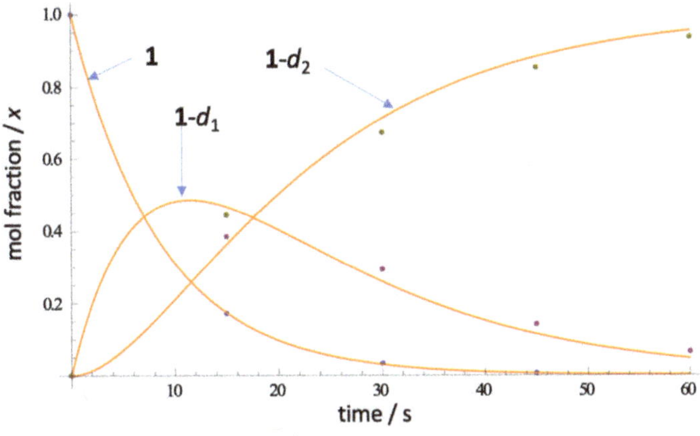

Figure 7. Representative ESIMS measurements of the rate of H-D exchange of petrosamine (**1** in CD_3OD, 23 °C) and fitted curves (non-linear regression). See S6–S10 for rate law and *k* determinations.

Table 2. Rate constants k_1 and k_2 for H-D exchange in **1** in CD_3OD (23 °C, see Figure 7) [1].

Reaction	k_1 (s^{-1})	k_2 (s^{-1})	95% Confidence Interval
1 –> **1**-d_1	0.1311(68)	–	{0.1176, 0.1447}
1-d_1 –> **1**-d_2	–	0.0755(26)	{0.0702, 0.0807}

[1] See S6–S10 for plotted raw ESIMS data and detailed kinetic analysis.

The exchange of **1** to **1**-d_2, in the absence of added acid, appears to be much faster than H-D exchange rates of other phenones, e.g., propiophenone (pK_a 24.4, DMSO [19] α-tetralone (pK_a 24.7 [20,21]. The rates of acid-promoted keto-enol equilibration of acetophenone (pK_a 18.4 ± 0.03) have been measured. For example, the rate of ketonization of acetophenone enol (1-phenylethen-1-ol) is linearly dependent upon [H_3O^+] with a catalytic efficiency determined to be k_{H+} = (1.25 ± 0.02) × 10^3 $M^{-1}s^{-1}$ [22]. We conclude that tautomerism of **1**, too, must be very fast, even in the absence of added acid. Either the enol **8a** or enolate **8b**, although undetectable as a discrete species in the time frame of ^{1}H NMR, present a pathway to the acid-base equilibria of **1** and **1**-d_2, but enolization of the keto form is likely dominant.

Some amount of discussion has been given on the keto-enol state of **1**. The NMR data for **1** reported by Suwanborirux and coworkers support the C5 keto form in DMSO-d_6 [5] and—in agreement with Molinski and Faulkner, the enol form in D_2O or CD_3OD [1]—but Quinn and coworkers find, *"no evidence for this keto-enol isomerism"* in **2**. Likely, the K_{eq} of keto-enol tautomerism lies on some continuum, dependent upon solvent dielectric and H-bond donor ability.

We find it unlikely that the position of the Br in ring *A*—the only difference between structures **1** and **2**—would exert a profound difference in physicochemical behavior. Quinn's argument—*"no involvement for this keto-enol isomerism"* in **2**—is confounded by two uncertainties: the composition of their NMR solvent (TFA, D_2O, DMSO-d_6) is not specified quantitatively, and facile interpretation of the ^{13}C NMR signals for the C8 signal: *"in contrast C-8 resonated at 155.8 ppm, supporting it as a phenolic resonance. It was, therefore, more likely that petrosamine also exists in the C-8 enol form and C-8* [Molinski and Faulkner [1] *was the carbon at 161 ppm"*, contrasts with those of Suwanborirux (δ 174.7 ppm, DMSO-d_6) and their observation of, *"the broad methylene carbon at δ_C 69.2"* (*DMSO-d_6, "100 μmol NaOD"* [5]). For comparative purposes, we synthesized a model quinoline **S1** (see Supplementary Material)—a merocyanine of class (n = 3, see labeling of **1a**,**b**) and found the ^{13}C NMR chemical shift of C8 in CD_3OD to be δ_C 172.2 ppm. Aside from these points, none of the three different ^{13}C NMR values for C8 in **1** and **2** are incompatible with the structures, both of which are not strictly aryl ketones but vinylogous amides, the dominant resonance form of which will be highly dependent upon solvent dielectric and H-bond donor properties.

The observed ^{1}H and ^{13}C NMR spectra of petrosamine (**1**), and the C6 exchange to CD_2 in CD_3OD/NaOD [5], support the enol form in hydroxylic solvents. It is likely fast exchange between the hybrid structure of **1** and **8a**,**b** with an equilibrium constant K_{eq1} largely in favor of **1** in DMSO-d_6, but moving to the dominant form **8a** in hydroxylic solvents. The enolate **8b** may be favored as the catalytically important intermediate under 'neutral' conditions for reasons of charge neutralization, but insufficient data preclude testing this hypothesis with more certainty. In either case—enol or enolate intermediate—we surmise the electron-withdrawing quaternized N^+Me_2 group in **1** and the related petrosamines, **2**, **3**, plays a significant role in accelerating the rates of H-D exchange and lowering the pK_a of the C-6 CH_2 group.

To the best of our knowledge, rapid H-D exchange of a substituted β-quaternary ammonium ketone within a natural product has been observed only in one other instance, coulteroberbinone (**7**, Figure 8 [23]), an *N*-quaternary ammonium isoquinoline alkaloid from the leaves of *Romneya trichocalyx*. The authors note the C-14 C-H in **7**, assigned to the α-proton between the carbonyl and quaternized nitrogen (δ_H 5.64, s), underwent rapid proton-deuterium exchange in D_2O or CD_3OD to C-D (**7**-d_1) under ambient conditions (supported by ^{1}H NMR and ESIMS data).

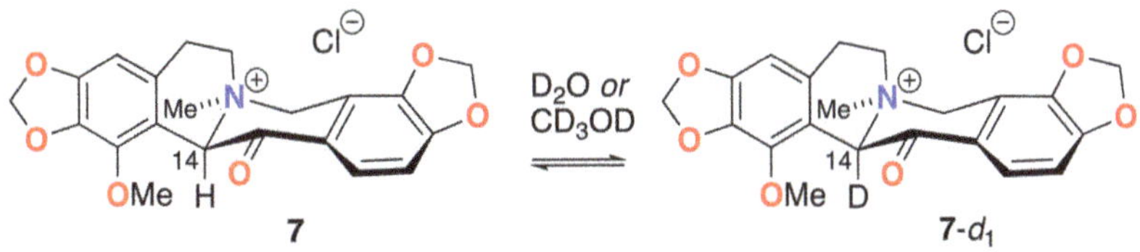

Figure 8. Rapid H-D exchange of coulterberbinone (**7**) in deuteric solvent under ambient conditions.

Two major factors most likely explain the relatively low pK_a of **1** and **7**: the electron-withdrawing (inductive) effect of the -N^+Me_2 quaternary ammonium group (N-7) and stabilization of enol **5** (or enolate **6**) through extensive conjugation of the heteroaromatic core, not unlike the stabilization of the enolates of alkyl phenones, e.g., acetophenone, determined by UV-vis (pK_a = 18.4). Indirect determinations of the pK_a of the enol of acetophenone have also been obtained from the kinetics of reactions of acetophenone: e.g., α-chlorination [24] and aminolysis of the corresponding enol acetate [25]. In contrast, **1** appears to undergo rapid enolization without added Brønsted acid, suggesting that this tautomerization may even be autocatalytic.

Is the enol **8a** of petrosamine (**1**) present in substantial concentrations? From the ^{1}H NMR spectrum of **1** (DMSO-d_6), we detect no signals attributable to **8a**. It can be ascertained from ^{1}H NMR that K_{eq1} (Figure 7) is very small (estimated from the limits of integration, $K_{eq1} < 0.05$), and the equilibrium of tautomers lies well towards the keto form **1**. As noted above, in highly-basic aqueous solutions of **1**, the charge-minimal enolate **8b** appears in substantial concentrations, placing a lower boundary of pK_a ~ 15 for **1**. In the absence of base, the mechanism of H-D exchange **1** to **1**-d_2 in CD_3OD most likely engages substantial equilibrium concentrations **8a** to allow rapid exchange of the CH_2 to CD_2 in deuteric solvents within less than 2 min at 23 °C. An accurate pK_A of **1** is not accessible from measurements in aqueous solvents, and securing an estimate will likely require measures in a suitable aprotic solvent (e.g., titration in DMSO [20,26]). Bordwell values of pK_a are conventionally obtained by titration of a solution of the weak Brønsted acid in DMSO with its non-nucleophilic conjugate base, 'dimsyl sodium' ($CH_3(S{=}O)CH_2^-$ Na^+) with Ph_3CH as an indicator [26] where a colored endpoint is presented by the deep-red Ph_3C^- anion. In the case of **1**, the high color of the substrate and its conjugate base should lend itself to 'self-indicating'.

3. Materials and Methods

3.1. General Experimental Procedures

Inverse detected 2D NMR spectra were measured on a Jeol ECA (500 MHz) spectrometer equipped with a 5 mm ^{1}H{^{13}C} 5 mm probe or a Bruker Avance III (600 MHz) NMR spectrometer with a 1.7 mm ^{1}H{^{13}C} microcryoprobe. ^{13}C NMR spectra were collected on a Varian NMR spectrometer (125 MHz) with a 5 mm Xsens ^{13}C{^{1}H} cryoprobe. NMR spectra were referenced to residual solvent signals, $(CD_3)_2CO$ (δ_H 2.05, δ_C 29.8). High-resolution ESITOF analyses were conducted on an Agilent 1200 HPLC coupled to an Agilent 6230 TOFMS. Low-resolution MS measurements were made on a Thermoelectron Surveyor UHPLC coupled to an MSD single-quadrupole detector. HPLC was performed on an Agilent 1200 HPLC. UV-vis spectra were measured on a Jasco V-630 spectrometer in quartz cells (1.00 cm path length, Helma). FTIR spectra were collected on thin film samples using a Jasco FTIR-4100 fitted with an ATR accessory (ZnSe plate). Optical densities (OD, λ nm) in microplate wells were measured using a Molecular Devices Spectramax 384 Plus. Measurements of pH were made with a digital pH meter (Denver Instrument, Arvada, CO, USA, Model 220), calibrated against standard solutions (NaH_2PO_4-Na_2HPO_4).

3.2. *UV-Vis Measurements*

Standard solutions of accurately-weighed **1** and **6b** were prepared in volumetric flasks and used for serial dilutions to the final working concentrations in specified media, either pure HPLC grade solvent (DCM, DMSO, acetone or DMF—see Supporting Information for **6b** in acetone) or mixtures of aqueous HPLC solvents of defined H_2O composition.

pH Dependence: Britton-Robinson buffer was prepared by dissolving 2.48 g of boric acid, 2.30 mL acetic acid and 2.72 mL phosphoric acid in 1 L of water and titrating in a 2.0 M solution of NaOH to the desired pH.

40 μL of a 2.37 mM of petrosamine (**1**) in MeOH was added to 960 μL of the desired buffer solution, and the UV-vis spectrum was measured from λ 190–700 nm.

Solvent-Dependence: Solutions of **1** in H_2O (95.0 μM) and acetone (95.0 μM) were mixed in the desired ratio, and the UV-vis spectrum was measured from λ 320–700 nm.

3.3. *DFT Calculations of* **1**

The X-ray coordinates of petrosamine (**1**) were used as a starting point for DFT calculations of the energy, geometry minimized structure, polar isosurfaces and LUMO-HOMO. Calculations were performed within the Spartan '20 package [27]. DFT energy minimized structure of **1** was obtained using B3LYP with the 6-31G* basis set under a polarization continuum model (=H_2O). Calculated minimized structure with overlaid frontier π-orbitals and electrostatic potential isosurfaces of petrosamine (**1**) are shown in Figures 4 and 5.

3.4. *Synthesis of Merocyanine Dye* (**6b**) *[16]*

Iodomethane (3.11 mL, 50 mmol) was slowly added to a cold mixture of 4-methylpyridine (4.86 mL, 50 mmol) and 2-propanol (5 mL). The stirred mixture was heated at reflux overnight. Removal of the solvent gave the crude 4-methylpyridinium methiodide (10.23 g) as an off-white solid. A portion of the latter salt (3.00 g, 12.8 mmol), 4-hydroxybenzaldehyde and piperidine (1.06 mL, 10.7 mmol) were dissolved in anhydrous EtOH (16 mL) and heated at reflux overnight, with stirring. Upon cooling, a red precipitate was deposited. The red solid was filtered (Büchner funnel), dissolved in KOH solution (0.2 M, 17 mL, 15 mmol), and heated with stirring for 30 min. Blue-red shiny crystals were recovered by filtration, washed with cold water, and dried to give **6b** (2.00 g, 65%). The ^{1}H NMR of the sample was consistent with the expected product.

3.5. *H-D Exchange Measurements of* **1** *by ESI-TOFMS*

H-D exchange measurements were made by ESI-TOFMS using the following conditions: VCap: 3500V; fragmentor voltage: 160 V; nozzle voltage: 500 V; drying gas temperature: 325 C, sheath gas temperature: 325 C, drying gas flow rate: 7.0 L/Min; sheath gas flow rate: 10 L/min; nebulizer pressure: 40 psi. A petrosamine (**1**) sample was rapidly dissolved in CD_3OD (23 °C). An aliquot of the solution was immediately introduced into the inlet of the TOF mass spectrometer, and the ratios of **1**-d_0, **1**-d_1 and **1**-d_2 were measured 'on the fly' from the corresponding molecular ion intensities $[M]^+$ normalized to the $[M]^+$ of a solution of **1** in CH_3OH at the same concentration. Subsequent measurements were made at 15 s intervals over a total reaction time of 90 s. HRMS data were acquired and stored in centroided mode. See Figure 7 and S6–S9.

4. Conclusions

Evidence supports that the major tautomer of petrosamine is the C5 and C8 di-keto form **1**, both in solution and solid states. While the enol form **8a** was not detectable by ^{1}H NMR in DMSO-d_6, it is nevertheless likely responsible for the H-D exchange of **1** in deuteric solvents. The exceptional solvatochromism of **1** is best attributable to charge delocalized resonance forms **1a** and **1d**—partial structures within **1** that are analogous to Brooker merocyanine dyes [11]. As with the latter, the hyposochromic solvent effects mostly correlate with polar-solvent stabilization of the ground state (HOMO) and attribution to the color changes of **1**. The complete isotopic exchange of the C6 CH_2 group in **1** to

CD_2 in a deuterated solvent under ambient conditions is supported by ^{1}H NMR and fast-sampled ESIMS measurements. The exchange mechanism likely proceeds through rapid acid-catalyzed keto-enol tautomerism.

Refinement of the molecular structure of **1** by QM methods maps the electron delocalization and accompanying charge distribution in **1**. These and other refinements complement computational docking studies that can lend a more precise definition of host-guest interactions of **1** in its cognate enzyme active site. In turn, these observations may assist in the design and synthesis of new AChE inhibitors based on the pyridoacridine skeleton: a privileged alkaloid molecular framework produced exclusively by marine invertebrates.

Supplementary Materials: The following supporting information can be downloaded at: https://www.mdpi.com/article/10.3390/md21080446/s1, Figure S1: UV-vis spectra of **1**. pH dependence in Britton-Robinson buffer (normalized); Figure S2: FTIR of **6b** (film, ATR); Figure S3: UV-vis spectra of **6b** in pure acetone; Figure S4: UV-vis spectra of **6b** in acetone-H_2O mixtures (normalized); Figure S5: In situ ESIMS measurements of the rate of H-D exchange of **1** in CD_3OD; Figure S6: Derivation of kinetic parameters of H-D exchange of **1** in CD_3OD and uncertainty analysis; Figure S7: HRMS. Time course of **1** in CD_3OD (triplicate runs) and Fitted Curves; Figure S8: ESI-TOFMS of **1** collected in (a) in CH3OH (b) CD_3OD at 30 s) (c) CD_3OD at 90 s; Figure S9: Time course of ESI-TOF-MS for petrosamine (**1**); Figure S10: ^{1}H NMR of **1** in a. CD_3OH and b. CD_3OD after 5 min; Figure S11: ^{1}H NMR temperature study of petrosamine (**1**) (CD_3OH, 500 MHz); Table S1: λ_{max} and absorbance of merocyanine dye **6b** in H_2O-acetone mixtures. Table S2: ^{1}H and ^{13}C NMR of **1** in CD_3OD and CD_3OH; Table S3: ^{1}H NMR data for compounds **S1** and **S6** (CD_3OD, 500 MHz); Scheme S1: Synthesis of compound **S1**: Experimental: Synthesis of Model Compound **S1**. References [28,29] are cited in the Supplementary Materials.

Author Contributions: C.J.G. carried out the photophysical measurements with consultation from T.F.M., X.W. and Y.S. and conducted the kinetic analysis of H-D exchange in **1**. The manuscript was co-written by C.J.G., X.W. and T.F.M. All authors have read and agreed to the published version of the manuscript.

Funding: The 500 MHz NMR spectrometer and the HPLC ESITOFMS instrument were purchased with funding from the NSF (Chemical Research Instrument Fund, CHE0741968) and the NIH Shared Instrument Grant (S10RR025636) programs, respectively. This research was supported by the Academic Senate of UC San Diego.

Institutional Review Board Statement: Not applicable.

Data Availability Statement: Original data will be made available upon reasonable request.

Acknowledgments: We are most grateful for a generous gift of petrosamine (**1**) from P. Stout (Sirenas Marine Discovery, LLC, San Diego, CA, USA) and to C. Perrin (UCSD) for helpful discussions and N. Goodyear for assistance with measurements

Conflicts of Interest: The authors declare no conflict of interest.

References and Notes

1. Molinski, T.F.; Faulkner, D.J. Petrosamine, a Novel Pigment from the Marine Sponge *Petrosia* sp. *J. Org. Chem.* **1988**, *53*, 1341–1343. [CrossRef]
2. Molinski, T.F. Marine Pyridoacridine Alkaloids: Structure, Synthesis, and Biological Chemistry. *Chem. Rev.* **1993**, *93*, 1825–1838. [CrossRef]
3. Schmitz, F.J.; Agarwal, S.K.; Gunasekera, S.P.; Schmidt, P.G.; Shoolery, J.N. Amphimedine, New Aromatic Alkaloid from a Pacific Sponge, *Amphimedon* sp. Carbon Connectivity Determination from Natural Abundance ^{13}C-^{13}C Coupling Constants. *J. Am. Chem. Soc.* **1983**, *105*, 4835–4836. [CrossRef]
4. Marshall, K.M.; Barrows, L.R. Biological activities of pyridoacridines. *Nat. Prod. Rep.* **2004**, *21*, 731–751. [CrossRef] [PubMed]
5. Nukoolkarn, V.S.; Saen-oon, S.; Rungrotmongkol, T.; Hannongbua, S.; Ingkaninan, K.; Suwanborirux, K. Petrosamine, a potent anticholinesterase pyridoacridine alkaloid from a Thai marine sponge *Petrosia* n. sp. *Bioorg. Med. Chem.* **2008**, *16*, 6560–6567. [CrossRef] [PubMed]

6. Carroll, A.R.; Ngo, A.; Quinn, R.J.; Redburn, J.; Hooper, J.N.A. Petrosamine B, an Inhibitor of the *Helicobacterpylori* Enzyme Aspartyl Semialdehyde Dehydrogenase from the Australian Sponge *Oceanapia* sp. *J. Nat. Prod.* **2005**, *68*, 804–806. [CrossRef] [PubMed]
7. Wei, X.; Bugni, T.S.; Harper, M.K.; Sandoval, I.T.; Manos, E.J.; Swift, J.; Van Wagoner, R.M.; Jones, D.A.; Ireland, C.M. Evaluation of Pyridoacridine Alkaloids in a Zebrafish Phenotypic Assay. *Mar. Drugs* **2010**, *8*, 1769–1778. [CrossRef]
8. Choi, D.-Y.; Choi, H. Natural products from marine organisms with neuroprotective activity in the experimental models of Alzheimer's disease, Parkinson's disease and ischemic brain stroke: Their molecular targets and action mechanisms. *Arch. Pharm. Res.* **2015**, *38*, 139–170. [CrossRef]
9. Britton, H.T.K.; Robinson, R.A. CXCVIII.—Universal buffer solutions and the dissociation constant of veronal. *J. Chem. Soc.* **1931**, 1456–1462. [CrossRef]
10. Exchange of ^1H-^{13}C for ^2H-^{13}C results in dramatic attenuation of the ^{13}C signal due to quadrupolar relaxation, and absence of the heteronuclear nOe, among other effects.
11. Brooker, L.G.S.; Heyes, G.H.; Sprague, R.H.; Van Dyke, R.H.; Van Lare, E.; Van Zandt, G.; White, F.L. Studies in the Cyanine Dye Series. XI. The Merocyanines. *J. Am. Chem. Soc.* **1951**, *73*, 5326–5332. [CrossRef]
12. Brooker, L.G.S.; Heyes, G.H.; Sprague, R.H.; Van Dyke, R.H.; Van Lare, E.; Van Zandt, G.; White, F.L.; Cressman, H.W.J.; Dent, S.G. Color and Constitution. X. Absorption of the Merocyanines. *J. Am. Chem. Soc.* **1951**, *73*, 5332–5350. [CrossRef]
13. Other possible resonance forms, for example invoking electron donation from Br, are considered only minor contributors.
14. Simpson, W.T. A Mathematical Treatment of the Color of the Merocyanine Dyes. *J. Am. Chem. Soc.* **1951**, *73*, 5359–5363. [CrossRef]
15. Morley, J.O.; Morley, R.M.; Docherty, R.; Charlton, M.H. Fundamental Studies on Brooker's Merocyanine. *J. Am. Chem. Soc.* **1997**, *119*, 10192–10202. [CrossRef]
16. Minch, M.J.; Sadiq Shah, S. A Merocyanin Dye Preparation for the Introductory Organic Laboratory. *J. Chem. Educ.* **1977**, *54*, 709. [CrossRef]
17. We confirm the solvatochromism of one example, ***i***, by synthesis and its solvent-dependent UV-vis properties. See Supporting Information.
18. Hasselbalch, K.A. Die Berechnung der Wasserstoffzahl des Blutes aus der freien und gebundenen Kohlensäure desselben, und die Sauerstoffbindung des Blutes als Funktion der Wasserstoffzahl. *Biochem. Z.* **1917**, *78*, 112–114.
19. Bordwell, F.K.; Harrelson, J.A., Jr. Acidities and homolytic bond dissociation energies of the αC—H bonds in ketones in DMSO. *Can. J. Chem.* **1990**, *68*, 1714–1718. [CrossRef]
20. Bordwell, F.G. Equilibrium acidities in dimethyl sulfoxide solution. *Acc. Chem. Res.* **1988**, *21*, 456–463. [CrossRef]
21. Available online: https://organicchemistrydata.org/hansreich/resources/pka/pka_data/pka-compilation-reich-bordwell.pdf (accessed on 8 July 2023).
22. Chiang, Y.; Kresge, J.; Wirz, J. Flash-Photolytic Generation of Acetophenone Enol. The Keto-Enol Equilibrium Constant and pKa of Acetophenone in Aqueous Solution. *J. Am. Chem. Soc.* **1984**, *106*, 6392–6395. [CrossRef]
23. Valpuesta, M.; Díaz, A.; Suau, R. Coulteroberbinone, a quaternary isoquinoline alkaloid from *Romneya coulteri*. *Phytochemistry* **1999**, *51*, 1157–1160. [CrossRef]
24. Guthrie, J.P.; Cossar, J.; Klym, A. p*K*a values for substituted acetophenones: Values determined by rates of halogenation. *Can. J. Chem.* **1987**, *65*, 2154–2159. [CrossRef]
25. Novak, M.; Marc Loudon, G. Aminolysis of Acetoxystyrenes. The pK_a of Acetophenones in Aqueous Solution. *J. Am. Chem. Soc.* **1976**, *98*, 3591–3597. [CrossRef]
26. Steiner, E.C.; Gilbert, J.M. The Acidities of Weak Acids in Dimethyl Sulfoxide. *J. Am. Chem. Soc.* **1963**, *85*, 3054–3055. [CrossRef]
27. Shao, Y.; Gan, Z.; Epifanovsky, E.; Gilbert, T.B.; Wormit, M.; Kussmann, J.; Lange, A.W.; Behn, A.; Deng, J.; Feng, X.; et al. Advances in molecular quantum chemistry contained in the Q-Chem 4 program package. *Mol. Phys.* **2015**, *113*, 184–215. [CrossRef]
28. Graulich, A.; Scuvée-Moreau, J.; Seutin, V.; Liégeois, J.-F. Synthesis and radioligand binding studies of C-5-and C-8-substituted 1-(3, 4-dimethoxybenzyl)-2, 2-dimethyl-1, 2, 3, 4-tetrahydroisoquinoliniums as SK channel blockers related to N-methyl-laudanosine and N-methyl-noscapine. *J. Med. Chem.* **2005**, *48*, 4972–4982. [CrossRef]
29. Molinski, T.F.; Faulkner, D.J. An antibacterial pigment from the sponge *Dendrilla membranosa*. *Tetrahedron Lett.* **1988**, *29*, 2137–2138. [CrossRef]

marine drugs

MDPI

Article

New Guanidine Alkaloids Batzelladines O and P from the Marine Sponge *Monanchora pulchra* Induce Apoptosis and Autophagy in Prostate Cancer Cells

Sergey A. Dyshlovoy [1,2,3,*,†], Larisa K. Shubina [4,†], Tatyana N. Makarieva [4,*], Alla G. Guzii [4], Jessica Hauschild [1,2], Nadja Strewinsky [1], Dmitrii V. Berdyshev [4], Ekaterina K. Kudryashova [4], Alexander S. Menshov [4], Roman S. Popov [4], Pavel S. Dmitrenok [4], Markus Graefen [2], Carsten Bokemeyer [1] and Gunhild von Amsberg [1,2]

1 Department of Oncology, Hematology and Bone Marrow Transplantation with Section Pneumology, Hubertus Wald-Tumorzentrum—University Cancer Center Hamburg, University Medical Center Hamburg-Eppendorf, 20251 Hamburg, Germany
2 Martini-Klinik, Prostate Cancer Center, University Medical Center Hamburg-Eppendorf, 20251 Hamburg, Germany
3 Institute of High Technologies and Advanced Materials, Far Eastern Federal University, 690922 Vladivostok, Russia
4 G.B. Elyakov Pacific Institute of Bioorganic Chemistry, Far Eastern Branch of the Russian Academy of Sciences, Pr. 100-let Vladivostoku 159, 690022 Vladivostok, Russia
* Correspondence: s.dyshlovoy@uke.de (S.A.D.); makarieva@piboc.dvo.ru (T.N.M.)
† These authors contributed equally to this work.

Citation: Dyshlovoy, S.A.; Shubina, L.K.; Makarieva, T.N.; Guzii, A.G.; Hauschild, J.; Strewinsky, N.; Berdyshev, D.V.; Kudryashova, E.K.; Menshov, A.S.; Popov, R.S.; et al. New Guanidine Alkaloids Batzelladines O and P from the Marine Sponge *Monanchora pulchra* Induce Apoptosis and Autophagy in Prostate Cancer Cells. *Mar. Drugs* **2022**, *20*, 738. https://doi.org/10.3390/md20120738

Academic Editor: William H. Fenical

Received: 18 October 2022
Accepted: 22 November 2022
Published: 25 November 2022

Abstract: Two new guanidine alkaloids, batzelladines O (**1**) and P (**2**), were isolated from the deep-water marine sponge *Monanchora pulchra*. The structures of these metabolites were determined by NMR spectroscopy, mass spectrometry, and ECD. The isolated compounds exhibited cytotoxic activity in human prostate cancer cells PC3, PC3-DR, and 22Rv1 at low micromolar concentrations and inhibited colony formation and survival of the cancer cells. Batzelladines O (**1**) and P (**2**) induced apoptosis, which was detected by Western blotting as caspase-3 and PARP cleavage. Additionally, induction of pro-survival autophagy indicated as upregulation of LC3B-II and suppression of mTOR was observed in the treated cells. In line with this, the combination with autophagy inhibitor 3-methyladenine synergistically increased the cytotoxic activity of batzelladines O (**1**) and P (**2**). Both compounds were equally active in docetaxel-sensitive and docetaxel-resistant prostate cancer cells, despite exhibiting a slight p-glycoprotein substrate-like activity. In combination with docetaxel, an additive effect was observed. In conclusion, the isolated new guanidine alkaloids are promising drug candidates for the treatment of taxane-resistant prostate cancer.

Keywords: batzelladines; sponge; *Monanchora pulchra*; anticancer activity; prostate cancer; apoptosis; autophagy; p-glycoprotein

1. Introduction

Guanidine alkaloids are a separate group of marine secondary metabolites with structural diversity and different biological activities. The largest number of natural guanidines has been isolated from marine sponges. Initially, they were suggested as chemotaxonomic markers of the marine sponges of *Ptilocaulis*, *Hemimycale*, *Crambe*, *Batzella*, *Clathria*, and *Monanchora* genera [1,2]. The guanidine alkaloids isolated from the sponges of the *Monanchora* genus of the Crambeidae family were reported to have unique structures and an impressive range of biological activities (reviewed in [1,3]). These compounds contain penta-, tri-, bicyclic, as well as acyclic guanidine moieties and were reported to be cytotoxic to various types of human cancer cells [4–10]. Recently, some aspects of the mechanism of action of the alkaloids have been reported, in particular, with regard to their anticancer activity. Thus, several guanidine-containing marine-derived metabolites could induce p53-independent cancer cell death [11] and the mechanism of this effect seems to be related to specific or unspecific induction of MAP kinases [11,12], induction of ROS, as well as induction of either intrinsic mitochondria-medicated apoptosis [13] or non-apoptotic cell death [12,14]. Moreover, some of these compounds could inhibit an EGF-induced malignant transformation of the cells in in vitro models [11,12]. However, the mechanism of action seems to significantly vary depending on the structure and the biological model used.

For example, for pentacyclic alkaloid monanchocidin A, an unusual mechanism of action cells has been reported in germ cell tumors (GCT) [5,7,14]. In a GCT model, this compound induced cytotoxic autophagy (type II programmed cell death) and lysosomal membrane permeabilization [14] as well as inhibition of cell migration [15]. Similar non-apoptotic cytotoxic effects were observed for other pentacyclic alkaloids, e.g., monanchoxymycalin C in prostate cancer cell models in vitro [12]. For another subgroup of pentacyclic guanidine alkaloids related to crambescidin, an inhibitory effect on cell migration was observed [16]. This was mediated by the suppression of tight junctions formation, cell–matrix and cell–cell adhesion, as well as the alteration of cytoskeleton dynamics [16]. In addition, ion channels were blocked [17,18]. Furthermore, the compound induced differentiation of chronic myelogenous leukemia cells [19].

Notably, for different guanidine-containing alkaloids, signs of caspase-dependent apoptosis were reported in various mammalian cancer cell models [11,13]. Thus, for cytotoxic bicyclic guanidine alkaloids urupocidin A and C, induction of intrinsic apoptosis was reported. These metabolites induced mitochondria membrane permeabilization and ROS upregulation, which consequently led to an activation of caspase-9 and -3, cleavage of PARP, DNA fragmentation, and apoptotic cell death [13]. However, the mechanism of action of different guanidine-containing alkaloids may vary.

Batzelladines represent a distinct class of guanidine-derived alkaloids that usually contain two main guanidinic moieties. The tricyclic guanidine core is connected with additional guanidine fragments of varying complexity via an ester linkage in a large number of these molecules. To date, batzelladines have been isolated only from the marine genera *Batzella* (family Chondropsidae), *Monanchora* (family Crambeidae), and *Clathria* (family Microcionidae) belonging to the order Poecilosclerida [20]. The unique and fascinating structures of these guanidines are coupled with a wide range of biological activities (reviewed in [21]), including cytotoxic [22–25], antiviral [10,26–28], and antiparasitic activities [29]. In addition, the anti-HIV activity of batzelladines was extensively studied. In fact, these molecules were found to mediate the inhibition of HIV gp120 binding to human CD4 cells [27,28]. Moreover, very recently, an anti-SARS-CoV-2 activity exerted via inhibition of the main virus M^{pro} protease has been predicted and is currently under investigation [30]. In addition, strong cytotoxic activity was observed in human cancer cells [22–25,27]. However, little is known about their mode of action in mammalian cells.

As part of our ongoing program in the search for bioactive compounds from the Northwestern Pacific marine invertebrates [31,32], we investigated a new collection of the sponge *Monanchora pulchra*, collected in the Okhotsk sea. Here, we report the isolation, structure elucidation, in vitro anticancer activity, and insights into the mode of action of two new guanidine alkaloids, batzelladines O (**1**) and P (**2**).

2. Results and Discussion

2.1. Isolation and Structure Elucidation of Batzelladines O and P

The crude EtOH extract of the marine sponge *Monanchora pulchra* was concentrated *in vacuo* and the obtained residue was fractionated by flash chromatography on a YMC*Gel ODS-A column. Further separation using reversed-phase HPLC resulted in the isolation of two new batzelladines O (**1**) and P (**2**) (Figure 1). The batzelladine **1** gave $[M]^+$ ion at *m/z* 487.3750 in the HRESIMS spectrum, appropriate for the molecular formula $C_{27}H_{47}N_6O_2$, and a peak at m/z 244.1921, corresponding to a doubly charged $[M + H]^{2+}$ ion. An intense fragment ion peak at *m/z* 374.2800 $[C_{22}H_{36}N_3O_2]^+$ in HRESIMS/MS showed a loss from the molecular ion of 113 Da, corresponding to the 4-guanidino-1-butyl, also observed in some batzelladines [26,33]. ^{13}C NMR data (Table 1) implied two C=C bonds (δ_C 129.2 and 132.2, and 103.7 and 151.3), one ester carbonyl (δ_C 166.2), and two guanidine carbons (δ_C 149.9 and 158.8). Detailed analysis of the ^1H, ^{13}C, COSY, HSQC, and HMBC NMR spectra of **1** (Table 1) revealed a close structural similarity between **1** and batzelladine C isolated from *Batzella* sp. [26], except that additional olefinic resonances were observed (δ_H 5.39, 5.45; δ_C 129.2, 132.2). In the COSY spectrum, there was presented a sequential spin system from the terminal methyl CH_3-27 (δ_H 0.92) to CH_2-25 (δ_H 2.06) through CH_2-26 (δ_H 1.40), while in the HMBC spectrum the methylene protons H_2-25 showed couplings to C-23 (δ_C 129.2) and C-24 (δ_C 132.2) (Figure 2), indicating a Δ^{23} unsaturation. The H_2-22 (δ_H 2.19) allylic protons were in turn coupled to signals at δ_H 1.75 and 1.64 in the COSY experiments and gave HMBC correlation with C-13 signal at δ_C 51.9. The ^{13}C chemical shifts of C-22 (δ_C 23.7) and C-25 (δ_C 30.4) suggested a Z configuration for this double bond, as in batzelladine E [26], proved by the synthesis [34]. This was further confirmed by an NOE correlation between H_2-22 and H_2-25 methylene protons (Figure 2).

Table 1. NMR Data for compounds **1** and **2** in CD_3OD.

	1		**2**	
position	δ_H (J in Hz)	δ_C [a] type	δ_H (J in Hz)	δ_C [a] type
1		158.8		158.8, C
2	3.22, t (6.8)	42.1, CH_2	3.23, t (6.8)	42.1, CH_2
3	1.67, m	26.6, CH_2	1.67, m	26.6, CH_2
4	1.76, m	27.1, CH_2	1.76, m	27.0, CH_2
5	4.23, m	65.2, CH_2	4.22, m	65.2, CH_2
6		166.2, C		166.2, C
7		103.7, C		103.7, C
8		151.3, C		151.4, C
9	3.43, m	30.8, CH_2	3.43, m	30.7, CH_2
	2.92, m		2.92, m	
10	2.46, m	30.3, CH_2	2.46, m	30.4, CH_2
	1.75, m		1.74, m	
11	4.10,m	58.7, CH	4.10, m	58.7, CH
12	2.46, m	34.2, CH_2	2.44, m	34.2, CH_2
12	1.40, m		1.40, m	
13	3.64, m	51.9, CH	3.63, m	51.9, CH
14		149.9, C		149.9, C
15	4.42, m	51.7, CH	4.42, m	51.7, CH
16	1.57, m	37.7, CH_2	1.57, m	37.7, CH_2
17	1.30, m	24.8, CH_2	1.30, m	25.1, CH_2
18	1.30, m	32.5, CH_2	1.30, m	32.7, CH_2
19	1.30, m	23.5, CH_2	1.30, m	23.5, CH_2
20	0.91, t (6.8)	14.3, CH_3	1.30, m	30.3, CH_2
21	1.75, m	35.7, CH_2	1.30, m	30.3, CH_2
	1.64, m			
22	2.19, m	23.7, CH_2	0.90, t (6.8)	14.3, CH_3
23	5.39, m	129.2, CH	1.75, m	35.8, CH_2
			1.64, m	
24	5.45, m	132.2, CH	2.18, m	23.7, CH_2
25	2.06, m	30.4, CH_2	5.39, m	129.0, CH
26	1.40, m	23.8, CH_2	5.45, m	132.2, CH
27	0.92, t (6.8)	14.0, CH_3	2.05, m	30.4, CH_2
28			1.40, m	23.8, CH_2
29			0.92, t (6.8)	14.0, CH_3

[a] ^{13}C NMR assignments supported by HSQC and HMBC data.

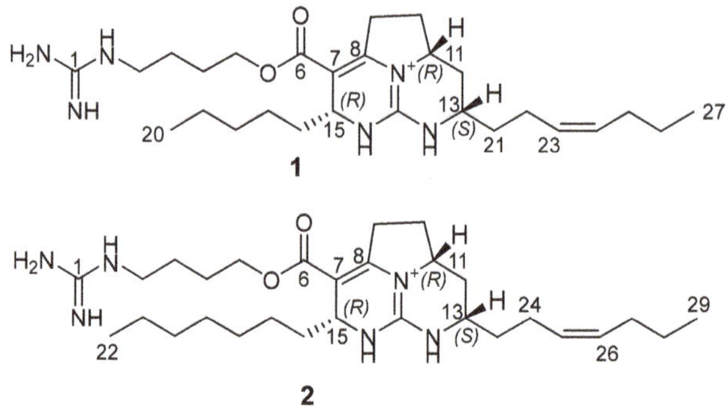

Figure 1. The structures of batzelladines O (**1**) and P (**2**).

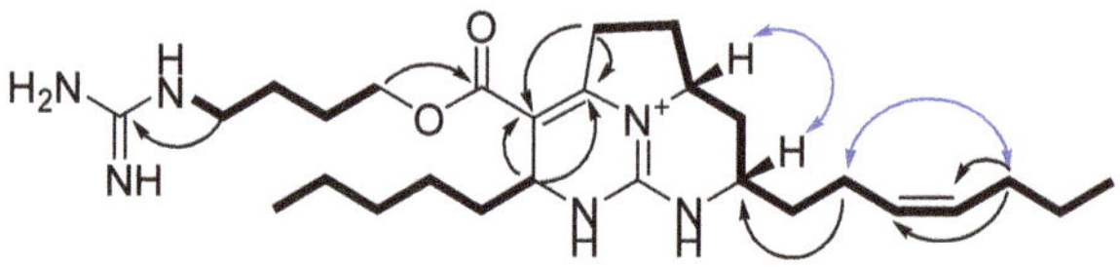

Figure 2. Key COSY (▬), HMBC (⌒), and ROESY (⌒) correlations for **1**.

The HMBC correlation of H_2-9 (δ_H 3.43/2.92) and H-15 (δ_H 4.42) with C-7 (δ_C 103.7) and C-8 (δ_C 151.3) (Figure 2) confirmed the placement of a tetrasubstituted double bond between the carbons C-7 and C-8.

The ROESY showed that hydrogen atoms at C-11 and C-13 are in a *cis* relationship.

The absolute configurations of the stereogenic centers of **1** were established by quantum-chemical calculations of ECD spectra (Figure 3 and Figure S15). The ECD spectra for four stereoisomers 11*R*,13*S*,15*R*-**1a,** 11*S*,13*R*,15*R*-**1b,** 11*R*,13*S*,15*S*-**1c,** and 11*S*,13*R*,15*S*-**1d** were calculated using GAUSSIAN_16 software (Gaussian Inc., Wallingford, USA). The main features of the experimental ECD spectrum of **1** are well reproduced with the ECD spectrum calculated for the 11*R*,13*S*,15*R*-**1a** stereoisomer. The spectrum of 11*S*,13*R*,15*R*-**1b** does not reproduce the features of a negative band in the $190 \leq \lambda \leq 235$ nm region; the ratio of intensities for two positive bands $\eta = I(\lambda = 295\ nm)/I(\lambda = 252\ nm)$ is also overestimated.

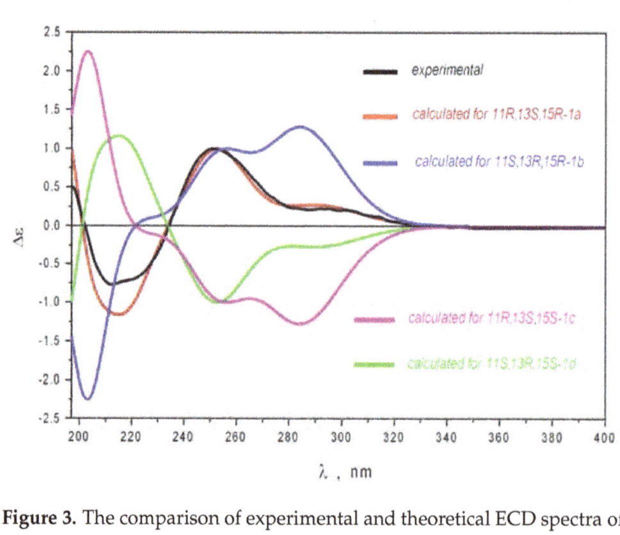

Figure 3. The comparison of experimental and theoretical ECD spectra of **1**.

The ECD spectra for 11*S*,13*R*,15*S*-**1d** and 11*R*,13*S*,15*S*-**1c** stereoisomers are mirror-imaged to those calculated for 11*R*,13*S*,15*R*-**1a** and 11*S*,13*R*,15*R*-**1b** stereoisomers, respectively, and completely disagree with the experimental spectrum. These results indicate the absolute configuration of compound **1** to be 11*R*,13*S*,15*R*.

Therefore, compound **1** differs from the known batzelladine C by an additional Δ^{23} double bond and was named batzelladine O.

Batzelladine P (**2**) was isolated as a colorless glassy solid and its molecular formula was determined to be $C_{29}H_{51}N_6O_2$ from the $[M]^+$ ion peak at *m/z* 515.4065 in the positive HRESIMS. The fragment ion peak at *m/z* 402.3117 in HRESIMS/MS corresponded to a loss of 4-guanidino-1-butyl from molecular ion. The ^1H and ^{13}C NMR spectra of **2** (Table 1) matched with those of batzelladine O. However, its molecular formula indicated the presence of two additional methylene groups. The HMBC correlations H_3-29/C-27, H_2-27/C-25, and H_2-24/C-13 of **2** together with ^1H, and ^{13}C NMR chemical shifts of H_3-29 (δ_H 0.92), H_2-27 (δ_H 2.05), C-25 (δ_c 129.0), and C-26 (δ_c 132.0) were consistent with the Δ^{23} seven-carbons chain at C-13 of batzelladine O. Therefore, two additional methylenes were located on the second alkyl side chain at C-15, thus including seven carbons. The configurations of the asymmetric centers in batzelladine P were the same as in batzelladine O.

2.2. Investigation of Cytotoxic Activity in Human Prostate Cancer Cells

Previously, we have shown that some spongian guanidine alkaloids reveal potent cytotoxic activity in human prostate cancer cells, including drug-resistant types [11,12]. Therefore, we tested the isolated batzelladines O and P in human prostate cancer 22Rv1, PC3, and PC3-DR cells. These cell lines represent human prostate cancer models bearing different levels of drug resistance. Prostate cells, both normal and malignant cells, are hormone-dependent. They require androgens for their growth and survival [35]. Androgens are the ligands for androgen receptor (AR) and their binding to AR leads to activation of the AR-mediated transcriptional pathway, which is essential for maintaining cell proliferation. Therefore, drug-mediated hormone deprivation (castration) is successfully applied as a therapeutic strategy to fight prostate cancer at its early stages [35]. However, practically all patients ultimately develop resistance to hormone-deprivation therapy and thus leading to so-called castration resistance. Here, more aggressive and less specific therapeutic approaches are required [36]. Thus, 22Rv1 cells express both androgen receptor full length (AR-FL) and androgen receptor splice variant V7 (AR-V7). AR-V7 maintains a constant autoactivation of the AR pathway even without androgens and makes 22Rv1 cells refractory to androgen deprivation including novel AR receptor targeting agents [36,37]. PC3 cells express neither AR-FL nor AR-V7 and are therefore completely androgen insensitive. PC3-DR cells were developed via the long-term treatment of parental PC3 cells with stepwise increasing concentrations of docetaxel [38]. PC3-DR cells are resistant to docetaxel and other taxanes via multiple mechanisms, with overexpression of p-glycoprotein (p-gp, MDR1) being the most relevant. P-gp is a cell membrane protein, which functions as a molecular pump responsible for the efflux of various small molecules, including different cytotoxic anticancer agents, out of the cells [39,40].

In our experiments, we applied a treatment time of 72 h, as this regime was previously shown to be the most suitable to maximize the effect of guanidine-containing alkaloids [14]. Compounds **1** and **2** exhibited strong cytotoxic activity in either cell line at low micromolar concentrations. Notably, both compounds **1** and **2** exhibited equal cytotoxicity in docetaxel-sensitive PC3 and docetaxel-resistant PC3-DR cells (Figure 4A), whereas docetaxel was 40-fold less active in the latter cell line [41]. For the further mechanistic studies, we selected 22Rv1 cells as they represent not a complete loss of AR, but rather a decreased sensitivity to AR-targeting agents, and this represents the most common situation in the patients. The examination of the long-term effects on cancer cell colony formation and survival did not reveal any significant changes in activity following an increased exposure time (Figure 4B).

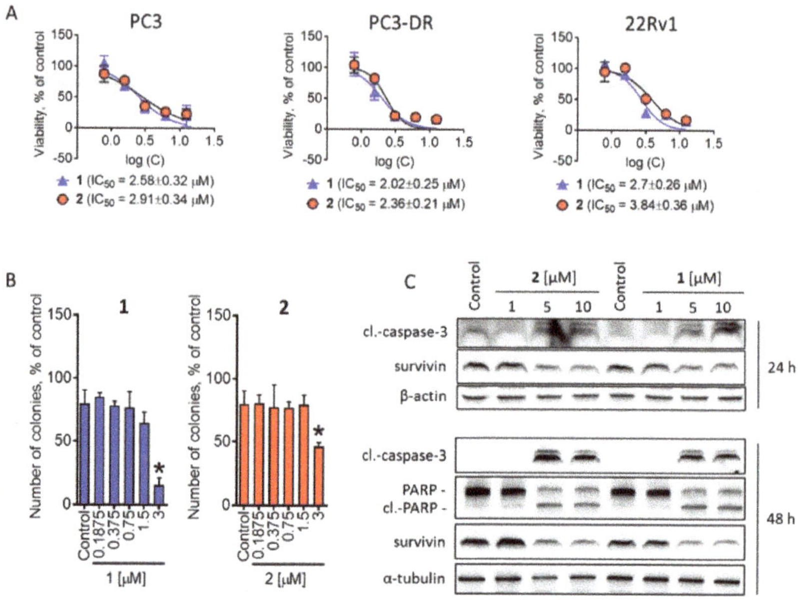

Figure 4. Cytotoxic activity in prostate cancer cells. (**A**), Cytotoxicity profiles of compounds **1** and **2** in PC3, PC3-DR, and 22Rv1 cells. The cells were treated with the investigated compounds for 72 h and the cell viability was measured using MTT assay. (**B**), Colony formation assay. 22Rv1 cells were treated with the indicated concentration of the investigated compounds for 48 h, and then the media was exchanged followed by incubation for additional 14 days. The colonies still alive were fixed, stained, and counted by naked eye. (**C**), Analysis of protein expression. 22Rv1 cells were treated with indicated concentration of the tested compounds for 24 h or 48 h and the protein expression was analyzed by Western blotting. Significant difference from control is indicated as * ($p < 0.05$, one-way ANOVA).

We further showed a time and dose-dependent cleavage of effector caspase-3 as well as of PARP, which suggests the induction of apoptosis-like or apoptosis-related processes (Figure 4C). In line with this, a down-regulation of antiapoptotic protein survivin was observed in the cells exposed to isolated compounds **1** and **2** (Figure 4C). Overall, **1** was more active against cancer cells and exhibited lower IC_{50} values in either experiment (Figure 4A–C).

2.3. Batzelladines O and P Induce Cytoprotective Autophagy in Prostate Cancer Cells

Previously, we have shown that the spongian guanidine-containing alkaloids, depending on the structure and model context, execute their cytotoxic activity via induction of classical caspase-dependent apoptosis [13] or activation of cytotoxic autophagy [14]. Therefore, we examined the effects of the isolated batzelladines O and P on the expression of LC3B-I and LC3B-II proteins (Figure 5A). LC3B-I is a soluble cytoplasmic protein; when autophagy is activated, LC3B-I undergoes conversion (via lipidation) to the LC3B-II protein, which is an essential component of the autophagosome membrane [42,43]. In addition, we evaluated the effect on SQSTM1 (sequestosome-1, p62), an important cargo protein, which binds to other proteins providing their selective delivery to autophagosomes for autophagic degradation [42]. Upregulation of LC3B-II in most cases indicates activation of autophagy, which can result either in cytotoxic or cytoprotective effects, whereas the pattern of SQSTM1 regulation may be significantly different depending on the stimulus and the model used [42]. In our experiments, we observed a pronounced up-regulation of LC3B-II proteins, suggesting activation of autophagy as it has been shown for monanchocidine A [14], while no effect on SQSTM1 has been detected (Figure 5A). Moreover, the downregulation of phospho-mTOR was observed under drug exposure (Figure 5A). It is well established that active (phosphorylated) mTOR suppresses autophagy, whereas mTOR inhibition, e.g., by rapamycin or siRNA, leads to autophagy activation [44]. Hence, the observed down-regulation of phospho-mTOR suggests induction of autophagy by compounds **1** and **2**. To distinguish between cytotoxic and cytoprotective autophagy we applied a combinational treatment with an established early steps autophagy inhibitor 3-methyladenine (3-MA). 3-MA inhibits PI3K kinase and therefore suppresses autophagosome formation ultimately inhibiting autophagy [45]. The combination of either compound **1** or **2** with 3-MA resulted in a synergistic effect, i.e., the investigated compounds were more cytotoxic when autophagy was blocked (Figure 5B,C). These results suggest batzelladines O and P induce cytoprotective autophagy in human cancer 22Rv1 cells.

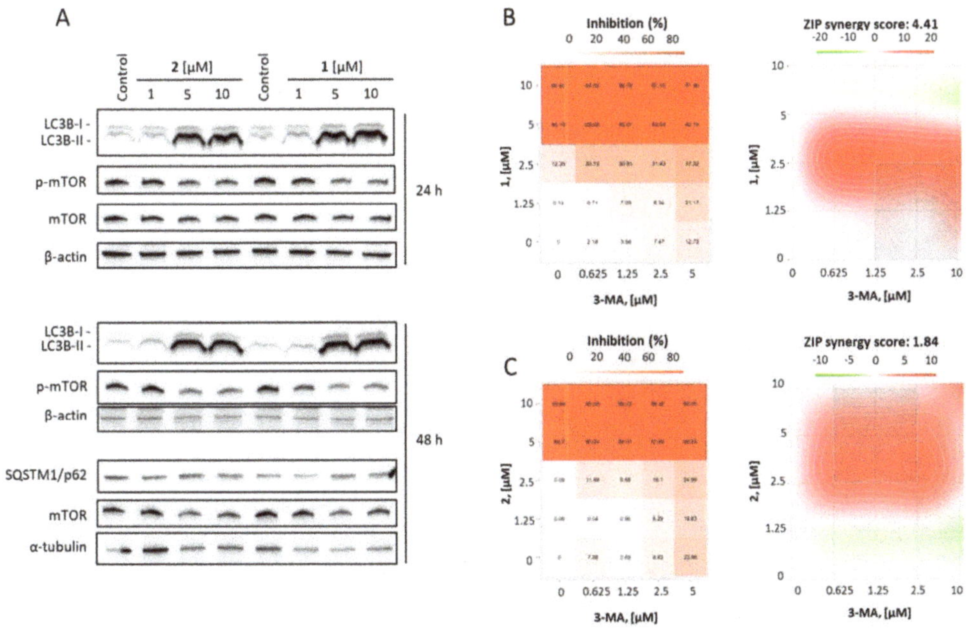

Figure 5. Analysis of the effect on autophagy. (**A**), Analysis of protein expression. 22Rv1 cells were treated with the indicated concentrations of the tested compounds for 24 h or 48 h and the protein expression was analyzed by Western blotting. (**B**,**C**), Effect of 3-methyladenine on cytotoxic activity of the compounds **1** (**B**) and **2** (**C**). 22Rv1 cells were treated with the indicated concentrations of the single drugs and their combinations for 72 h and viability was measured using MTT assay. The cytotoxic heat maps (left panels) and synergy maps (right panels) were generated using SynergyFinder 2.0 software. Red areas indicate synergistic effects of the combinations of specific concentrations of the drugs, green areas indicate antagonistic effects (right panels).

Cytoprotective autophagy is a well-established survival mechanism that helps the cells to overcome unfavorable conditions and in the case of cancer cells helps to survive radio- or chemotherapy [46,47]. Remarkably, another guanidine alkaloid monanchocidine A induced cytotoxic autophagy in GCT cells, and its activity could be antagonized by 3-MA [14]. Therefore, this effect seems to be depending on the drug structure and the model used. Based on our findings in prostate cancer, a combination with established pharmacological autophagy inhibitors should be considered as a strategy for further development of batzelladines O and P as well as related compounds as anticancer agents, in order to prevent resistance mediating, cytoprotecting autophagy.

2.4. Examination of the Effect on P-Glycoprotein Activity

An interesting observation was the equal cytotoxicity of the isolated compounds in PC3 and docetaxel-resistant PC3-DR cells (Figure 4A). Previously, we and others have shown a strong overexpression of p-glycoprotein (p-gp) in PC3-DR cells to be responsible for the resistance to docetaxel. Of note, docetaxel is a well-known substrate of p-gp [32,38–41]. To examine whether either compound is a substrate or inhibitor of p-gp we applied a calcen-AM excretion assay. Calcein-AM is a fluorescence dye that can passively diffuse into cells where it is metabolized by esterases into a fluorescent calcein, which can be further detected. However, in cells overexpressing active p-gp, calcen-AM is rapidly evacuated out to the extracellular space, which results in a decrease or lack of fluorescence. In the case of p-gp-overexpressing cells, inhibition of p-gp activity results in an increase in fluorescence. In our experiments, treatment of the PC3-DR cells with the investigated drugs induced green fluorescence of the cells (Figure 6A). However, the control of the cell viability performed under the identical treatment regime indicated a viability drop-down at the same concentrations at which an increase in fluorescence was observed (Figure 6A). These results indicate that the observed intracellular accumulation of calcian was due to cellular membrane permeabilization, where a disrupted membrane facilitates a passive calcian diffusion inside of the cell and makes a p-gp-medicated calcein efflux ineffective. Interestingly, a pretreatment of PC3-DR cells with a well-established p-gp inhibitor tariquidar (TQD) resulted in a slight but significant increase in cytotoxicity of both compounds **1** and **2** (Figure 6B), indicating that batzelladines O and P exhibit slight p-gp substrate-like properties and can be at least partially excreted from the cells via the p-gp system (Figure 6B). However, no reduction of cytotoxic activity was found in PC3-DR cells when compared to PC3 cells (Figure 4A). This may be due to a multitarget mode of action of compounds **1** and **2**. The details of the aforementioned effect are to be further elucidated.

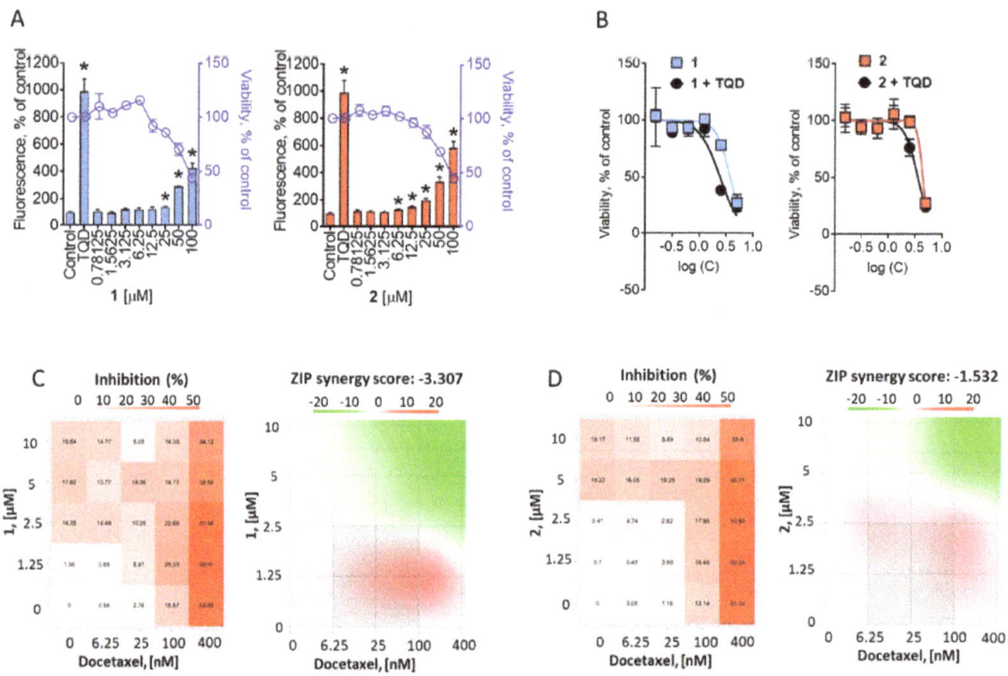

Figure 6. P-gp-independent effect of the compounds. (**A**), Effect on p-glycoprotein activity. PC3-DR cells were treated with indicated concentration of the drugs for 30 min and then incubated with

calcein-AM solution for 15 min. The green fluorescence was measured using Infinite F200PRO TECAN plate spectrophotometer. Tariquidar (TQD, 50 nM) was used as a positive control. The effect of the drugs on cell viability was measured using MTT assay in the same experimental settings, the percentage of viable cells is indicated as a blue curve. (**B**), Effect of tariquidar (TQD) on cytotoxic activity of the compounds. PC3-DR cells were pre-treated with 50 nM TQD for 1 h and then co-treated with indicated concentration of the compounds for additional 72 h. The cellular viability was measured using MTT assay. (**C**,**D**), Effect of the compounds **1** (**C**) and **2** (**D**) in combination with docetaxel. PC3-DR cells were co-treated with indicated concentration of the single drugs and their combination for 48 h and viability was measured using MTT assay. The cytotoxic heat maps (left panels) and synergy maps (right panels) were constructed using SynergyFinder 2.0 software. Red areas indicate synergistic effects of the combination of specific concentrations of the drugs, green areas indicate antagonistic effects (right panels). Significant difference from control is indicated as * ($p < 0.05$, one-way ANOVA).

Finally, we investigated the effects of the isolated metabolites in combination with docetaxel. In line with the previous results, a combination with docetaxel revealed a slight synergistic effect at lower concentrations of compounds **1** and **2**; however, an overall ZIP synergy score (δ) indicated a rather additive effect (Figure 6C,D).

3. Materials and Methods

3.1. General Procedures

Optical rotations were measured using a PerkinElmer 343 polarimeter (Waltham, MA, USA). ECD spectra were recorded with an Chirascan Plus spectropolarimeter (Applied Photophysics , Leatherhead, UK). The ^{1}H and ^{13}C NMR spectra were obtained using Bruker Avance III-700 spectrometer (Bruker, Ettlingen, Germany). Chemical shifts were referenced to the corresponding residual solvent signal (δ_H 3.31/δ_C 49.0 for CD_3OD). HRESIMS were measured using Bruker maXis Impact II mass spectrometer (Bruker Daltonics, Bremen, Germany). Low-pressure column liquid chromatography was performed using YMC*Gel ODS-A (YMC Co., Ltd., Kyoto, Japan). HPLC was performed using Shimadzu Instrument equipped with RID-10A refractive index detector (Shimadzu Corporation, Kyoto, Japan) and YMC-Pack ODS-A (250 × 10 mm) column (YMC Co., Ltd., Kyoto, Japan).

3.2. Animal Material

Specimens of *M. pulchra* were collected in Okhotsk sea, near Iturup Island (45°21.4 N; 148°23.5 E) by dredging at 94 m depth on July 2015, and identified by Grebnev B. B. A voucher specimen was deposited under registration number O47-002 in the collection of marine invertebrates of the G.B. Elyakov Pacific Institute of Bioorganic Chemistry (Vladivostok, Russia).

3.3. Extraction and Isolation

The freshly collected specimens of *M. pulchra* (dry weight 10 g) were extracted with EtOH (2 × 0.2 L). The EtOH extract after evaporation in vacuo was fractioned by flash column chromatography on YMC*Gel ODS-A (75 μm), eluting with a step gradient of H_2O, EtOH:H_2O (40:60, v/v), and EtOH:H_2O (65:35, v/v + 0.1% TFA) with monitoring by HPLC. The fractions that eluted with 65% EtOH+ 0.1% TFA were further purified by repeated reversed-phase HPLC (YMC-Pack ODS-A column (250 × 10 mm), 1.6 mL/min, EtOH:H_2O (65:35, v/v + 0.01% TFA)) to afford batzelladines O (**1**, 6 mg) and P (**2,** 8 mg).

3.4. Compound Characterization Data

Batzelladine O (**1**): colorless glassy solid; $[\alpha]_D^{20}$ +12 (c 0.1, MeOH); UV (MeOH) λmax (log ε) 289 (4.38) nm; ECD (7.9 × 10^{-4}, MeOH) λmax (Δε) 255 (1.49), 299 (0.31) nm; ^{1}H and ^{13}C NMR data, Table; HRESIMS *m/z* 487.3750 $[M]^+$ (calcd for $C_{27}H_{47}N_6O_2$, 487.3755), *m/z* 244.1921 $[M^+ + H]^{2+}$ (calcd for $C_{27}H_{48}N_6O_2$, 244.1914).

Batzelladine P (**2**): colorless glassy solid; $[\alpha]_D^{20}$ +15 (*c* 0.1, MeOH); UV (MeOH) λmax (log ε) 290 (4.42) nm; ECD (10.2 × 10^{-4}, MeOH) λmax (Δε) 254 (1.30), 299 (0.29) nm; ^{1}H and ^{13}C NMR data, Table; HRESIMS *m/z* 515.4065 $[M]^+$ (calcd for $C_{29}H_{51}N_6O_2$, 515.4068), *m/z* 258.2075 $[M^+ + H]^{2+}$ (calcd for $C_{29}H_{51}N_6O_2$, 258.2070).

3.5. Quantum Chemical Modeling

Theoretical modeling of ECD spectra for compound **1** was performed using GAUSSIAN_16 software. The conformational analysis was performed at B3LYP/6-31G(d)_PCM level of theory with CH_3OH as a solvent. Conformations, in which electronic energies are in diapason $\Delta E \leq 5$ kcal/mol were then chosen for calculation of vertical electronic transitions at TDDFT_cam-B3LYP/6-311G(d)_PCM//B3LYP/6-31G(d)_PCM level of theory. ECD spectra for each conformation were simulated as a superposition of bands, generated by individual transitions, using GAUSS band shapes. The used bandwidth, taken at 1/e of peak height, is $\sigma = 0.34$ eV. The UV shift was taken as $\Delta\lambda = +18$ nm.

3.6. Reagents and Antibodies for Bioactivity Assay

Docetaxel was purchased from Pharmacy of the University Hospital Hamburg-Eppendorf (Hamburg, Germany); PhosphoSTOP™ *EASY*packs phosphotase inhibitors cocktail and cOmplete™ *EASY*packs protease inhibitors cocktail were purchased from Roche (Mannheim, Germany); MTT (3-(4,5-dimethylthiazol-2-yl)-2,5-diphenyltetrazolium bromide) was purchased from Sigma (Taufkirchen, Germany); 3-Methyladenine was purchased from Enzo Life Sciences (Farmingdale, NY, USA); Tariquidar was purchased from MedChemExpress (Monmouth Junction, NJ, USA); Primary and secondary antibodies used for Western blotting are listed in Table 2.

Table 2. List of antibodies used.

Antibodies	Clonality	Source	Cat.-No.	Dilution	Manufacturer
anti-SQSTM/p62	pAb	rabbit	#5114	1:1000	Cell Signaling
anti-phospho-mTOR	mAb	rabbit	#5536	1:1000	Cell Signaling
anti-mTOR	mAb	rabbit	#2983	1:1000	Cell Signaling
anti-LC3B-I/II	pAb	rabbit	#2775	1:1000	Cell Signaling
anti-β-Actin-HRP	pAb	goat	sc-1616	1:10,000	Santa Cruz
anti-α-Tubulin	mAb	mouse	T5168	1:5000	Sigma-Aldrich
anti-LC3B-I/II	pAb	rabbit	#2775	1:1000	Cell Signaling
anti-cleaved Caspase-3	mAb	rabbit	#9664	1:1000	Cell Signaling
anti-PARP	pAb	rabbit	#9542	1:1000	Cell Signaling
anti-Survivin	pAb	rabbit	NB500-201	1:1000	Novus
anti-mouse IgG-HRP		sheep	NXA931	1:10,000	GE Healthcare
anti-rabbit IgG-HRP		goat	#7074	1:5000	Cell Signaling

3.7. Cell Lines and Culture Conditions

The human prostate cancer cell lines PC3 and 22Rv1 cells were purchased from ATCC (Manassas, VA, USA). Docetaxel-resistant cells PC3-DR were generated from PC3 cells via long-term treatment with stepwise increasing concentrations of docetaxel and kindly provided by Dr. S. J. Oh-Hohenhorst and Prof. Z. Culig [38]. The cells were regularly tested for mycoplasma infection, checked microscopically for stable phenotype, and were kept in culture for maximum of 2 months. The passage number of either cell line was <30. Cells were cultured in a humidified 5% (*v*/*v*) CO_2 atmosphere at 37°C as a monolayer. For PC3 and 22Rv1 cells 10% FBS/RPMI medium was used (RPMI medium supplemented with Glutamax^{TM}-I (gibco® Life technologies^{TM}, Paisley, UK), 10% fetal bovine serum (gibco® Life technologies™), and 1% penicillin/streptomycin (gibco® Life technologies™). PC3-DR cells were cultured in 10% FBS/RPMI medium containing 12.5 nM of docetaxel. All the experiments with PC3-DR cells were performed in docetaxel-free 10% FBS/RPMI culture media unless docetaxel was applied as drug for combinational treatment.

3.8. MTT Assay

The MTT assay was used to estimate cell viability. The cells were seeded in 96-well plates, 6×10^3 cells/well in 100 µL/well. Cells were incubated overnight and the medium was exchanged with 100 µL/well of fresh culture medium containing drugs or vehicle at the indicated concentrations. Cells treated with vehicle were used as a negative control. The plates were further incubated for indicated time and 10 µL/well of 5 mg/mL MTT solution in PBS (3-(4,5-dimethylthiazol-2-yl)-2,5-diphenyltetrazolium bromide) was added. After 2–4 h of additional incubation, the media was aspirated, and the plates were dried overnight. The 50 µL/well of DMSO was added to dissolve the formazan crystals and absorbance was measured using Infinite F200PRO reader (TECAN, Männedorf, Switzerland). The results were proceeded with, and IC_{50}s calculated, using GraphPad Prism software v.9.1.1 (GraphPad Software, San Diego, CA, USA).

3.9. Western Blotting

The Western blot was used to evaluate the protein expression and was executed as previously described [48]. Cells were seeded in ø 6 cm Petri dishes (1×10^6 cells/dish in 5 mL/dish) and incubated overnight. Then, the cells were treated with the compounds in fresh culture media (5 mL/dish) for 48 h, unless otherwise stated. The cells were harvested and lysed in the RIPA buffer containing a cocktail of protease and phosphatase inhibitors (Roche, Mannheim, Germany). Nonelysed cell particles were separated by centrifugation. Afterward, the total protein lysates were separated using a ready-made gradient Mini-PROTEAN® TGX Stain-Free™ gels (Bio-Rad, Hercules, CA, USA) by SDS-PAGE. The proteins were transferred onto ø 0.2 µm pore PVDF membrane, which was further blocked and incubated with primary and secondary antibodies, listed in Table 2. The protein signals were developed using the ECL chemiluminescence system (Thermo Scientific, Rockford, IL, USA). The original Western blotting pictures are presented in the Supplementary information.

3.10. Drug Combination Studies

To evaluate a possible synergistic effect of combinations with docetaxel we used a Zero interaction potency (ZIP) reference model [49] and the SynergyFinder 2.0 software (https://synergyfinder.fimm.fi [50], accessed on 2 October 2022) as previously described [11]. The 22Rv1 or PC3-DR cells were seeded in 96-well plates and treated with individual drugs or their combinations in 100 µL/well of culture media, as described for MTT assay. Following 48 h of incubation (unless otherwise stated) the cellular viability was estimated using MTT assay, as described above. The difference between expected and detected effects of the drug combinations was analyzed and visualized using SynergyFinder 2.0 tool. positive δ-values (red areas) indicate synergistic effects and negative δ-values (green areas) indicate antagonistic effects. The effects having $-10 \leq \delta \leq 10$ are considered additive.

3.11. P-Glycoprotein Activity Analysis

The evolution of p-glycoprotein activity was performed in p-gp overexpressing PC3-DR cells, as previously reported [32]. PC3-DR cells (6×10^3 cells/well) were seeded in 96-well plates (black, clear bottom) in 100 µL/well of the culture medium and the plates were incubated overnight. The culture media was substituted with 50 µL/well of DPBS, containing investigated compounds at the indicated concentrations. The plates were incubated for 30 min, followed by addition of 50 µL of calcein-AM solution (1 µM in DPBS) and incubation for additional 15 min. The green fluorescence of free calcein produced by cellular esterases via removal of the acetoxymethyl (AM) was measured with Infinite

F200PRO reader (TECAN, Männedorf, Switzerland). The values were normalized to the possible background autofluorescence of the drugs' solutions. The viability of the cells was simultaneously measured by MTT assay using the same treatment regime.

3.12. Data and Statistical Analysis

The experiments were performed in biological triplicates ($n = 3$). Cells treated with vehicle were used as a negative control in all the experiments. IC_{50}s values and statistical analysis were performed using GraphPad Prism v.9.1.1 software (GraphPad Software, San Diego, CA, USA). Data are shown as mean $\pm$ standard deviation (SD). The one-way ANOVA in combination with Dunnett's post-hoc tests were used for multiple group comparisons. The Student's t-test was used for comparison of the two groups. Statistically significant difference is indicated as: * if $p < 0.05$ (ANOVA or Student's *t*-test).

4. Conclusions

In conclusion; two new batzelladines O and P were isolated from the deep-water marine sponge *Monanchora pulchra*. Both compounds exhibited a pronounced cytotoxic activity in human prostate cancer cells executed via induction of apoptosis and could inhibit colony formation and survival of the cancer cells. The compounds induced pro-survival autophagy. Batzelladines O and P were equally active in docetaxel-sensitive and -resistant prostate cancer cells, despite the slight p-glycoprotein substrate-like activity, and exhibited an additive effect in combination with docetaxel. To the best of our knowledge, this is the very first study reporting insights into the mechanisms of the cytotoxic action of batzelladines in human cancer cells.

Supplementary Materials: The following are available online at https://www.mdpi.com/article/10.3390/md20120738/s1, Copies of CD, HRESIMS, 1D-, and 2D-NMR spectra of compounds **1, 2**. Theoretical modeling of ECD spectra for compound 1 was done using GAUSSIAN_16 software [51].

Author Contributions: S.A.D., J.H., and N.S. performed the bioactivity assays, L.K.S. and E.K.K. isolated the metabolites, T.N.M. and L.K.S. elucidated structures; S.A.D., L.K.S., D.V.B., and A.G.G. wrote the manuscript which was revised and approved by all the authors, A.S.M. performed the NMR spectra; R.S.P. and P.S.D. performed the mass spectra; D.V.B. calculated ECD spectra; S.A.D., T.N.M., M.G., C.B., and G.v.A. analyzed and discussed the results; M.G., C.B., and G.v.A. provided essential materials and resources for this study. All authors have read and agreed to the published version of the manuscript.

Funding: This scientific work was carried out with the support of the Grant of the Ministry of Science and Higher Education, Russian Federation 13.1902.21.0012 (contract No 075-15-2020-796).

Institutional Review Board Statement: Not applicable.

Data Availability Statement: Not applicable.

Acknowledgments: We devote this article to the 80th anniversary of the birth of an outstanding scientist, Valentin A. Stonik, who made a great contribution to this study of various natural products. For many years Stonik has been leading a worldwide known scientific school working on marine and terrestrial natural compounds from diverse chemical classes such as steroids, terpenoids, alkaloids, glycosides, peptides, and lipids as well as products of mixed biosynthesis. The authors would like to thank Zoran Culig (Innsbruck Medical University, Innsbruck, Austria) and Su Jung Oh-Hohenhorst (University Medical Center Hamburg-Eppendorf) for providing the docetaxel-resistant PC3-DR cells. This study was carried out on the equipment of the Collective Facilities Center "The Far Eastern Center for Structural Molecular Research (NMR/MS) PIBOC FEB RAS." The authors are thankful to Natalya Yu. Kim for providing UV and CD spectroscopic data.

Conflicts of Interest: The authors declare no conflict of interest.

References

1. Berlinck, R.G.S.; Bertonha, A.F.; Takaki, M.; Rodriguez, J.P.G. The chemistry and biology of guanidine natural products. *Nat. Prod. Rep.* **2017**, *34*, 1264–1301. [CrossRef] [PubMed]
2. Carroll, A.R.; Copp, B.R.; Davis, R.A.; Keyzers, R.A.; Prinsep, M.R. Marine natural products. *Nat. Prod. Rep.* **2019**, *36*, 122–173. [CrossRef]
3. Shi, Y.; Moazami, Y.; Pierce, J.G. Structure, synthesis and biological properties of the pentacyclic guanidinium alkaloids. *Bioorg. Med. Chem.* **2017**, *25*, 2817–2824. [CrossRef] [PubMed]
4. Hamann, M.T.; Gul, W.; Hammond, N.L.; Yousaf, M.; Bowling, J.J.; Schinazi, R.F.; Wirtz, S.S.; Andrews, G.D.C.; Cuevas, C. Modification at the C9 position of the marine natural product isoaaptamine and the impact on HIV-1, mycobacterial, and tumor cell activity. *Bioorg. Med. Chem.* **2006**, *14*, 8495–8505.
5. Guzii, A.G.; Makarieva, T.N.; Denisenko, V.A.; Dmitrenok, P.S.; Kuzmich, A.S.; Dyshlovoy, S.A.; Krasokhin, V.B.; Stonik, V.A. Monanchocidin: A new apoptosis-inducing polycyclic guanidine alkaloid from the marine sponge *Monanchora pulchra*. *Org. Lett.* **2010**, *12*, 4292–4295. [CrossRef] [PubMed]
6. Makarieva, T.N.; Tabakmaher, K.M.; Guzii, A.G.; Denisenko, V.A.; Dmitrenok, P.S.; Kuzmich, A.S.; Lee, H.-S.; Stonik, V.A. Monanchomycalins A and B, unusual guanidine alkaloids from the sponge *Monanchora pulchra*. *Tetrahedron Lett.* **2012**, *53*, 4228–4231. [CrossRef]
7. Makarieva, T.N.; Tabakmaher, K.M.; Guzii, A.G.; Denisenko, V.A.; Dmitrenok, P.S.; Shubina, L.K.; Kuzmich, A.S.; Lee, H.S.; Stonik, V.A. Monanchocidins B-E: Polycyclic guanidine alkaloids with potent antileukemic activities from the sponge *Monanchora pulchra*. *J. Nat. Prod.* **2011**, *74*, 1952–1958. [CrossRef]
8. Guzii, A.G.; Makarieva, T.N.; Korolkova, Y.V.; Andreev, Y.A.; Mosharova, I.V.; Tabakmaher, K.M.; Denisenko, V.A.; Dmitrenok, P.S.; Ogurtsova, E.K.; Antonov, A.S.; et al. Pulchranin A, isolated from the Far-Eastern marine sponge, *Monanchora pulchra*: The first marine non-peptide inhibitor of TRPV-1 channels. *Tetrahedron Lett.* **2013**, *54*, 1247–1250. [CrossRef]
9. Bensemhoun, J.; Bombarda, I.; Aknin, M.; Vacelet, J.; Gaydou, E.M. Ptilomycalin D, a Polycyclic Guanidine Alkaloid from the Marine Sponge *Monanchora dianchora*. *J. Nat. Prod.* **2007**, *70*, 2033–2035. [CrossRef]
10. Gallimore, W.A.; Kelly, M.; Scheuer, P.J. Alkaloids from the sponge *Monanchora unguifera*. *J. Nat.Prod.* **2005**, *68*, 1420–1423. [CrossRef]
11. Dyshlovoy, S.A.; Tabakmakher, K.M.; Hauschild, J.; Shchekaleva, R.K.; Otte, K.; Guzii, A.G.; Makarieva, T.N.; Kudryashova, E.K.; Fedorov, S.N.; Shubina, L.K.; et al. Guanidine alkaloids from the marine sponge *Monanchora pulchra* show cytotoxic properties and prevent EGF-induced neoplastic transformation in vitro. *Mar. Drugs* **2016**, *14*, 133. [CrossRef] [PubMed]
12. Dyshlovoy, S.A.; Kaune, M.; Kriegs, M.; Hauschild, J.; Busenbender, T.; Shubina, L.K.; Makarieva, T.N.; Hoffer, K.; Bokemeyer, C.; Graefen, M.; et al. Marine alkaloid monanchoxymycalin C: A new specific activator of JNK1/2 kinase with anticancer properties. *Sci. Rep.* **2020**, *10*, 13178. [CrossRef]
13. Dyshlovoy, S.A.; Kudryashova, E.K.; Kaune, M.; Makarieva, T.N.; Shubina, L.K.; Busenbender, T.; Denisenko, V.A.; Popov, R.S.; Hauschild, J.; Fedorov, S.N.; et al. Urupocidin C: A new marine guanidine alkaloid which selectively kills prostate cancer cells via mitochondria targeting. *Sci. Rep.* **2020**, *10*, 9764. [CrossRef] [PubMed]
14. Dyshlovoy, S.A.; Hauschild, J.; Amann, K.; Tabakmakher, K.M.; Venz, S.; Walther, R.; Guzii, A.G.; Makarieva, T.N.; Shubina, L.K.; Fedorov, S.N.; et al. Marine alkaloid monanchocidin A overcomes drug resistance by induction of autophagy and lysosomal membrane permeabilization. *Oncotarget* **2015**, *6*, 17328–17341. [CrossRef] [PubMed]
15. Dyshlovoy, S.A.; Venz, S.; Hauschild, J.; Tabakmakher, K.M.; Otte, K.; Madanchi, R.; Walther, R.; Guzii, A.G.; Makarieva, T.N.; Shubina, L.K.; et al. Anti-migratory activity of marine alkaloid monanchocidin A—proteomics-based discovery and confirmation. *Proteomics* **2016**, *16*, 1590–1603. [CrossRef]
16. Rubiolo, J.A.; Lopez-Alonso, H.; Roel, M.; Vieytes, M.R.; Thomas, O.; Ternon, E.; Vega, F.V.; Botana, L.M. Mechanism of cytotoxic action of crambescidin-816 on human liver-derived tumour cells. *Br. J. Pharmacol.* **2014**, *171*, 1655–1667. [CrossRef]
17. Berlinck, R.G.; Braekman, J.C.; Daloze, D.; Bruno, I.; Riccio, R.; Ferri, S.; Spampinato, S.; Speroni, E. Polycyclic guanidine alkaloids from the marine sponge *Crambe crambe* and Ca++ channel blocker activity of crambescidin 816. *J. Nat. Prod.* **1993**, *56*, 1007–1015. [CrossRef] [PubMed]
18. Martin, V.; Vale, C.; Bondu, S.; Thomas, O.P.; Vieytes, M.R.; Botana, L.M. Differential effects of crambescins and crambescidin 816 in voltage-gated sodium, potassium and calcium channels in neurons. *Chem. Res. Toxicol.* **2013**, *26*, 169–178. [CrossRef]
19. Aoki, S.; Kong, D.; Matsui, K.; Kobayashi, M. Erythroid differentiation in K562 chronic myelogenous cells induced by crambescidin 800, a pentacyclic guanidine alkaloid. *Anticancer Res.* **2004**, *24*, 2325–2330.
20. Sfecci, E.; Lacour, T.; Amade, P.; Mehiri, M. Polycyclic Guanidine Alkaloids from Poecilosclerida Marine Sponges. *Mar. Drugs* **2016**, *14*, 77. [CrossRef]
21. El-Demerdash, A.; Atanasov, A.G.; Bishayee, A.; Abdel-Mogib, M.; Hooper, J.N.A.; Al-Mourabit, A. *Batzella, Crambe* and *Monanchora:* Highly Prolific Marine Sponge Genera Yielding Compounds with Potential Applications for Cancer and Other Therapeutic Areas. *Nutrients* **2018**, *10*, 33. [CrossRef] [PubMed]
22. Sun, H.H.; Sakemi, S.; Burres, N.; McCarthy, P. Isobatzellines A, B, C, and D. Cytotoxic and antifungal pyrroloquinoline alkaloids from the marine sponge *Batzella* sp. *J. Org. Chem.* **1990**, *55*, 4964–4966. [CrossRef]
23. Laville, R.; Thomas, O.P.; Berrué, F.; Marquez, D.; Vacelet, J.; Amade, P. Bioactive Guanidine Alkaloids from Two Caribbean Marine Sponges. *J. Nat. Prod.* **2009**, *72*, 1589–1594. [CrossRef] [PubMed]

24. Kohn, L.K.; Porto, P.S.S.; Bianchi, B.R.; Santos, M.F.C.; Berlinck, R.G.S.; Arns, C.W. NOR-Batzelladine L from the sponge *Monanchora* sp. displays antiviral acyivity against Herpes Simplex virus type 1. *Planta Med.* **2012**, *78*, CL27. [CrossRef]
25. Pessoa, C.; dos Santos, M.F.C.; Berlinck, R.G.S.; Ferreira, P.M.P.; Cavalcanti, B.C. Cytotoxic Batzelladine L from the Brazilian Marine Sponge *Monanchora arbuscula*. *Planta Med.* **2013**, *79*, 844. [CrossRef]
26. Zhou, X.; Liu, J.; Yang, B.; Lin, X.; Yang, X.W.; Liu, Y. Marine natural products with anti-HIV activities in the last decade. *Curr. Med. Chem.* **2013**, *20*, 953–973.
27. Patil, A.D.; Kumar, N.V.; Kokke, W.C.; Bean, M.F.; Freyer, A.J.; Brosse, C.D.; Mai, S.; Truneh, A.; Carte, B. Novel Alkaloids from the Sponge *Batzella* sp.: Inhibitors of HIV gp120-Human CD4 Binding. *J. Org. Chem.* **1995**, *60*, 1182–1188.
28. Hua, H.-M.; Peng, J.; Dunbar, D.C.; Schinazi, R.F.; de Castro Andrews, A.G.; Cuevas, C.; Garcia-Fernandez, L.F.; Kelly, M.; Hamann, M.T. Batzelladine alkaloids from the caribbean sponge *Monanchora unguifera* and the significant activities against HIV-1 and AIDS opportunistic infectious pathogens. *Tetrahedron* **2007**, *63*, 11179–11188. [CrossRef]
29. Santos, M.F.C.; Harper, P.M.; Williams, D.E.; Mesquita, J.T.; Pinto, É.G.; da Costa-Silva, T.A.; Hajdu, E.; Ferreira, A.G.; Santos, R.A.; Murphy, P.J.; et al. Anti-parasitic Guanidine and Pyrimidine Alkaloids from the Marine Sponge *Monanchora arbuscula*. *J. Nat. Prod.* **2015**, *78*, 1101–1112. [CrossRef] [PubMed]
30. Elgohary, A.M.; Elfiky, A.A.; Pereira, F.; Abd El-Aziz, T.M.; Sobeh, M.; Arafa, R.K.; El-Demerdash, A. Investigating the structure-activity relationship of marine polycyclic batzelladine alkaloids as promising inhibitors for SARS-CoV-2 main protease (Mpro). *Comput. Biol. Med.* **2022**, *147*, 105738. [CrossRef] [PubMed]
31. Shubina, L.K.; Makarieva, T.N.; Denisenko, V.A.; Popov, R.S.; Dyshlovoy, S.A.; Grebnev, B.B.; Dmitrenok, P.S.; von Amsberg, G.; Stonik, V.A. Gracilosulfates A-G, Monosulfated Polyoxygenated Steroids from the Marine Sponge *Haliclona gracilis*. *Mar. Drugs* **2020**, *18*, 454. [CrossRef] [PubMed]
32. Dyshlovoy, S.A.; Shubina, L.K.; Makarieva, T.N.; Hauschild, J.; Strewinsky, N.; Guzii, A.G.; Menshov, A.S.; Popov, R.S.; Grebnev, B.B.; Busenbender, T.; et al. New diterpenes from the marine sponge *Spongionella* sp. overcome drug resistance in prostate cancer by inhibition of P-glycoprotein. *Sci. Rep.* **2022**, *12*, 13570. [CrossRef] [PubMed]
33. Braekman, J.C.; Daloze, D.; Tavares, R.; Hajdu, E.; Van Soest, R.W.M. Novel Polycyclic Guanidine Alkaloids from Two Marine Sponges of the Genus *Monanchora*. *J. Nat. Prod.* **2000**, *63*, 193–196. [CrossRef] [PubMed]
34. Snider, B.B.; Chen, J. Synthesis of batzelladine E and its E isomer. *Tetrahedron Lett.* **1998**, *39*, 5697–5700. [CrossRef]
35. Sandhu, S.; Moore, C.M.; Chiong, E.; Beltran, H.; Bristow, R.G.; Williams, S.G. Prostate cancer. *Lancet* **2021**, *398*, 1075–1090. [CrossRef] [PubMed]
36. Nelson, P.S. Targeting the androgen receptor in prostate cancer—A resilient foe. *N. Engl. J. Med.* **2014**, *371*, 1067–1069. [CrossRef] [PubMed]
37. Sampson, N.; Neuwirt, H.; Puhr, M.; Klocker, H.; Eder, I.E. In vitro model systems to study androgen receptor signaling in prostate cancer. *Endocr. Relat. Cancer* **2013**, *20*, R49–R64. [CrossRef] [PubMed]
38. Puhr, M.; Hoefer, J.; Schäfer, G.; Erb, H.H.H.; Oh, S.J.; Klocker, H.; Heidegger, I.; Neuwirt, H.; Culig, Z. Epithelial-to-mesenchymal transition leads to docetaxel resistance in prostate cancer and is mediated by reduced expression of miR-200c and miR-205. *Am. J. Pathol.* **2012**, *181*, 2188–2201. [CrossRef]
39. Gjyrezi, A.; Xie, F.; Voznesensky, O.; Khanna, P.; Calagua, C.; Bai, Y.; Kung, J.; Wu, J.; Corey, E.; Montgomery, B.; et al. Taxane resistance in prostate cancer is mediated by decreased drug-target engagement. *J. Clin. Investig.* **2020**, *130*, 3287–3298. [CrossRef] [PubMed]
40. Zhu, Y.; Liu, C.; Nadiminty, N.; Lou, W.; Tummala, R.; Evans, C.P.; Gao, A.C. Inhibition of ABCB1 expression overcomes acquired docetaxel resistance in prostate cancer. *Mol. Cancer Ther.* **2013**, *12*, 1829–1836. [CrossRef] [PubMed]
41. Dyshlovoy, S.A.; Kaune, M.; Hauschild, J.; Kriegs, M.; Hoffer, K.; Busenbender, T.; Smirnova, P.A.; Zhidkov, M.E.; Poverennaya, E.V.; Oh-Hohenhorst, S.J.; et al. Efficacy and Mechanism of Action of Marine Alkaloid 3,10-Dibromofascaplysin in Drug-Resistant Prostate Cancer Cells. *Mar. Drugs* **2020**, *18*, 609. [CrossRef] [PubMed]
42. Klionsky, D.J.; Abdel-Aziz, A.K.; Abdelfatah, S.; Abdellatif, M.; Abdoli, A.; Abel, S.; Abeliovich, H.; Abildgaard, M.H.; Abudu, Y.P.; Acevedo-Arozena, A.; et al. Guidelines for the use and interpretation of assays for monitoring autophagy (4th edition). *Autophagy* **2021**, *17*, 1–382. [PubMed]
43. Tanida, I.; Ueno, T.; Kominami, E. LC3 and Autophagy. *Methods Mol. Biol.* **2008**, *445*, 77–88. [PubMed]
44. Dossou, A.S.; Basu, A. The Emerging Roles of mTORC1 in Macromanaging Autophagy. *Cancers* **2019**, *11*, 1422. [CrossRef]
45. Wu, Y.-T.; Tan, H.-L.; Shui, G.; Bauvy, C.; Huang, Q.; Wenk, M.R.; Ong, C.-N.; Codogno, P.; Shen, H.-M. Dual Role of 3-Methyladenine in Modulation of Autophagy via Different Temporal Patterns of Inhibition on Class I and III Phosphoinositide 3-Kinase. *J. Biol. Chem.* **2010**, *285*, 10850–10861. [CrossRef] [PubMed]
46. Hippert, M.M.; O'Toole, P.S.; Thorburn, A. Autophagy in cancer: Good, bad, or both? *Cancer Res.* **2006**, *66*, 9349–9351. [CrossRef]
47. Mathew, R.; Karantza-Wadsworth, V.; White, E. Role of autophagy in cancer. *Nat. Rev. Cancer* **2007**, *7*, 961–967. [CrossRef]
48. Dyshlovoy, S.A.; Pelageev, D.N.; Hauschild, J.; Sabutskii, Y.E.; Khmelevskaya, E.A.; Krisp, C.; Kaune, M.; Venz, S.; Borisova, K.L.; Busenbender, T.; et al. Inspired by sea urchins: Warburg effect mediated selectivity of novel synthetic non-glycoside 1,4-naphthoquinone-6S-glucose conjugates in prostate cancer. *Mar. Drugs* **2020**, *18*, 251. [CrossRef]
49. Yadav, B.; Wennerberg, K.; Aittokallio, T.; Tang, J. Searching for Drug Synergy in Complex Dose–Response Landscapes Using an Interaction Potency Model. *Comput. Struct. Biotechnol. J.* **2015**, *13*, 504–513. [CrossRef]

50. Ianevski, A.; Giri, A.K.; Aittokallio, T. SynergyFinder 2.0: Visual analytics of multi-drug combination synergies. *Nucleic Acids Res.* **2020**, *48*, W488–W493. [CrossRef] [PubMed]
51. Frisch, M.J.; Trucks, G.W.; Schlegel, H.B.; Scuseria, G.E.; Robb, M.A.; Cheeseman, J.R.; Scalmani, G.; Barone, V.; Petersson, G.A.; Nakatsuji, H.; et al. *Gaussian 16 Software, Revision A.01*; Gaussian, Inc.: Wallingford. CT, USA, 2016.

marine drugs

Article

Analogs of 6-Bromohypaphorine with Increased Agonist Potency for α7 Nicotinic Receptor as Anti-Inflammatory Analgesic Agents

Igor A. Ivanov [1], Andrei E. Siniavin [1,2], Victor A. Palikov [3], Dmitry A. Senko [4], Irina V. Shelukhina [1], Lyubov A. Epifanova [1], Lucy O. Ojomoko [1], Svetlana Y. Belukhina [5], Nikita A. Prokopev [6], Mariia A. Landau [7], Yulia A. Palikova [3], Vitaly A. Kazakov [3], Natalia A. Borozdina [3], Arina V. Bervinova [3], Igor A. Dyachenko [3], Igor E. Kasheverov [1], Victor I. Tsetlin [1] and Denis S. Kudryavtsev [1,*]

1 Shemyakin-Ovchinnikov Institute of Bioorganic Chemistry, Russian Academy of Sciences, 117997 Moscow, Russia; victortsetlin3f@gmail.com (V.I.T.)
2 N.F. Gamaleya National Research Center for Epidemiology and Microbiology, Ivanovsky Institute of Virology, Ministry of Health of the Russian Federation, 123098 Moscow, Russia
3 The Branch of the Shemyakin-Ovchinnikov Institute of Bioorganic Chemistry, Russian Academy of Sciences, 142290 Pushchino, Russia
4 Center Neurobiology and Brain Restoration, Skolkovo Institute of Science and Technology, 121205 Moscow, Russia
5 Center of Life Sciences, Skolkovo Institute of Science and Technology, 121205 Moscow, Russia
6 Department of Biology, M.V. Lomonosov Moscow State University, 119991 Moscow, Russia
7 Moscow Institute of Physics and Technology, 141701 Dolgoprudny, Russia
* Correspondence: kudryavtsev@ibch.ru

Abstract: Hypaphorines, tryptophan derivatives, have anti-inflammatory activity, but their mechanism of action was largely unknown. Marine alkaloid L-6-bromohypaphorine with EC_{50} of 80 μM acts as an agonist of α7 nicotinic acetylcholine receptor (nAChR) involved in anti-inflammatory regulation. We designed the 6-substituted hypaphorine analogs with increased potency using virtual screening of their binding to the α7 nAChR molecular model. Fourteen designed analogs were synthesized and tested in vitro by calcium fluorescence assay on the α7 nAChR expressed in neuro 2a cells, methoxy ester of D-6-iodohypaphorine (6ID) showing the highest potency (EC_{50} 610 nM), being almost inactive toward α9α10 nAChR. The macrophages cytometry revealed an anti-inflammatory activity, decreasing the expression of TLR4 and increasing CD86, similarly to the action of PNU282987, a selective α7 nAChR agonist. 6ID administration in doses 0.1 and 0.5 mg/kg decreased carrageenan-induced allodynia and hyperalgesia in rodents, in accord with its anti-inflammatory action. Methoxy ester of D-6-nitrohypaphorine demonstrated anti-oedemic and analgesic effects in arthritis rat model at i.p. doses 0.05–0.26 mg/kg. Tested compounds showed excellent tolerability with no acute in vivo toxicity in dosages up to 100 mg/kg i.p. Thus, combining molecular modelling and natural product-inspired drug design improved the desired activity of the chosen nAChR ligand.

Keywords: α7 nicotinic acetylcholine receptor; nAChR; agonist; ligand; inflammation; potency; hypaphorines

Citation: Ivanov, I.A.; Siniavin, A.E.; Palikov, V.A.; Senko, D.A.; Shelukhina, I.V.; Epifanova, L.A.; Ojomoko, L.O.; Belukhina, S.Y.; Prokopev, N.A.; Landau, M.A.; et al. Analogs of 6-Bromohypaphorine with Increased Agonist Potency for α7 Nicotinic Receptor as Anti-Inflammatory Analgesic Agents. *Mar. Drugs* **2023**, *21*, 368. https://doi.org/10.3390/md21060368

Academic Editors: Vladimir Kalinin, Pavel Dmitrenok and Natalia Ivanchina

Received: 20 April 2023
Revised: 3 June 2023
Accepted: 14 June 2023
Published: 20 June 2023

1. Introduction

Homopentameric α7 nicotinic acetylcholine receptors (nAChRs) are ligand-gated cationic ion channels with high permeability for calcium ions and a fast desensitization (see reviews [1,2]). They are widespread in the nervous system, mainly in areas responsible for cognitive function and memory—such as the hippocampus, cerebral cortex, and several subcortical structures of the limbic system [3]. Dysfunction of α7 nAChR is associated with neuropsychiatric and neurological disorders such as schizophrenia and Alzheimer's disease [4,5]. α7 nAChRs are also expressed in non-neuronal cells, in particular, in immune and endothelial cells where they play a decisive role in at least some types of inflammation [6]. These receptors are also involved in neuroprotection in astrocytes, oligodendrocyte precursor cells, and microglia [7–9].

In non-neuronal cells, including lymphocytes, dendritic cells, and macrophages, α7 nAChR is an essential participant in the cholinergic anti-inflammatory pathway, which is the link between the innate immune system and efferent nerves [10]. As well, α7 nAChR mediates the cholinergic pathway in brain, which is crucial in neuroprotection and, probably, in Parkinson's disease, oxygen deprivation, and global ischemia [9,11].

Intracellular signaling, mediated by the activation of α7 nAChR, is based on conducting short-term currents of Na^+, K^+, and Ca^{2+} ions, the mechanism of calcium-induced calcium release through the IP3 receptors being one of the essential ones. In certain cases, α7 nAChRs function as metabotropic receptors, mediating intracellular signals by binding to G proteins (Gα and Gβγ) [12,13]. In immune cells, α7 nAChR is involved in several intracellular pathways resulting in the anti-inflammatory effects [14]. For example, α7 nAChR has been shown to activate JAK2/STAT3 cascade, which leads to inhibition of the transcription factor NF-κB and to the production of anti-inflammatory cytokines [15,16]. It has also been shown that α7 nAChR activates the PI3K/Akt pathway, which promotes Nrf-2 translocation into the nucleus and an increase in heme oxygenase (HO-1) expression, which leads to potent anti-inflammatory effects [11,17–19].

Thus, the activation of α7 nAChR is considered a potential therapeutic strategy against neurological and inflammatory disorders [14,20]. The search for new agonists, especially selective ones, is especially perspective. The α7 nAChR agonist EVP-6124 has shown positive results in clinical trials for the treatment of conditions such as schizophrenia [21], but trials were discontinued due to side effects from the digestive system; ABT-594 is a less toxic analog of nicotine and has analgesic properties [22]. Anabasein analog GTS-21 is another example of analgesic and anti-inflammatory activity of α7 nAChR agonists (see Section 2.4 in review [23]). Unfortunately, no α7 nAChR agonists have passed phase III of clinical trials as analgesic, anti-inflammatory, or anti-psychotic means.

Several years ago, we found that 6-bromohypaphorine (6BHP) from the nudibranch Hermissenda carassicornis acts as an agonist of human α7 nAChRs [24]. Currently, anti-inflammatory activity of its parent compound hypaphorine is described in detail [25–28]. Interestingly, acetylcholine esterase-blocking activity has been demonstrated recently for L- and D-hypaphorines [29]. Moreover, intracranial hypaphorine induces sleep in mice [30], and cholinergic system is known to be involved in the sleep–wake cycle regulation [31]. Thus, we decided to check the anti-inflammatory and analgesic activities of 6BHP and to try to increase its affinity for the α7 nAChRs by designing its new analogs.

In the present communication, we explored diverse 6BHP synthetic analogs described by the general formula depicted on Scheme 1, where X = NO_2, NH_2, OH, OCH_3, Cl, Br, I; Q = OH, OCH_3, OC_2H_5, NH_2, $NHCH_3$; R = H, CH_3.

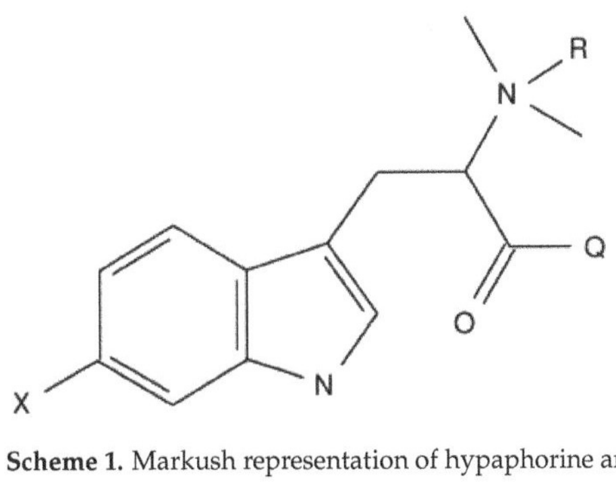

Scheme 1. Markush representation of hypaphorine analogs structures.

2. Results

2.1. Rational Design and Virtual Screening

The potency of naturally occurring 6-bromohypaphorine (~90 μM) is considerably lower than that of the traditional α7 nAChR agonists acetylcholine (8 μM) and epibatidine (15 μM). However, it is selective for α7 nAChR [24], which makes it a promising candidate for the design of structurally similar ligands with a higher affinity for α7 nAChRs. The original compound has a relatively simple structure, a small number of rotatable bonds and, in general, similar features with other agonists of nicotinic receptors: quaternary amine and aromatic moiety. Thus, it was possible to develop trustworthy models of complexes of 6-bromohypaphorine and analogs thereof with α7 nAChR using OpenBabel script to generate SMILES [32,33]. Indeed, docking of 6-bromohypaphorine to AChBP/α7 nAChR chimera (PDB 3SQ6) and the recently published cryo-EM structure of α7 nAChR in complex with epibatidine (PDB 7KOX) show binding at the orthosteric site under the loop C (Figure 1a–c). Moreover, 6BHP analogs and epibatidine appear to occupy similar positions in the orthosteric site showing remarkable similarity in the binding modes (Figure 1d,e).

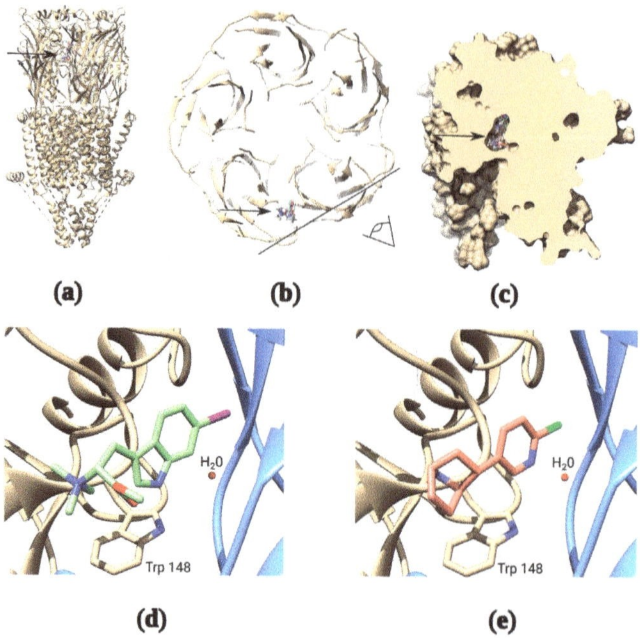

Figure 1. Molecular docking of synthetic hypaphorines. The receptor has five identical binding sites, but docking of ligand is shown to only one of them for clarity (**a**): Binding positions of several synthetic hypaphorines identified under the loop C of the receptor (arrow); (**b**): Top view from the extracellular space of the α7 nAChR cross-section showing binding positions of the major synthetic hypaphorines studied in this paper. The black line and an eye symbol designate the section plane used in panel "c"; (**c**): Space-filling model of two adjacent extracellular domains of α7 nAChR demonstrating inclusion of synthetic hypaphorines into the binding pocket under the loop C; (**d**): Example of (D)-6-iodohypaphorine methyl ester (6ID) docking pose showing the close proximity of quaternary ammonium nitrogen to W148 in the orthosteric binding site that is known to form pi–cation interaction with amine groups of various nAChR agonists. Note, that water molecule observed in many X-ray and cryo-EM structures is conserved in the binding site to accurately mimic binding conditions; (**e**): Structure of the epibatidine complex with α7 nAChR from recently published PDB 7KOX structure [34] by Noviello et al. Note, that putative structure of synthetic hypaphorine (6ID) complex shows remarkable similarity to the structure of complex with epibatidine.

It was previously found that hydrophilic amino acid residues in orthosteric binding site influence agonist binding by interaction with aromatic moiety of the later [35]. Interestingly, carboxylic group of hypaphorine analogs might fit in this interaction and add some subtype selectivity.

2.2. Synthesis of a Hypaphorine Analog Series

Modification of indole core was performed by Sandmeyer reaction in Filimonov modification [36]. See Scheme 2 for the illustration. The key intermediate protected 6-aminoindoline derivate was easily synthesized by tryptophan's selective nitration [37], followed by total protection and reduction. After introducing an appropriate substitution at the 6th position, the indolyl ring was quantitatively reoxidized by 2,3-dichloro-5,6-dicyanobenzoquinone (DDQ). The obtained protected tryptophan derivate was easily converted to the corresponding hypaphorine derivate.

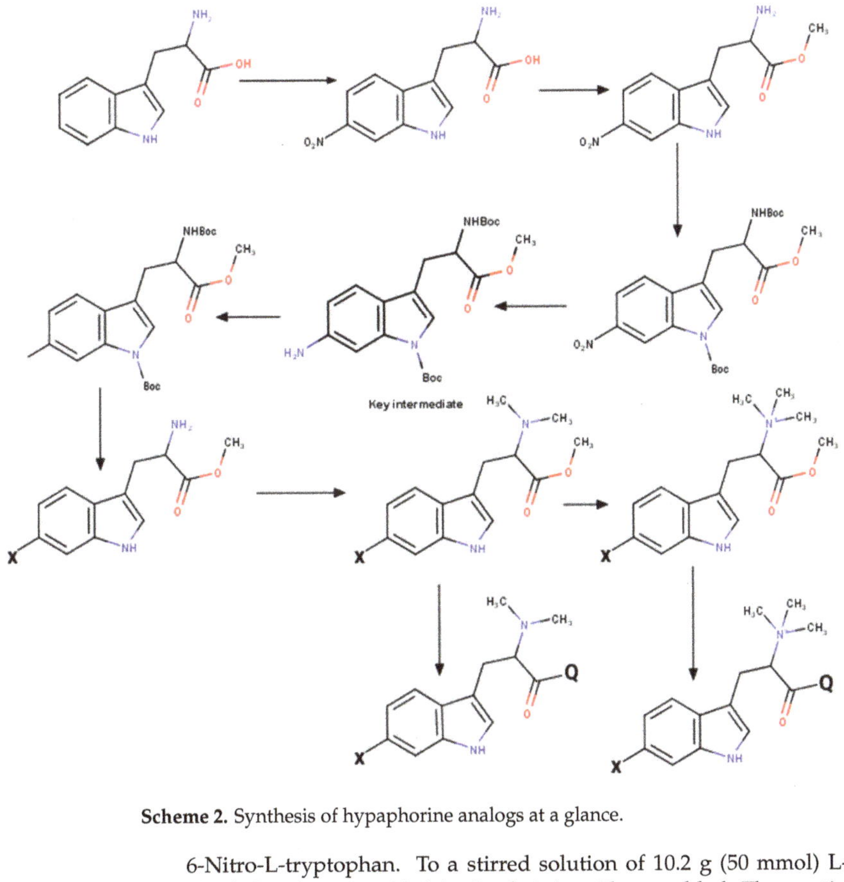

Scheme 2. Synthesis of hypaphorine analogs at a glance.

6-Nitro-L-tryptophan. To a stirred solution of 10.2 g (50 mmol) L-tryptophan in 100 mL of glacial acetic acid, 1.9 mL of nitric acid was added. The reaction mixture was cooled to 5 °C, followed by dropwise addition of 4.4 mL nitric acid solution in the 20 mL of glacial acetic acid. After 48 h stirring at room temperature (RT), the precipitate was filtered and washed by cold acetic acid, followed by thorough washings with acetone and drying. The resultant solid was dissolved in 15 mL of saturated sodium acetate solution, and the precipitate was washed and dried under P_2O_5. As a result, 2.6 g (21%) of 6-nitro-L-tryptophan was obtained as a yellow fluffy powder. 1H NMR (500 MHz, Deuterium Oxide, pH 12) δ 7.69 (d, J = 7.9 Hz, 1H), 7.60 (d, J = 8.0 Hz, 1H), 7.27 (s, 1H), 7.05 (t, J = 8.0 Hz, 1H), 3.16 (t, J = 7.0 Hz, 1H), 3.03 dd, J = 14.3, 6.6 Hz, 1H), 2.86 (dd, J = 14.3, 7.4 Hz, 1H).

6-Nitro-L-tryptophan methyl ester hydrochloride. To a cooled to −15 °C methanol, 2.5 mL of thionyl chloride was slowly added, and after that, a 1 g of 6-nitro-L-tryptophan was added, mixed to a complete dissolution of solids and warmed to RT. After that, the mixture was heated to the reflux for 1 h and subsequently evaporated under vacuum. The residue was suspended in diethyl ether and filtered. The solid was dried under vacuum to receive 1.1 g (yield 91%) 6-nitro-L-tryptophan methyl ester hydrochloride.

Nα,Nin-di-Boc-6-nitro-L-tryptophan methyl ester. To a suspension of 8.0 g 6-nitro-L-tryptophan methyl ester hydrochloride in 200 mL acetonitrile, 6.2 g of Boc anhydride was added under stirring, followed by portion-wise N,N-Diisopropylethylamine (DIPEA) (5 mL) addition. After 15 min stirring, the second equivalent of Boc anhydride, followed by 350 mg 4-Dimethylaminopyridine (DMAP) was added, and the reaction mixture was stirred overnight and then evaporated. The residue was dissolved in ethyl acetate, washed triple with the citric acid solution, then twice with brine, dried over anhydrous sodium sulphate and evaporated to afford 11 g (89%) of Nα,Nin-di-Boc-6-nitro-L-tryptophan methyl ester. 1H NMR (700 MHz, Chloroform-d) δ 9.11 (s, 1H), 8.19 (dd, J = 8.7, 2.1 Hz, 1H), 7.72 (s, 1H), 7.65 (d, J = 8.7 Hz, 1H), 5.22 (d, J = 7.8 Hz, 1H), 4.71 (d, J = 7.8 Hz, 1H), 3.78 (s, 3H), 3.42–3.34 (m, 1H), 3.24 (d, J = 13.5 Hz, 1H), 1.77 (s, 9H), 1.56–1.42 (m, 9H).

The solution of 6.0 g Nα,Nin-di-Boc-6-nitro-L-tryptophan methyl ester in 150 mL of methanol was hydrogenated for 12 h over 800 mg of 10% Pd/C under the pressure of hydrogen 1 MPa, after that the catalyst was filtered with celite, and the filtrate was evaporated under vacuum to afford the 5.9 g of Nα,Nin-di-Boc-6-nitro-L-dihydrotryptophan methyl ester as a cream foam. 1H NMR (700 MHz, Chloroform-d) δ 7.13 (dd, J = 7.9, 1.7 Hz, 0H), 7.02 (d, J = 7.9 Hz, 0H), 5.16 (s, 1H), 4.25–4.14 (m, 1H), 3.82 (d, J = 2.3 Hz, 3H), 3.78 (s, 1H), 3.73 (ddd, J = 14.4, 9.7, 5.6 Hz, 1H), 3.41 (tt, J = 9.8, 5.4 Hz, 1H), 2.28 (dt, J = 13.1, 5.6 Hz, 1H), 2.12–2.00 (m, 1H), 1.91 (dt, J = 14.0, 8.4 Hz, 1H), 1.73 (s, 2H), 1.64 (d, J = 5.1 Hz, 9H), 1.53 (d, J = 4.4 Hz, 9H).

2.3. Functional Assay on α7 nAChR

A panel of the synthesized compounds was tested to determine their activity on α7 nAChR by fluorescence detection of cytoplasmic calcium rise (Figure 2a). All modified hypaphorines (i.e., N,N,N-trimethyl tryptophan derivatives) have shown agonistic activity in the presence of the positive allosteric modulator PNU 120,596 in the range from sub-micromolar to high micromolar concentrations. L-hypaphorine without additional modifications in the indole ring failed to activate α7 nAChR in these conditions at concentrations up to 1 mM, whereas methyl ester of L-hypaphorine showed agonistic activity with EC_{50} between 18 and 37 μM. 6-bromohypaphorine as previously reported shows agonistic activity in high micromolar concentrations [24]. 6-nitrohypaphorine does not show any agonistic activity at concentrations up to 1 mM, however methyl ester of 6-nitrohypaphorine activates α7 nAChR with EC_{50} of 14 μM. These observations suggest that the presence of indole ring substitutions as well as esterification of the carboxy group are crucial to develop α7 nAChR agonist on the basis of hypaphorine structure (Figure 2b). None of N,N-dimethyltryptophan derivatives have shown α7 nAChR agonistic activity (Table 1).

To explore the space of possible substitutions in the hypaphorine molecule (Figure 2b), a systematic approach has been used. Table 1 summarizes the findings of in vitro screening based on fluorescent calcium detection. The most striking effect on the activity showed a variation of carboxylic group modification. Substitutions in the indole ring (denoted as "X" on Figure 2b) consisted of halogens, hydrogen bond acceptors, or hydrogen bond donors. Halogens showed the most positive impact on the tested compound potency, the iodine-substituted one being the most potent (EC_{50} 611 nM). Contrary to that, hydrogen bond donor groups (NH_2 and OH), when placed in the sixth position of the hypaphorine indole ring, resulted in very low potency with EC_{50} from 52 to 599 μM, respectively. Hydrogen bond acceptors in the indole ring sixth position showed intermediate results. Thus, using rational design methods to get new ligands for α7 nAChR, it was possible to significantly improve their potency for this receptor.

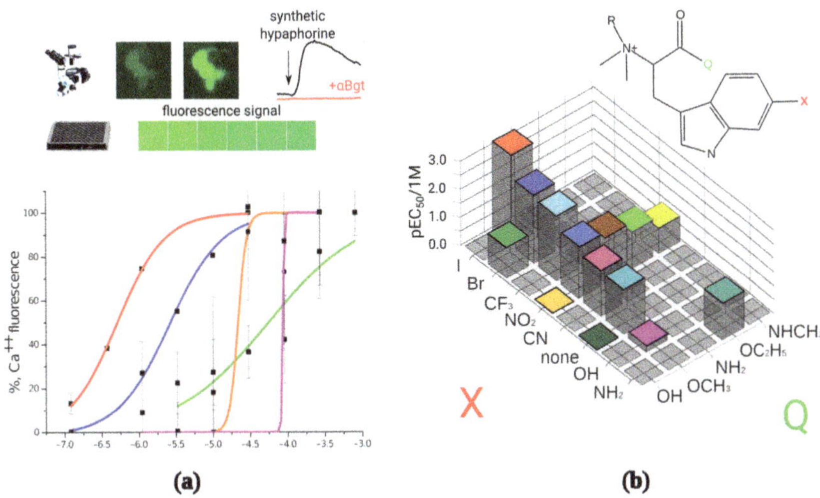

Figure 2. In vitro activity determination of the synthesized compounds. (**a**): Synthetic hypaphorines activate human α7 nAChRs expressed on transiently transfected neuro 2a cell. The receptor activation was detected as an increase in fluorescent signal in cell cytoplasm by fluorescence microscopy or as total fluorescence of the well in 96-well plate reader setup. Normalized fluorescence amplitude was plotted against base 10 logarithm of mole per liter concentration of the tested agonist and EC_{50} values were than calculated using dose-response equation. Numbering corresponds to the Table 1; (**b**): Structure–activity landscape explored in calcium fluorescence assay identifies modifications of the hypaphorine structure most strongly influencing the activity of the α7 nAChR. Since N,N-dimethyl variants (R = H) did not show the agonistic activity (see Table 1), only quaternary amines (R = CH_3) are shown on the figure. Substitutions in the 6th position of the indole ring (designated as X) were halogens (green shadow), hydrogen bond acceptors (yellow shadow) or hydrogen bond donors (blue shadow). Carboxy group modifications were comprised by free acid (Q = OH), ethers (Q = OCH_3, OC_2H_5), and amides (Q = NH_2, $NHCH_3$). The most active compound identified is (D)-6-iodohypaphorine methyl ester.

Table 1. The α7 nAChR potency of the synthetic hypaphorines. All compounds were in L form, unless stated otherwise.

95% CI	EC_{50}, μM	Code	No.
(0.13, 2.93)	0.61	D-isomer of X = I, Q = OCH_3, R = CH_3 (6ID)	1
(2.47, 9.02)	4.72	L-isomer of X = I, Q = OCH_3, R = CH_3	2
(0.08, 6139.56)	22.0	X = CN, Q = OCH_3, R = CH_3	3
(18.63, 37.87)	27.0	X = H, Q = OCH_3, R = CH_3	4
(10.32, 18.75)	14.0	X = NO_2, Q = OCH_3, R = CH_3 (6ND)	5
(43.38, 61.98)	52.0	X = NH_2, Q = OC_2H_5, R = CH_3	6
(48.35, 52.47)	50.0	X = NO_2, Q = OC_2H_5, R = CH_3	7
(1.16, 12.42)	4.0	X = Br, Q = OCH_3, R = CH_3	8
(21.82, 24.17)	23.0	X = NO_2, Q = NH_2, R = CH_3	9
(105.81, 154.79)	128	X = NO_2, Q = $NHCH_3$, R = CH_3 (6NAM)	10
(536.64, 667.88)	599	X = OH, Q = OCH_3, R = CH_3	11
	no agonistic activity	X = Br, Q = OCH_3, R = H	12
	no agonistic activity	X = NO_2, Q = OC_2H_5, R = H	13
	no agonistic activity	X = NO_2, Q = OH, R = CH_3	14
(4.40, 6.00)	5.14	X = CF_3, Q = OCH_3, R = CH_3 (6CF)	15
	no agonistic activity	X = H, Q = OH, R = CH_3 (hypaphorine)	16

2.4. Inhibition of α3-Containing (α3*) nAChRs by Hypaphorine Derivatives

Two of the studied compounds with high and medium agonist potency for α7 nAChR (6CF and 6ID, Table 1) were tested for activity toward α3* nAChR by calcium imaging. Neither 6ID, which was most active agonist of α7 nAChR, nor 6CF showed detectable agonism on α3* nAChR at concentrations up to 30 μM. Test results showed their antagonistic effect on human α3-containing nAChRs (Figure 3). The most effective was (D)-6-trifluoromethylhypaphorine methyl ester (6CF, IC_{50} = 2.4 ± 1.2 μM), least active was (D)-6-iodohypaphorine methyl ester (6ID, IC_{50} = 4.4 ± 1.0 μM).

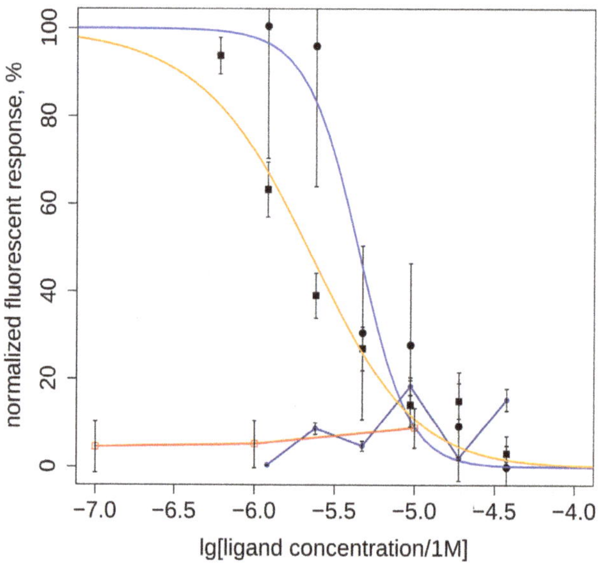

Figure 3. Inhibitory action of methylamide (D)-6-trifluoromethylhypaphorine methyl ester (6CF3, IC_{50} = 2.4 ± 1.2 μM, orange line and squares) and (D)-6-iodohypaphorine methyl ester (6ID, IC_{50} = 4.4 ± 1.0 μM, blue line and circles) on nicotine (100 μM)-induced increase in intracellular calcium ion concentration mediated by activation of α3-containing nAChRs in human neuroblastoma SH-SY5Y cells. (D)-6-iodohypaphorine methyl ester (6ID, dark blue line and circles) does not evoke noticeable calcium response in range of 3–30 μM (black line and circles). (D)-6-trifluoromethylhypaphorine methyl ester (6CF3, brown line and squares)) does not evoke noticeable calcium response in concentrations 10, 1, and 0.1 μM. Normalized data are presented as mean ± standard error. n = 6–7.

It should be noted that fluorescent calcium assay results cannot exclude the possibility of partial agonism of the tested compounds. One can imagine such weak activation of the α3* nAChR that does not produce calcium response strong enough to be detected by the plate reader, but will desensitize the receptor (thus, diminishing cellular response to nicotine). Moreover, partial agonists can occupy orthosteric sites and prevent nicotine binding, which would appear as antagonism under these conditions. Further investigation using electrophysiology might resolve this question.

2.5. Effects of PNU 282987, Hypaphorine Methyl Ester, and D-6-Iodohypaphorine Methyl Ester (6ID) on the Expression of Macrophage Markers

Basing on our recent work [14], where for the first time activation of α7 nAChR on macrophages by selective agonist PNU 282,987 was shown to affect the expression of various membrane proteins, here, the action of the α7 nAChR agonists was compared with that of parental hypaphorine (not interacting with nAChRs) and with 6ID (our analog with the highest affinity for α7 nAChR) on the expression of macrophage membrane markers

CD11b and CD11c. The selective agonist PNU 282,987 increased CD11b and CD11c protein expression by an average of 10% and 26%, respectively (Figure 4a,b). There was some tendency towards an increase in CD11c membrane protein expression under the action of 6ID, however, not statistically significant at $p < 0.05$ (one-way ANOVA, Tukey post hoc test).

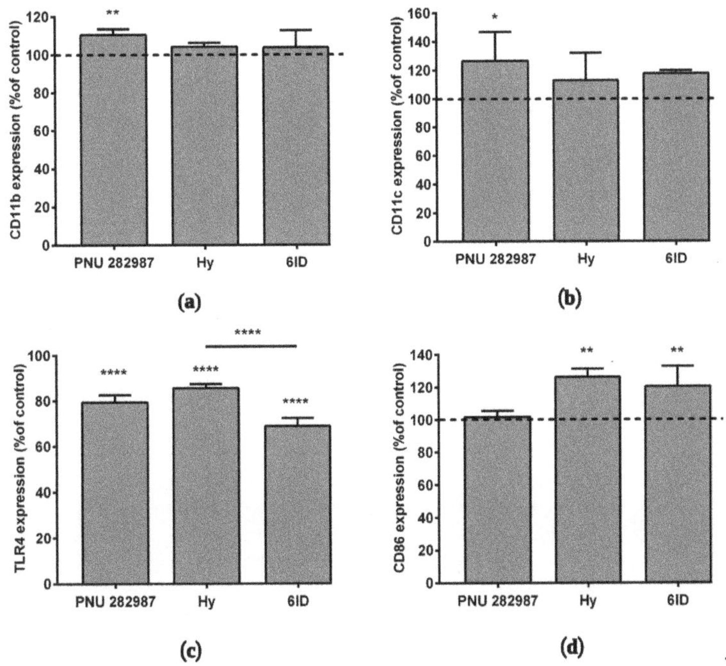

Figure 4. The effects of PNU 282987, hypaphorine, and D-6-iodohypaphorine methyl ester (6ID) on macrophage membrane proteins expression. Cells were treated with these compounds for 48 h and then analyzed by flow cytometry. PNU 282,987 led to an increase in the (**a**) CD11b and (**b**) CD11c expression. (**c**): All tested compounds led to a decrease in TLR4 expression. (**d**): Macrophages stimulation with hypaphorine and 6ID promoted an increase in CD86 expression. Results obtained in three independent experiments on macrophages from different donors and presented as mean ± SD. One-way ANOVA tests with Tukey post hoc test: * $p < 0.05$, ** $p < 0.01$, **** $p < 0.0001$ in comparison with untreated cells.

The effect of the above-listed compounds on the expression of the TLR4 receptor and the CD86 protein was studied by flow cytometry after 48 h incubation with macrophages. The interaction of bacterial endotoxins with the TLR4 receptor is known to activate the expression of several pro-inflammatory cytokine genes [38], while the co-stimulatory molecule CD86 plays an important role in suppressing inflammatory responses [39]. A decrease in the expression of the TLR4 receptor was observed under the action of all tested compounds (Figure 4c,e), the rank order of these TLR4-suppressing effects was as follows: hypaphorine (14%) < PNU 282,987 (21%) < 6ID (31%).

In addition, we observed an increase in the expression of CD86 under the action of hypaphorine methyl ester (26%) and 6ID (21%), but not PNU 282,987 (Figure 4d,f).

2.6. *Involvement of ERK and STAT3 in the Protective Role of PNU 282987, Hypaphorine, and 6ID in LPS-Mediated Inflammation in Macrophages*

To gain an insight into the signaling induced by the tested compounds, ERK1/2 and STAT3 phosphorylation was checked. Macrophages were treated with tested compounds for 1 h, after which they were stimulated with LPS for 3 h.

Using flow cytometry, it was determined that application of all compounds resulted in a marked increase in the ERK 1/2 phosphorylation (Figure 5a). ERK 1/2 phosphorylation peaked upon 6ID stimulation (MFI cntr vs. 6ID: 194 vs. 274; $p < 0.001$, see Figure 5c). Activation of the STAT3 pathway was less pronounced. However, stimulation with all tested compounds led to a statistically significant increase in the phosphorylation of STAT3 (Figure 5b). 6ID led to the maximum level of STAT3 phosphorylation in macrophages (MFI cntr vs. 6ID: 49 vs. 58; $p < 0.01$).

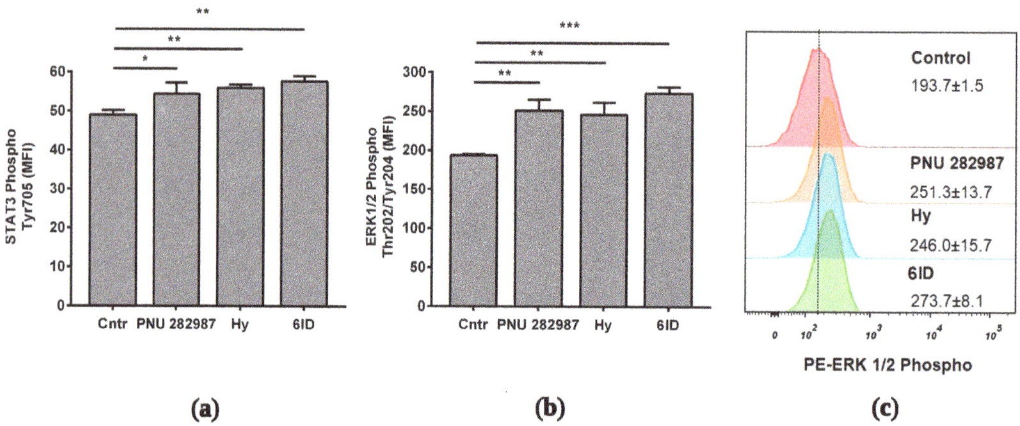

Figure 5. Effects of the analyzed compounds on signaling pathways in LPS-stimulated macrophages. Cells were treated with PNU 282987, hypaphorine, or 6-iodohypaphorine methyl ester (6ID) for 1 h and then stimulated with LPS for 3 h. Phosphorylated ERK1/2 and STAT3 were quantified by flow cytometry with specific antibodies. (**a**): All studied compounds led to an increase in the phosphorylation of ERK 1/2; (**b**): All studied compounds led to an increase in the phosphorylation of STAT3. (**c**): Representative histogram showing the increase in ERK 1/2 phosphorylation in comparison to control. The numbers in the histograms show the MFI ± SD ($n = 3$). MFI—geometrical mean fluorescence intensity. One-way ANOVA tests with Tukey post hoc test: * $p < 0.05$, ** $p < 0.01$, *** $p < 0.001$ in comparison with untreated cells.

2.7. CFA-Induced Inflammation Test

Anti-inflammatory effect of two analogs of 6-substituted hypaphorine (6ND and 6ID) was tested in a model with subplantar administration of complete Freund's adjuvant (CFA) in mice. CFA was injected once into the footpad to induce oedema in 24 h. Oedema was measured with an electronic calliper. Immediately after paw measurement, 6ND or 6ID was injected intramuscularly in doses 1, 0.5, and 0.1 mg/kg. Paws were measured at 6, 12, 24, and 48 h after administration of 6ND or 6ID. Only the dose 0.5 mg/kg of 6ND significantly reduced oedema (Figure 6a), whereas 6ID was effective against oedema at doses 0.5 and 1 mg/kg (Figure 6b). Hypaphorine, which did not show α7 nAChR agonistic activity in concentrations up to 1 mM in fluorescent Ca2+ detection (Table 1), has nonetheless demonstrated anti-oedemic properties (Figure 6c). End points at 48 h of drug treatments at dose 0.5 mg/mL have been analyzed in terms of paw diameter normalized to the maximal oedemic levels in each group (Figure 6d). Hypaphorine showed markedly less effective reduction of oedema than 6ND and 6ID.

Additionally, after 2 h following the drug injection, the analgesia test was conducted on the "hot plate" preheated to 53 °C. The time of withdrawal or licking of the affected paw was assessed for each animal. According to the hot plate test results, it was found that 6ND at a dose of 0.1 mg/kg has a pronounced analgesic effect, but higher doses did not show a difference from the control group (Figure 6c). 6ID at doses 0.5 and 1 mg/kg showed significant analgesia (Figure 6d).

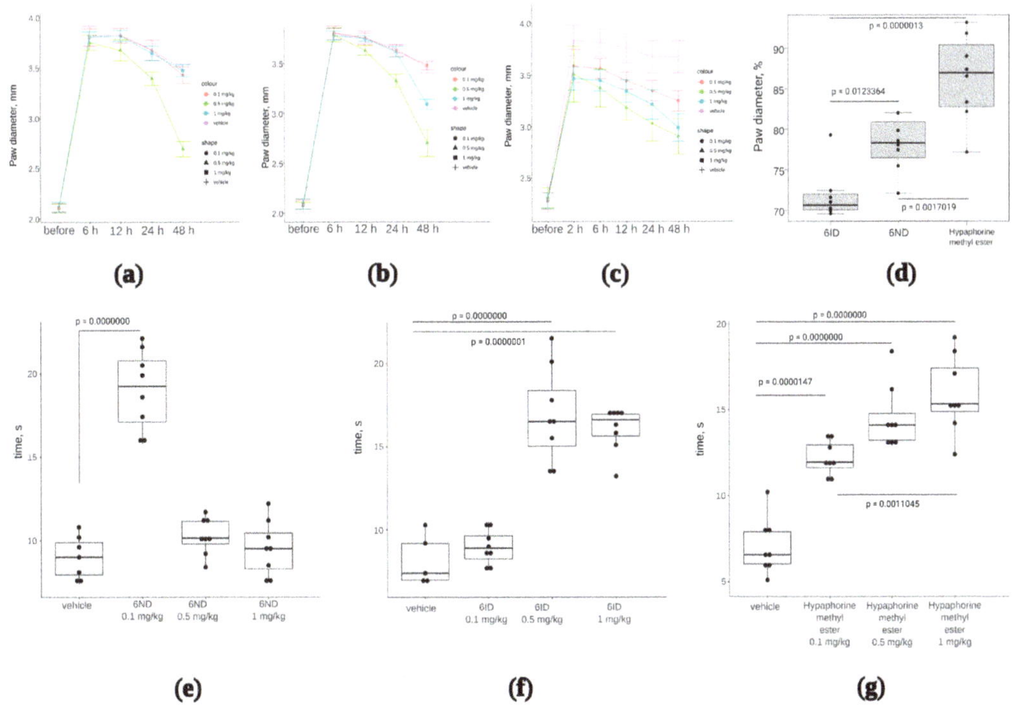

Figure 6. The effect of synthetic 6ND, 6ID, and hypaphorine on CFA-induced inflammation oedema and hyperalgesia. One-way ANOVA with Tukey HSD test were utilized to calculate the p-values. (**a**): Joint oedema, measured as paw diameter, was significantly reduced by 6ID at 0.5 mg/kg starting from 12 h of treatment; (**b**): 6ND significantly reduced joint edema at 0.5 mg/kg starting from 12 h of treatment and at 1 mg/kg 48 h after the initial treatment; (**c**): Unmodified hypaphorine also significantly reduced edema in a concentration-dependent manner at doses 0.1–1 mg/kg. Unlike α7 nAChR agonists 6ID and 6ND, higher doses of unmodified hypaphorine did not show a decrease in anti-oedemic effect; (**d**): Analysis of 48 h end points of CFA-induced inflammation treatment (0.5 mg/kg) in terms of oedema reduction expressed as percents of the maximal measured paw diameter; (**e**): 6ND showed significant analgesia in hot plate test only at 0.1 mg/kg, whereas higher doses were not efficient; (**f**): 6ID was effective in the hot plate test at 0.5 as well as at 1 mg/kg; (**g**): Hypaphorine showed analgesia in hot plate test at all tested doses from 0.1 to 1 mg/kg.

2.8. Carrageenan-Induced Inflammation Test

Subplantar 1% carrageenan injections in a volume of 30 μL into the right hind paw were carried out to generate an inflammation model in mice. The paw volume was measured beforehand using an electronic calliper. The maximum effect of the introduction of carrageenan was observed after 3 h. One hour after carrageenan injection, the drugs (hypaphorine, 6ND, or 6ID) were administered at a dose of 0.5 mg/kg. Hypaphorine and 6ID were also administered at 1 mg/kg. The control group was treated with saline (0.9% NaCl). Two hours after the drug administration (total 3 h from carrageenan injection), the size of the pad of the right hind paw was measured, the von Frey test and paw pressure analgesimeter pain sensitivity assessment were performed.

Carrageenan injections provoked marked oedema in the absence of any treatment (Figure 7a–c, "control" box plots). In all treatment groups, oedema was reduced, although not to a baseline (Figure 7a–c). However, data normalization to the average paw diameter in each treatment group revealed that effects of 6ND (Figure 7d) and 6ID (Figure 7e) were statistically significant, whereas hypaphorine showed only a tendency to decrease oedema, not statistically significant (Figure 7f, $p = 0.131$, one-way ANOVA, $n = 8$).

All three compounds showed statistically significant analgesia in paw pressure tests (Figure 7g), but no statistically significant difference between tested compounds were detected. Similarly, von Frey test revealed significant analgesia in the case of 6ND, 6ID, and hypaphorine, compared to vehicle, but differences between the compounds were not significant (Figure 7h). Thus, anti-oedemic effect seems to depend on α7 nAChR activity to a greater extent than the analgesic activity of the compounds.

2.9. Monosodium Iodoacetate-Induced Arthritis Model

Osteoarthritis is a condition characterized by articular cartilage degradation and activation of inflammation molecular signaling pathways [40]. One of the well-established osteoarthritis animal models is monosodium iodoacetate (MIA) intra-articular injection [41], which was used in the present study to evaluate the perspectives of hypaphorine analogs in anti-inflammatory osteoarthritis treatment.

We previously have published the results of 6ID and 6ND tests in myocardial infarction model, showing significant positive effect of 6ND [42]. In the current communication, we chose 6ND for the test on osteoarthritis rat model.

By the 8th day of the experiment, the maximum inflammation of the knee joint was noted. The intramuscular administration of compound 6ND in the studied doses was carried out from the 8th day of the study to the 16th day. On the 8th and 16th days, the effect of the compound was assessed using a set of functional tests: weight distribution between the healthy and inflamed limbs immediately after the first 6ND injection (Figure 8a) and after eight days of 6ND administration (Figure 8b), muscle strength grasping the hind limbs after the first 6ND injection (Figure 8c) and after eight days of 6ND administration (Figure 8d), mechanical allodynia after the start of 6ND administration and after eight days of 6ND administration (Figure 8e,f, respectively), thermal hypersensitivity 2 h after administration of 6ND (Figure 7g), and measuring the diameter of the knee joint (Figure 8h).

According to the tests, 6ND did not influence body weight distribution between healthy and affected limbs, but significantly improved grip strength of affected limb after eight days of *i.p.* administration (Figure 8d). 6ND exhibited analgesic (anti-allodynic) effect (Figure 8f) and anti-oedemic effect on the MIA-injected joint (Figure 8g), but did not influence thermal hyperalgesia in hot plate test in this model.

2.10. Histological Study

After the experiment termination and euthanasia of the experimental animals, biomaterials were taken for histological examination. Soft tissues around the right knee joint were excised as much as possible. Femur, tibia, and fibula were cut across in the middle of the diaphysis closer to the corresponding articular surfaces. The joint was fixed in a 10% solution of neutral formalin for seven days, washed in running water and then decalcified in Trilon B for 7–14 days. After satisfactory decalcification of the bone and cartilage tissue, the joint was cut in the sagittal plane. The diaphysis of the tubular bones was shortened to the border of 2–3 mm from the metaepiphyseal cartilage. In this form, the biomaterial was repeatedly washed in running water, dehydrated in alcohols of ascending concentration, and embedded in paraffin. Paraffin sections 5–7 μm thick were stained with hematoxylin and eosin and examined using conventional light microscopy. During histological analysis, the following morphological signs were assessed: inflammatory infiltration of the synovial membrane (synovitis), synovial hyperplasia, destructive changes in the articular cartilage, and destructive changes in bone tissue (if any). The following scale was used to assess the severity of a particular morphological trait: 0 points—within normal limits, 1—minimal severity, 2—weak, 3—medium (moderate), 4—strong, 5—very strong [43].

After intra-articular administration of MIA, all animals on the 15th day showed characteristic signs of arthritis—inflammatory infiltration of the synovial membrane (synovitis) with signs of its hyperplasia, destructive changes in the articular cartilage of the femur and tibia, as well as destructive and necrotic changes in the menisci. Thus, to the described picture on the 15th day of the experiment, the term "MIA-induced arthritis" can be fully used.

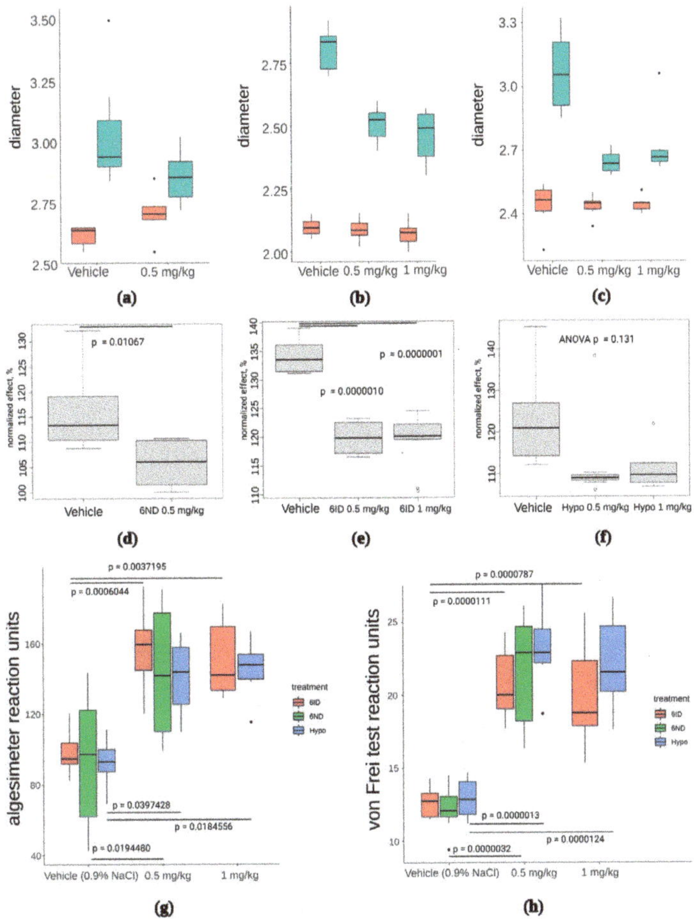

Figure 7. The effect of 6ND, 6ID, and synthetic hypaphorine on carrageenan-induced inflammation edema and allodynia. (**a**): 6ND demonstrated moderate anti-oedemic effect at a dose of 0.5 mg/kg. Red boxes represent distribution of paw diameters before the carrageenan injection, blue boxes reflect the increase in paw diameter after carrageenan injection; (**b**): 6ID showed marked decrease in paw diameter at doses 0.5 and 1 mg/kg. Red boxes represent distribution of paw diameters before the carrageenan injection, blue boxes reflect the increase in paw diameter after carrageenan injection; (**c**): A tendency to oedema reduction was observed under 0.5 and 1 mg/kg administration of synthetic hypaphorine. Red boxes represent distribution of paw diameters before the carrageenan injection, blue boxes reflect the increase in paw diameter after carrageenan injection; (**d**): Paw diameters normalized to pre-injection values reveal anti-oedemic effect of 6ND significant at $p < 0.05$ level in comparison to vehicle-treated animals (students *t*-test); €: Analysis of normalized paw diameters show significant reduction of oedema under administration of 0.5 and 1 mg/kg 6ID (one-way ANOVA and Tukey HSD test); (**f**): No significant effect of synthetic hypaphorine was detected in terms of normalized paw diameter at doses up to 1 mg/kg (one-way ANOVA and Tukey HSD test); (**g**): All tested compounds were significantly ($p < 0.05$, one-way ANOVA and Tukey HSD test) effective analgesics in mechanical hyperalgesia test performed with filament algesimeter at all doses tested. However, no significant difference between different drugs was discovered; (**h**): Von Frey type of mechanical hyperalgesia test revealed significant efficacy of all three tested compounds but did not detect significant differences between drugs at all doses ($p < 0.05$, one-way ANOVA and Tukey HSD test).

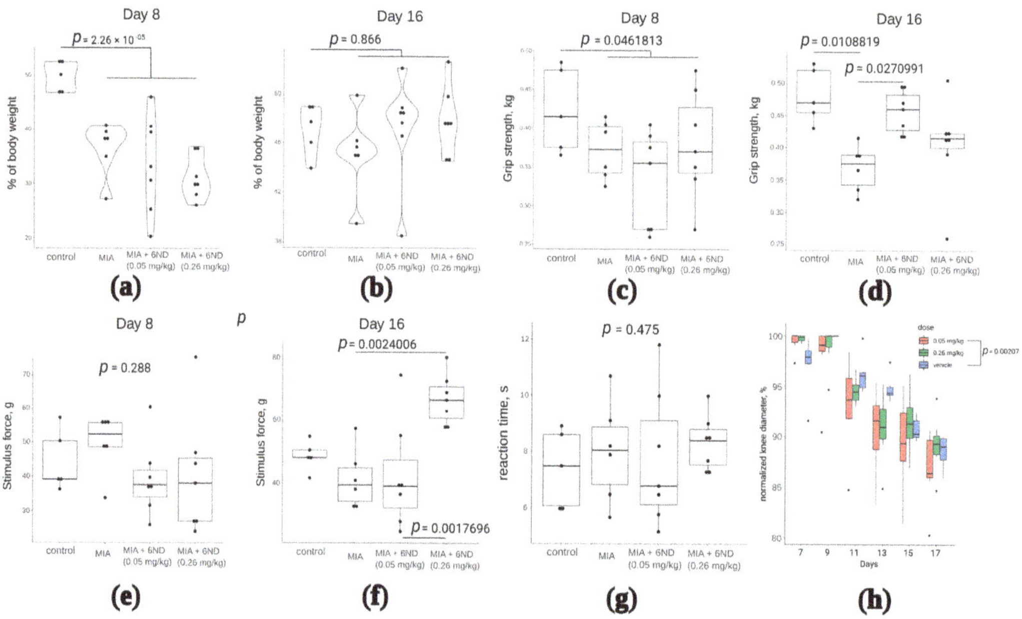

Figure 8. Investigation of 6ND effects in MIA-induced arthritis model rats. (**a**): MIA significantly shifts the body weight distribution between affected and healthy limbs after the intra-synovial injection. 6ND did not immediately show significant improvement in terms of body weight distribution between affected and healthy limbs (two-way ANOVA and Tukey HSD test); (**b**): On the 16th day, there remained no statistically significant differences between control and MIA-injected rats in terms of invalidation. No significant influence of 6ND on recovery was detected (two-way ANOVA and Tukey HSD test); (**c**): On 8th day after MIA injection, a significant decrease in grip strength was observed in comparison to control group. No significant immediate effects of 6ND at doses 0.05 and 0.26 mg/kg were observed (two-way ANOVA and Tukey HSD test); (**d**): 6ND significantly improved grip strength in MIA-induced arthritis in rats at a dose of 0.05 but not 0.26 mg/kg (two-way ANOVA and Tukey HSD test); (**e**): No signs of mechanical allodynia were detected in rats on day 8 after the MIA injection and 6ND showed no influence, irrespective of the dose (two-way ANOVA and Tukey HSD test); (**f**): 6ND shows significant (two-way ANOVA and Tukey HSD test) dose-dependent improvement in allodynia test 16 days after MIA injection (eight days of 6ND administration); (**g**): No significant effects of 6ND in hot plate test on MIA-induced arthritis rats were detected (two-way ANOVA and Tukey HSD test); (**h**): Significant effect in dynamics of oedema was detected in rats treated with 0.05 mg/kg 6ND comparing to vehicle-treated rats (repeated measures ANOVA).

In the group of animals that received saline after intra-articular administration of MIA, a pronounced inflammatory infiltration of the synovial membrane was observed on the 15th day (mean score 3.67), accompanied by synovial hyperplasia (2.33). Destructive changes in the epiphyseal articular cartilage of the femur (3.0) and tibia (3.33) were observed (Figure 9a,b).

The administration of 6ND at a dose of 0.05 mg/kg did not change the inflammatory infiltration of the synovial membrane of the right knee joint of female rats (Figure 9c,d) compared to the control group on the 15th day of observation (average score 3.75, versus 3.67 in control). The mean score for synovial hyperplasia among the animals of this group was estimated as 2.25. Destructive changes in the epiphyseal articular cartilage of the femur and tibia (3.0 and 3.5, respectively) were comparable to those in the group of animals with MIA-induced arthritis treated with saline in a volume of 2 mL/kg.

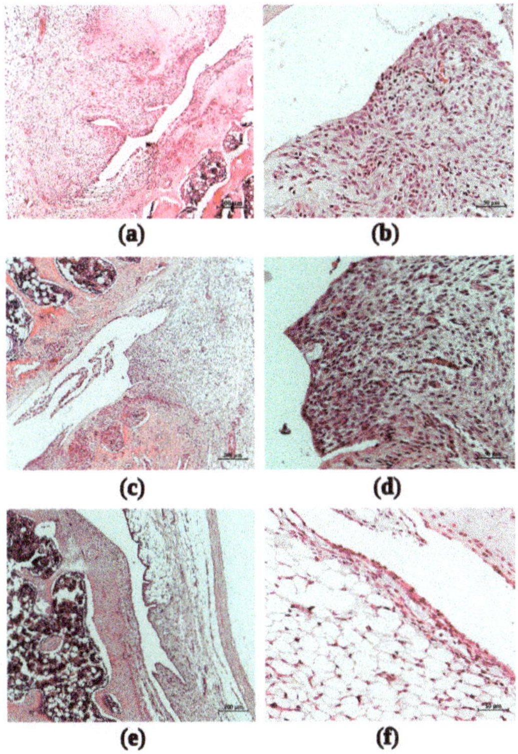

Figure 9. Fragments of the synovial membrane of the right knee joint of female rats with MIA-induced arthritis during the administration of saline (**a**,**b**); 6ND at dose 0.05 mg/kg (**c**,**d**); 6ND at dose 0.26 mg/kg (**e**,**f**). Staining with hematoxylin and eosin. Magnification: 50× (left, ruler size corresponds to 200 μm) and 200 × (right, ruler size corresponds to 50 μm).

The administration of 6ND in the model of MIA-induced arthritis at a dose of 0.26 mg/kg contributed to a significant reduction in the morphological manifestations of synovitis and synovial hyperplasia of the right knee joint of female rats: the average score of inflammatory infiltration of the synovial membrane corresponded to 2.5, synovial hyperplasia–1.5 (see Figure 9e,f).

The minimum, average, and maximum estimated scores of synovitis, synovial hyperplasia, and destructive changes in the epiphyseal articular cartilage of the femur and tibia for each experimental group of animals are presented in Tables 2 and 3.

Table 2. Estimated on the 15th day of observation points of synovitis and synovial hyperplasia of the right knee joint in female rats with MIA-induced arthritis on the background of intramuscular injection of saline and 6ND at doses of 0.05 and 0.26 mg/kg.

Synovial Hyperplasia			Inflammatory Infiltration of the Synovium			
Min	**Me**	**Max**	**Min**	**Me**	**Max**	**Evaluation Score**
1.0	2.33	3.0	3.0	3.67	4.0	MIA + saline (n = 3)
2.0	2.75	3.0	3.0	3.75	4.0	MIA + 6ND (0.05 mg/kg) (n = 4)
1.0	1.5	2.0	2.0	2.5	3.0	MIA + 6ND (0.26 mg/kg) (n = 4)

Table 3. Estimated on the 15th day of observation scores of destructive changes in the distal epiphyseal cartilage of the femur and tibia of the right knee joint in female rats with MIA-induced arthritis on the background of intramuscular injection of saline and 6ND at doses of 0.05 and 0.26 mg/kg.

Destructive Changes in the Distal Epiphyseal Cartilage of the Femur			Destructive Changes in the Proximal Epiphyseal Cartilage of the Tibia			
Min	Me	Max	Min	Me	Max	Evaluation Score
3.0	3.0	3.0	3.0	3.33	4.0	MIA + saline (n = 3)
3.0	3.0	3.0	3.0	3.5	4.0	MIA + 6ND (0.05 mg/kg) (n = 4)
3.0	3.0	3.0	3.0	3.5	4.0	MIA + 6ND (0.26 mg/kg) (n = 4)

3. Discussion

6-Bromohypaphorine (6BHP) isolated from the marine mollusk *Hermissenda* sp. acts as a silent or partial agonist of α7 nAChR [24]. It was previously shown that activation of α7 nAChR by its selective agonist PNU 282,987 leads to an increase in the expression of macrophage membrane proteins, including HLA-DR, CD11b, and CD54, but reduces the expression of the CD14 receptor [14]. These macrophage markers play a major role in the complex pathogenesis of sepsis, inflammation, and immunosuppression. In current study, using computer-aided design, molecular modelling, and synthesis of fourteen 6-bromohypaphorine analogs, we approached the new class of α7 nAChR agonists and explored their anti-inflammatory and analgesic properties. Unmodified hypaphorine did not show agonistic properties at α7 nAChR (Table 1) but was described previously as anti-inflammatory agent [26] possessing anticholinesterase activity [29]. In the present paper, it was used as a source of complementary information regarding different aspects of 6ID and 6ND activity, which might not be related directly to α7 nAChR activity.

Synthetic hypaphorines inhibit heteromeric α3* nAChRs at micromolar and α9/α10 nAChRs at sub-millimolar concentrations, which might explain some of the effects outside the scope of the α7 nAChR-attributed anti-inflammatory activity (Figure 3 and Figure S1). It should be stressed, that calcium response detected in the assay described in this paper is an indirect measure of the receptor activity. Thus, the real affinity toward α7 nAChR of the synthesized compounds might deviate from the measured efficiency (potency). However, the activity of the synthesized compounds was also determined in competition with radioiodinated α-Bgt for α7 nAChR binding site (Figure S2) and it can be ranked in a good agreement with the calcium imaging results: compound 1 > 4 > 5, 9 > 6, 7, 10, 12 vs. 1 > 4, 5 > 9 > 6, 7 > 10 > 12, respectively (numbering corresponds to the Table 1).

The analysis of biological activity of 6ID, a new α7 nAChR agonist, in comparison with PNU 282987, a well-established selective agonist of this receptor subtype, was performed in vitro on human primary macrophages. It was found that PNU 282987, but not 6ID or hypaphorine, leads to a significant increase in the surface expression of CD11b and CD11c proteins (Figure 4a,b). Complement receptors CR3 (CD11b/CD18) and CR4 (CD11c/CD18) belong to the β2-integrin family and play an important role in cell adhesion and migration, as well as in phagocytosis [44].

On the other hand, PNU 282987, 6ID, and hypaphorine significantly decreased the toll-like receptor 4 (TLR4) surface expression (Figure 4c,e). TLR4 is a member of the TLR family that is recognized and activated by bacterial lipopolysaccharide (LPS), a major component of the cell wall of gram-negative bacteria. Molecular recognition of LPS by the TLR4 receptor system triggers a cascade leading to the production of pro-inflammatory cytokines initiating the inflammatory responses [45]. Reduction of cell surface TLR4 expression under the influence of PNU 282987, 6ID, and hypaphorine may alleviate inflammatory response. The most pronounced inhibition of TLR4 receptor was observed when macrophages were treated with 6ID. Similar data were published earlier, showing that treatment of HMEC-1 endothelial cells by hypaphorine from Erythrina velutina resulted in the inhibition of TLR4 expression [27]. Thus, the hyperinflammatory response can be partially blocked by these compounds due to suppressing the expression of TLR4 receptors.

Complementary to that, 6ID and hypaphorine significantly raised the surface expression of CD86, a known regulator of IL-10 anti-inflammatory response [46]. CD86 is a co-stimulatory molecule for the priming and activation of T cells. At an early stage of the immune response, CD86 is expressed in cells of primary lymphoid tissue and is constitutively expressed in the antigen-presenting cells [46]. CD86 promotes antigen presentation to T cells and regulates the anti-inflammatory response. In addition, CD86 is essential for Th2 response [47]. There are reports that acute inflammation during sepsis is associated with suppression of constitutive expression of CD86 [46,48]. In addition, some studies show that CD86 plays a role in the regulation of the inflammatory response in vivo in diseases regulated by the adaptive immune response [49]. We found that treatment of macrophages with hypaphorine and 6ID resulted in the increased expression of CD86 while PNU 282,987 had no effect, suggesting different molecular mechanisms.

Signaling pathways upon activation of macrophages by LPS were also explored in the presence of the compounds described in this communication. Previous studies have shown that PNU 282,987 [50] and nicotine [51,52] increase phosphorylation of ERK 1/2 in different types of cells. In addition, activation of the STAT3 pathway through α7 nAChR has been shown to play a critical role in the prevention of inflammation [16,53]. In our study, it is shown that the treatment of LPS-stimulated macrophages with PNU282987, hypaphorine, or 6ID leads to an increase in the phosphorylation of ERK 1/2 and STAT3 (Figure 5). Interestingly, treatment of murine macrophages with hypaphorine was reported to decrease EPK 1/2 phosphorylation [26]. These results suggest that modulation of EPK 1/2 by hypaphorines obtained from different sources may proceed differently in different types of cells, thereby regulating various kinds of biological functions.

Anti-inflammatory properties of new hypaphorine analogs were confirmed in several in vivo models. In a mice model with subplantar administration of complete Freund's adjuvant (CFA), the compounds 6ID, 6ND, and hypaphorine demonstrated anti-oedemic properties and an analgesia in a hot plate test (Figure 6). Hypaphorine, lacking α7 nAChR-agonistic properties, had significantly lower anti-oedemic activity (Figure 6d). The latter conclusion has been confirmed in test on mice inflammation model with subplantar 1% carrageenan injections: 6ND and 6ID demonstrated significant (at $p < 0.05$ level) anti-oedemic effect in comparison to vehicle-treated animals, whereas hypaphorine was ineffective ($p = 0.131$, one-way ANOVA). Despite being ineffective against oedema, hypaphorine showed strong evidence of analgesia in algesimeter and von Frey tests (Figures 7g and 7h, respectively). These results suggest that analgesic effects of hypaphorine and its analogs 6ND and 6ID depend on different molecular mechanisms, some of them not involving α7 nAChR.

Modelling of arthritis-like inflammation by introducing monoiodoacetate (MIA) through the patellar ligament into the intra-articular space of the right knee revealed anti-invalidation effects of the 6ND hypaphorine analog (improved grip strength, Figure 8d). 6ND also showed analgesic (in mechanical allodynia test, Figure 8f) and anti-oedemic (Figure 8h) effects in MIA-induced arthritis rat model.

The administration of 6ND in the model of MIA-induced arthritis at a dose of 0.26 mg/kg demonstrated a significant reduction in the morphological manifestations of synovitis and synovial hyperplasia of the right knee joint of female rats.

Thus, 6-substituted esterified hypaphorine analogs show promising results as a potential new class of anti-inflammatory agents, having an increased affinity for α7 nAChR.

4. Materials and Methods

4.1. Rational Design and Virtual Screening

A virtual library of fifty-six hypaphorine analogs was constructed using the GNU/Bash script to automatically generate simplified molecular-input line-entry system (SMILES) format [32] of all possible combinations of hypaphorine modifications from the following set: X = NO_2, NH_2, OH, OCH_3, Cl, Br, I; Q = OH, OCH_3, OC_2H_5, NH_2, $NHCH_3$; R = H, CH_3, where X denotes indole ring substitution in 6th position, Q is either carboxy, amide, or ester group, and R encodes the methylation state of the ammonium nitrogen (see Figure 2b). Open Babel [33] was used to convert generated SMILES to MOL2 files and UCSF Chimera was utilized to minimize internal energy of the generated structures. Autodock Tools were used to generate PDBQT files for the designed panel of molecular structures and prepare the receptor structure (PDB 7KOX [34]) for virtual screening. In brief, water molecules except the conservative water in the orthosteric binding site have been removed, Gasteiger charges were added, and all hydrogens, with the exception of polar ones, were removed. Grid box was centered at Trp 148 located in the orthosteric binding site and playing the crucial role in the receptor activation. Grid box dimensions were set to 22 × 22 × 22 points (at spacing 0.375Å) to accommodate all possible conformations of hypaphorine analogs. Final docking results were visually inspected in UCSF Chimera (Figure 2) and the ligands which produced meaningful conformations of the complexes were selected for chemical synthesis.

4.2. Calcium Imaging

Mouse neuroblastoma Neuro2a cells were purchased from the Russian collection of cell cultures (Institute of Cytology, Russian Academy of Sciences, Saint Petersburg, Russia). Cells were cultured in Dulbecco's modified Eagle's essential medium (Paneco, Moscow, Russia) supplemented with 10% fetal bovine serum (Sigma, St. Louis, MO, USA). The cells one day before transfection were subcultured and plated at a density of 10,000 cells per well on a 96-well plate. The next day, Neuro2a cells were transiently transfected with plasmids coding α7 nAChR (human α7 nAChR-pCEP4) or its mutants, the chaperone Ric-3 (Ric3-pCMV6-XL5, OriGene, USA) and a fluorescent calcium sensor Case12 (pCase12-cyto vector, Evrogen, Russia) in molar ratio 4:1:1. Lipofectamine transfection protocol (Invitrogen, Waltham, MA, USA) was performed as recommended by the manufacturer. Transfected cells were grown at 37 °C in a CO_2 incubator for 48 h.

Calcium immobilization assay was performed as described previously [35]. Briefly, Transfected Neuro2a cells were grown on black 96-well plates (Corning, Somerville, MA, USA) at 37 °C in a CO_2 incubator for 72 h, then growth medium was substituted with buffer containing 140 mM NaCl, 2 mM $CaCl_2$, 2.8 mM KCl, 4 mM $MgCl_2$, 20 mM HEPES, 10 mM glucose; pH 7.4.

Cells were incubated with the α7 nAChR positive allosteric modulator PNU120596 (10 μM, Tocris, Bristol, UK) for 20 min at room temperature before the addition of tested compounds. Plates were measured in microplate reader Hidex Sence (Hidex, Turku, Finland) using excitation at 485 nm and emission at 535 ± 10 nm. Fluorescence peak intensity in each well was expressed as a percentage of the maximal obtained response. Data files were analysed using Hidex Sence software (Hidex, Turku, Finland). Controls were run in the presence of 4 μM α-cobratoxin.

4.3. Flow Cytometry

Peripheral blood mononuclear cells (PBMCs) were separated by Ficoll-Paque PLUS density gradient centrifugation. PBMCs were placed in a sterile Petri dish and incubated at 37 °C for 2 h. Unattached cells were then removed by washing with PBS and substituted with fresh complete RPMI 1640 medium. To generate monocyte-derived macrophages (MDMs), 50 ng/mL GM-CSF was added to the isolated monocytes and cultured for 6 days to differentiate them into non-polarized MDMs.

4.4. In Vivo Anti-Inflammatory Activity

4.4.1. Animals

Specific pathogen-free outbred ICR male mice (6 to 8 weeks old, weighing 29 to 33 g) and Wistar female rats (8–9 weeks old 250 ± 25 g) were obtained from the Animal Breeding Facility of the Branch of the Shemyakin–Ovchinnikov Institute of Bioorganic Chemistry of the Russian Academy of Sciences (Pushchino). The animals were acclimatized for 2 weeks before the experimental procedures and were kept in two corridor barrier rooms under a controlled environment: a temperature of 20 to 24 °C, a relative humidity of 30% to 60%, and a 12 h light cycle. The animals were housed in Type 3 standard polycarbonate cages (820 cm2) on bedding (LIGNOCEL BK 8/15, JRS, Rosenberg, Germany), with ad libitum access to feed (SSNIFF V1534-300, Spezialdiaeten, GmbH, Soest, Germany) and filtered tap water. The mouse cages were also supplied with material for environmental enrichment, i.e., Mouse House (Techniplast, Buguggiate, Italy).

The study was conducted in AAALAC (Association for Assessment and Accreditation of Laboratory Animal Care International) accredited facility in compliance with the standards of the Guide for Care and Use of Laboratory Animals (8th edition, Institute for Laboratory Animal Research). Animal treatment procedures were approved by the Institutional Animal Care and Use Committee (IACUC) of the Branch of the Shemyakin–Ovchinnikov Institute of Bioorganic Chemistry, Russian Academy of Sciences, the experimental protocol code is no. 688/19 (date of approval: 10 January 2019).

4.4.2. CFA-Induced Inflammation Test

The development of the inflammation and thermal hyperalgesia of the paw was induced by the injection of the oil/saline (1:1) CFA emulsion (Sigma-Aldrich, St. Louis, MO, lot # WH327536) into the subplantar surface of the hind paw of the ICR mice (20 µL/paw) 24 h before the measurement. Oedema was measured with an electronic calliper. Immediately after paw measurement, 6ND or 6ID were injected intramuscularly in doses: 1, 0.5, and 0.1 mg/kg. Paws were measured at 6, 12, 24, and 48 h after administration of 6ND or 6ID. After 2 h following the drug injection, the inflamed paw withdrawal or licking latencies to thermal stimulation were measured on a hot plate device (Hot Plate Analgesia Meter, Columbus Instruments) with a set temperature of 53 ± 0.1 °C and a cut-off time of 60 s.

4.4.3. Carrageenan-Induced Inflammation Test

Carrageenan inflammation was induced by injecting 30 µL of a 1% carrageenan suspension (Sigma-Aldrich, lot #SLBK3896V) into the subplantar surface of the ICR mice hind paw. The oedema degree of the carrageenan-induced paw was evaluated using an electronic calliper by the volume difference of the animal's right hind paw after (3 h from carrageenan injection) and before being hurt by carrageenan.

One hour after carrageenan injection, the drugs (hypaphorine, 6ND, or 6ID) were administered at a dose of 0.5 mg/kg. Hypaphorine and 6ID were also administered at 1 mg/kg. The control group was treated with saline (0.9% NaCl).

Antinociceptive testing was performed 2 h after the drug administration (3 h after carrageenan injection).

Nociceptive testing was performed by utilizing the von Frey test and Paw pressure test.

4.4.4. Von Frey Test

The animals were placed in 20 cm × 20 cm Plexiglas boxes equipped with a metallic mesh floor. Animals were allowed to habituate themselves to their environment for 15 min before the test. The electronic von Frey instrument (model BIO-EVF4; Bioseb, Vitrolles, France) was used to vertically stimulate the center of the rat hind paw with increasing intensity until the hind paw was lifted, the withdrawal threshold was automatically displayed on the screen. The paw sensitivity threshold was defined as the minimum force required to elicit a robust and immediate withdrawal reflex of the paw. Spontaneous movements associated with locomotion were not considered as a withdrawal response. Measurements were repeated 3 times and the final value was obtained by averaging the 3 measurements.

4.4.5. Paw Pressure Test

Mechanical hyperalgesia was measured as a paw withdrawal response to a gradual increase of mechanical pressure applied by the Rodent pinchers—analgesia meter (model BIO-RP-M; BioSeb, Vitrolles, France). The influence of stimulation on each hind paw was recorded three times. The maximum force applied to the paw was recorded as the grams (g) of force on the dynamometer.

4.4.6. Monoiodoacetate-Induced Arthritis Model

Before the study, 26 female Wistar rats (250 $\pm$ 25 g), 8–9 weeks old, were divided into cages of four so that the average body weight did not differ between groups. Four groups of animals were formed: control and three groups with modelling of inflammation by introducing monoiodoacetate (MIA). Three groups of animals were injected with 3 mg MIA in 50 μL of 0.9% sodium chloride through the patellar ligament into the intra-articular space of the right knee using a 29-G needle. Control rats received an equivalent volume of 0.9% sodium chloride.

Assessments of inflammation in vivo and functional tests were conducted on days 8 and 16 after arthritis model induction. Thermal and mechanical hypersensitivity and pain-induced articular disability were tested in this model. Body weight gain and paw oedema were also monitored during the experiment.

The inflamed paw withdrawal or licking latencies to thermal stimulation were measured on a hot plate device (Hot Plate Analgesia Meter, Columbus Instruments) with a set temperature of 53 $\pm$ 0.1 °C and a cut-off time of 60 s.

Mechanical hypersensitivity was measured by the electronic von Frey instrument (model BIO-EVF4; Bioseb, Vitrolles, France) used to vertically stimulate the center of the rat hind paw with increasing intensity. Measurements were repeated 3 times and the final value was obtained by averaging the 3 measurements.

Behavioral assessment of movement-caused pain was carried out in a hind limb grip strength test using a Grip Strength Meter (Columbus Instruments, Columbus, OH, USA) consisting of a wire mesh frame connected to the gauge. Rats were restrained and allowed to grasp the wire mesh frame with their hind paws and then were pulled backwards until the grip released. Three measurements were conducted with an interval of 30 s for the averaged grip strength calculation.

4.5. Statistical Analysis

The R statistical programming environment "https://www.r-project.org/" (accessed on 16 June 2023) was used to analyze the results. Base R graphics and ggplot2 [54] graphical library were used for the data visualisation. Dose-response curves were fitted using "nlm" function and the following equation: response ~ $(100/1 + (EC_{50}/x)\hat{}n)$, where x denotes base 10 logarithm of ligand concentration in moles per liter (normalized to 1M) and n is the Hill coefficient. Libraries "rgl" "https://CRAN.R-project.org/package=rgl" (accessed on 16 May 2023) and "barplot3d" "https://cran.r-project.org/src/contrib/Archive/barplot3d/" (accessed on 16 May 2023) were used to generate the 3D bar chart shown on Figure 2b.

One- and two-way ANOVA with Tukey post hoc test were used to examine the significance of the differences between groups. Repeated measures ANOVA was utilized to explain longitudinal data. In all cases $p < 0.05$ was interpreted as significant.

4.6. Two-Electrode Voltage-Clamp

Recordings were performed using Axopatch 200 amplifier in two-electrode voltage-clamp setup (Molecular Devices, LLC, San Jose, CA, USA) on oocytes removed from mature *Xenopus* frogs. Two to three days before taking recordings, oocytes were injected with in vitro transcribed RNA, coding α9 and α10 nAChR and kept at 18 °C in ND96 electrophysiology buffer solution (5 mM HEPES/NaOH at pH 7.6 and 18 °C, 96 mM NaCl, 2 mM KCl, 1.8 mM $CaCl_2$, 2 mM $MgCl_2$).

4.7. Radioligand Competition

Tested compounds were incubated with GH4C1 cells expressing α7 nAChR in 50 μL of 20 mM Tris-HCl buffer pH 8.0, containing 1 mg/mL BSA (binding buffer) for 30 min. Then, 0.5–0.9 nM of mono-iodinated 125I-αBgt was added for 5 min and cells suspensions were applied to GF/C glass filters (Whatman, Maidstone, UK) presoaked in 0.1 mg/mL BSA, and unbound radioactivity was removed from the filter by washing (3 × 3 mL) with 20 mM Tris-HCl buffer, pH 8.0 containing 0.1 mg/mL BSA (washing buffer). Nonspecific binding was determined by preliminary incubation of GH4C1 cells with 30 μM α-cobratoxin. The bound radioactivity was determined using a Wizard 1470 Automatic Gamma Counter.

5. Conclusions

In this paper, we describe a panel of synthetic analogs of natural marine product 6-bromohypaphorine with elevated potency toward α7 nAChR (as can be deduced from intracellular calcium rise and competition with radioactive α-bungarotoxin), demonstrating the pronounced anti-inflammatory and analgesic activities. Introduction of iodine instead of bromine, esterification of carboxy group, and changing chirality from L isomers to D was found to be the most productive strategy.

Supplementary Materials: The following supporting information can be downloaded at: https://www.mdpi.com/article/10.3390/md21060368/s1, Figure S1: Two-electrode voltage-clamp recording of alpha9/alpha10 nAChR inhibition by 6ND; Figure S2: Inhibition of radioiodinated α-butgarotoxin binding by 10 μM synthetic hypaphorine analogs (numbering corresponds to the Table 1).

Author Contributions: Conceptualization, I.A.I. and D.S.K.; Investigation, I.A.I., A.E.S., V.A.P., D.A.S., I.V.S., L.A.E., L.O.O., S.Y.B., N.A.P., M.A.L., Y.A.P., V.A.K., N.A.B., A.V.B., I.A.D. and D.S.K.; Project administration, D.S.K.; Writing—original draft, D.S.K.; Writing—review & editing, I.E.K. and V.I.T. All authors have read and agreed to the published version of the manuscript.

Funding: The work of I.A.I., D.S.K. and L.A.E. was supported by the RSF grant No. 21-74-10092, "https://rscf.ru/project/21-74-10092/" (accessed on 16 June 2023).

Institutional Review Board Statement: The study was conducted in an AAALAC (Association for Assessment and Accreditation of Laboratory Animal Care International) accredited facility in compliance with the standards of the Guide for Care and Use of Laboratory Animals (8th edition, Institute for Laboratory Animal Research). Animal treatment procedures were approved by the Institutional Animal Care and Use Committee (IACUC) of the Branch of the Shemyakin–Ovchinnikov Institute of Bioorganic Chemistry, Russian Academy of Sciences, the experimental protocol code is no. 688/19 (date of approval: 10 January 2019).

Data Availability Statement: The data presented in this study are available on request from the corresponding author.

Conflicts of Interest: The authors declare no conflict of interest.

References

1. Papke, R.L.; Horenstein, N.A. Therapeutic Targeting of A7 Nicotinic Acetylcholine Receptors. *Pharmacol. Rev.* **2021**, *73*, 1118–1149. [CrossRef] [PubMed]
2. Borroni, V.; Barrantes, F.J. Homomeric and Heteromeric A7 Nicotinic Acetylcholine Receptors in Health and Some Central Nervous System Diseases. *Membranes* **2021**, *11*, 664. [CrossRef] [PubMed]
3. Lendvai, B.; Kassai, F.; Szájli, Á.; Némethy, Z. A7 Nicotinic Acetylcholine Receptors and Their Role in Cognition. *Brain Res. Bull.* **2013**, *93*, 86–96. [CrossRef] [PubMed]
4. Tregellas, J.R.; Wylie, K.P. Alpha7 Nicotinic Receptors as Therapeutic Targets in Schizophrenia. *Nicotine Tob. Res.* **2018**, *21*, 349–356. [CrossRef]
5. Roberts, J.P.; Stokoe, S.A.; Sathler, M.F.; Nichols, R.A.; Kim, S. Selective Coactivation of A7- and A4β2-Nicotinic Acetylcholine Receptors Reverses Beta-Amyloid–Induced Synaptic Dysfunction. *J. Biol. Chem.* **2021**, *296*, 100402. [CrossRef] [PubMed]
6. Martín, A.; Domercq, M.; Matute, C. Inflammation in Stroke: The Role of Cholinergic, Purinergic and Glutamatergic Signaling. *Ther. Adv. Neurol. Disord.* **2018**, *11*, 175628641877426. [CrossRef] [PubMed]
7. Patel, H.; McIntire, J.; Ryan, S.; Dunah, A.; Loring, R. Anti-Inflammatory Effects of Astroglial A7 Nicotinic Acetylcholine Receptors Are Mediated by Inhibition of the NF-κB Pathway and Activation of the Nrf2 Pathway. *J. Neuroinflamm.* **2017**, *14*, 192. [CrossRef]

8. Vélez-Fort, M.; Audinat, E.; Angulo, M.C. Functional A7-Containing Nicotinic Receptors of NG2-Expressing Cells in the Hippocampus. *Glia* **2009**, *57*, 1104–1114. [CrossRef]
9. Shytle, R.D.; Mori, T.; Townsend, K.; Vendrame, M.; Sun, N.; Zeng, J.; Ehrhart, J.; Silver, A.A.; Sanberg, P.R.; Tan, J. Cholinergic Modulation of Microglial Activation by A7 Nicotinic Receptors. *J. Neurochem.* **2004**, *89*, 337–343. [CrossRef]
10. Martelli, D.; McKinley, M.J.; McAllen, R.M. The Cholinergic Anti-Inflammatory Pathway: A Critical Review. *Auton. Neurosci.* **2014**, *182*, 65–69. [CrossRef]
11. Corradi, J.; Bouzat, C. Understanding the Bases of Function and Modulation of A7 Nicotinic Receptors: Implications for Drug Discovery. *Mol. Pharmacol.* **2016**, *90*, 288–299. [CrossRef] [PubMed]
12. King, J.R.; Gillevet, T.C.; Kabbani, N. A G Protein-coupled A7 Nicotinic Receptor Regulates Signaling and TNF-α Release in Microglia. *FEBS Open Bio* **2017**, *7*, 1350–1361. [CrossRef] [PubMed]
13. King, J.R.; Kabbani, N. A7 Nicotinic Receptor Coupling to Heterotrimeric G Proteins Modulates RhoA Activation, Cytoskeletal Motility, and Structural Growth. *J. Neurochem.* **2016**, *138*, 532–545. [CrossRef] [PubMed]
14. Siniavin, A.; Streltsova, M.; Kudryavtsev, D.; Shelukhina, I.; Utkin, Y.; Tsetlin, V. Activation of A7 Nicotinic Acetylcholine Receptor Upregulates HLA-DR and Macrophage Receptors: Potential Role in Adaptive Immunity and in Preventing Immunosuppression. *Biomolecules* **2020**, *10*, 507. [CrossRef] [PubMed]
15. He, P.-F.; A-Ru-Na; Chen, H.; Wei, H.-Y.; Cao, J.-S. Activation of Alpha7 Nicotinic Acetylcholine Receptor Protects Bovine Endometrial Tissue against LPS-Induced Inflammatory Injury via JAK2/STAT3 Pathway and COX-2 Derived Prostaglandin E2. *Eur. J. Pharmacol.* **2021**, *900*, 174067. [CrossRef]
16. Marrero, M.B.; Bencherif, M. Convergence of Alpha 7 Nicotinic Acetylcholine Receptor-Activated Pathways for Anti-Apoptosis and Anti-Inflammation: Central Role for JAK2 Activation of STAT3 and NF-κB. *Brain Res.* **2009**, *1256*, 1–7. [CrossRef]
17. Báez-Pagán, C.A.; Delgado-Vélez, M.; Lasalde-Dominicci, J.A. Activation of the Macrophage A7 Nicotinic Acetylcholine Receptor and Control of Inflammation. *J. Neuroimmune Pharmacol.* **2015**, *10*, 468–476. [CrossRef]
18. Egea, J.; Buendia, I.; Parada, E.; Navarro, E.; León, R.; Lopez, M.G. Anti-Inflammatory Role of Microglial Alpha7 nAChRs and Its Role in Neuroprotection. *Biochem. Pharmacol.* **2015**, *97*, 463–472. [CrossRef]
19. De Jonge, W.J.; Ulloa, L. The Alpha7 Nicotinic Acetylcholine Receptor as a Pharmacological Target for Inflammation. *Br. J. Pharmacol.* **2007**, *151*, 915–929. [CrossRef]
20. Nurkhametova, D.; Siniavin, A.; Streltsova, M.; Kudryavtsev, D.; Kudryavtsev, I.; Giniatullina, R.; Tsetlin, V.; Malm, T.; Giniatullin, R. Does Cholinergic Stimulation Affect the P2X7 Receptor-Mediated Dye Uptake in Mast Cells and Macrophages? *Front. Cell. Neurosci.* **2020**, *14*, 548376. [CrossRef]
21. Barbier, A.J.; Hilhorst, M.; Vliet, A.V.; Snyder, P.; Palfreyman, M.G.; Gawryl, M.; Dgetluck, N.; Massaro, M.; Tiessen, R.; Timmerman, W.; et al. Pharmacodynamics, Pharmacokinetics, Safety, and Tolerability of Encenicline, a Selective A7 Nicotinic Receptor Partial Agonist, in Single Ascending-Dose and Bioavailability Studies. *Clin. Ther.* **2015**, *37*, 311–324. [CrossRef] [PubMed]
22. Dutta, S.; Hosmane, B.S.; Awni, W.M. Population Analyses of Efficacy and Safety of ABT-594 in Subjects with Diabetic Peripheral Neuropathic Pain. *AAPS J.* **2012**, *14*, 168–175. [CrossRef] [PubMed]
23. Kasheverov, I.; Kudryavtsev, D.; Shelukhina, I.; Nikolaev, G.; Utkin, Y.; Tsetlin, V. Marine Origin Ligands of Nicotinic Receptors: Low Molecular Compounds, Peptides and Proteins for Fundamental Research and Practical Applications. *Biomolecules* **2022**, *12*, 189. [CrossRef] [PubMed]
24. Kasheverov, I.; Shelukhina, I.; Kudryavtsev, D.; Makarieva, T.; Spirova, E.; Guzii, A.; Stonik, V.; Tsetlin, V. 6-Bromohypaphorine from Marine Nudibranch Mollusk Hermissenda Crassicornis Is an Agonist of Human A7 Nicotinic Acetylcholine Receptor. *Mar. Drugs* **2015**, *13*, 1255–1266. [CrossRef]
25. Ding, Y.; Miao, R.; Zhang, Q. Hypaphorine Exerts Anti-inflammatory Effects in Sepsis Induced Acute Lung Injury via modulatingDUSP1/P38/JNKpathway. *Kaohsiung J. Med. Sci.* **2021**, *37*, 883–893. [CrossRef]
26. Sun, H.; Cai, W.; Wang, X.; Liu, Y.; Hou, B.; Zhu, X.; Qiu, L. Vaccaria Hypaphorine Alleviates Lipopolysaccharide-Induced Inflammation via Inactivation of NFκB and ERK Pathways in Raw 264.7 Cells. *BMC Complement. Altern. Med.* **2017**, *17*, 120. [CrossRef]
27. Sun, H.; Zhu, X.; Cai, W.; Qiu, L. Hypaphorine Attenuates Lipopolysaccharide-Induced Endothelial Inflammation via Regulation of TLR4 and PPAR-γ Dependent on PI3K/Akt/mTOR Signal Pathway. *Int. J. Mol. Sci.* **2017**, *18*, 844. [CrossRef]
28. Sun, H.; Zhu, X.; Lin, W.; Zhou, Y.; Cai, W.; Qiu, L. Interactions of TLR4 and PPARγ, Dependent on AMPK Signalling Pathway Contribute to Anti-Inflammatory Effects of Vaccariae Hypaphorine in Endothelial Cells. *Cell. Physiol. Biochem.* **2017**, *42*, 1227–1239. [CrossRef]
29. Yonekawa, M.K.A.; Penteado, B.d.B.; Dal'Ongaro Rodrigues, A.; Lourenço, E.M.G.; Barbosa, E.G.; das Neves, S.C.; de Oliveira, R.J.; Marques, M.R.; Silva, D.B.; de Lima, D.P.; et al. L-Hypaphorine and d-Hypaphorine: Specific Antiacetylcholinesterase Activity in Rat Brain Tissue. *Bioorganic Med. Chem. Lett.* **2021**, *47*, 128206. [CrossRef]
30. Ozawa, M.; Honda, K.; Nakai, I.; Kishida, A.; Ohsaki, A. Hypaphorine, an Indole Alkaloid from Erythrina Velutina, Induced Sleep on Normal Mice. *Bioorganic Med. Chem. Lett.* **2008**, *18*, 3992–3994. [CrossRef]
31. Jouvet, M. The Role of Monoamines and Acetylcholine-Containing Neurons in the Regulation of the Sleep-Waking Cycle. In *Neurophysiology and Neurochemistry of Sleep and Wakefulness*; Springer: Berlin/Heidelberg, Germany, 1972; pp. 166–307.

32. Weininger, D. SMILES, a Chemical Language and Information System. 1. Introduction to Methodology and Encoding Rules. *J. Chem. Inf. Comput. Sci.* **1988**, *28*, 31–36. [CrossRef]
33. O'Boyle, N.M.; Banck, M.; James, C.A.; Morley, C.; Vandermeersch, T.; Hutchison, G.R. Open Babel: An Open Chemical Toolbox. *J. Cheminform.* **2011**, *3*, 33. [CrossRef] [PubMed]
34. Noviello, C.M.; Gharpure, A.; Mukhtasimova, N.; Cabuco, R.; Baxter, L.; Borek, D.; Sine, S.M.; Hibbs, R.E. Structure and Gating Mechanism of the A7 Nicotinic Acetylcholine Receptor. *Cell* **2021**, *184*, 2121–2134.e13. [CrossRef] [PubMed]
35. Shelukhina, I.; Spirova, E.; Kudryavtsev, D.; Ojomoko, L.; Werner, M.; Methfessel, C.; Hollmann, M.; Tsetlin, V. Calcium Imaging with Genetically Encoded Sensor Case12: Facile Analysis of A7/A9 nAChR Mutants. *PLoS ONE* **2017**, *12*, e0181936. [CrossRef] [PubMed]
36. Filimonov, V.D.; Trusova, M.; Postnikov, P.; Krasnokutskaya, E.A.; Lee, Y.M.; Hwang, H.Y.; Kim, H.; Chi, K.-W. Unusually Stable, Versatile, and Pure Arenediazonium Tosylates: Their Preparation, Structures, and Synthetic Applicability. *Org. Lett.* **2008**, *10*, 3961–3964. [CrossRef]
37. Moriya, T.; Hagio, K.; Yoneda, N. A Facile Synthesis of 6-Chloro-D-Tryptophan. *Bull. Chem. Soc. Jpn.* **1975**, *48*, 2217–2218. [CrossRef]
38. Molteni, M.; Gemma, S.; Rossetti, C. The Role of Toll-Like Receptor 4 in Infectious and Noninfectious Inflammation. *Mediat. Inflamm.* **2016**, *2016*, 1–9. [CrossRef]
39. Pinto, B.F.; Medeiros, N.I.; Teixeira-Carvalho, A.; Eloi-Santos, S.M.; Fontes-Cal, T.C.M.; Rocha, D.A.; Dutra, W.O.; Correa-Oliveira, R.; Gomes, J.A.S. CD86 Expression by Monocytes Influences an Immunomodulatory Profile in Asymptomatic Patients with Chronic Chagas Disease. *Front. Immunol.* **2018**, *9*, 454. [CrossRef]
40. Lu, K.; Ma, F.; Yi, D.; Yu, H.; Tong, L.; Chen, D. Molecular Signaling in Temporomandibular Joint Osteoarthritis. *J. Orthop. Transl.* **2022**, *32*, 21–27. [CrossRef]
41. Bove, S.E.; Calcaterra, S.L.; Brooker, R.M.; Huber, C.M.; Guzman, R.E.; Juneau, P.L.; Schrier, D.J.; Kilgore, K.S. Weight Bearing as a Measure of Disease Progression and Efficacy of Anti-Inflammatory Compounds in a Model of Monosodium Iodoacetate-Induced Osteoarthritis. *Osteoarthr. Cartil.* **2003**, *11*, 821–830. [CrossRef]
42. Shaykhutdinova, E.R.; Kondrakhina, A.E.; Ivanov, I.A.; Kudryavtsev, D.S.; Dyachenko, I.A.; Murashev, A.N.; Tsetlin, V.I.; Utkin, Y.N. Synthetic Analogs of 6-Bromohypaphorine, a Natural Agonist of Nicotinic Acetylcholine Receptors, Reduce Cardiac Reperfusion Injury in a Rat Model of Myocardial Ischemia. *Dokl. Biochem. Biophys.* **2022**, *503*, 47–51. [CrossRef] [PubMed]
43. Mann, P.C.; Vahle, J.; Keenan, C.M.; Baker, J.F.; Bradley, A.E.; Goodman, D.G.; Harada, T.; Herbert, R.; Kaufmann, W.; Kellner, R.; et al. International Harmonization of Toxicologic Pathology Nomenclature. *Toxicol. Pathol.* **2012**, *40*, 7S–13S. [CrossRef]
44. Lukácsi, S.; Gerecsei, T.; Balázs, K.; Francz, B.; Szabó, B.; Erdei, A.; Bajtay, Z. The Differential Role of CR3 (CD11b/CD18) and CR4 (CD11c/CD18) in the Adherence, Migration and Podosome Formation of Human Macrophages and Dendritic Cells under Inflammatory Conditions. *PLoS ONE* **2020**, *15*, e0232432. [CrossRef] [PubMed]
45. Kuzmich, N.; Sivak, K.; Chubarev, V.; Porozov, Y.; Savateeva-Lyubimova, T.; Peri, F. TLR4 Signaling Pathway Modulators as Potential Therapeutics in Inflammation and Sepsis. *Vaccines* **2017**, *5*, 34. [CrossRef]
46. Newton, S.; Ding, Y.; Chung, C.-S.; Chen, Y.; Lomas-Neira, J.L.; Ayala, A. Sepsis-Induced Changes in Macrophage Co-Stimulatory Molecule Expression: CD86 as a Regulator of Anti-Inflammatory IL-10 Response. *Surg. Infect.* **2004**, *5*, 375–383. [CrossRef] [PubMed]
47. Maj, T.; Slawek, A.; Chelmonska-Soyta, A. CD80 and CD86 Costimulatory Molecules Differentially Regulate OT-II CD4+T Lymphocyte Proliferation and Cytokine Response in Cocultures with Antigen-Presenting Cells Derived from Pregnant and Pseudopregnant Mice. *Mediat. Inflamm.* **2014**, *2014*, 1–8. [CrossRef] [PubMed]
48. Nolan, A.; Kobayashi, H.; Naveed, B.; Kelly, A.; Hoshino, Y.; Hoshino, S.; Karulf, M.R.; Rom, W.N.; Weiden, M.D.; Gold, J.A. Differential Role for CD80 and CD86 in the Regulation of the Innate Immune Response in Murine Polymicrobial Sepsis. *PLoS ONE* **2009**, *4*, e6600. [CrossRef] [PubMed]
49. Hosiawa, K.A.; Wang, H.; DeVries, M.E.; Garcia, B.; Liu, W.; Zhou, D.; Akram, A.; Jiang, J.; Sun, H.; Cameron, M.J.; et al. CD80/CD86 Costimulation Regulates Acute Vascular Rejection. *J. Immunol.* **2005**, *175*, 6197–6204. [CrossRef] [PubMed]
50. Kouhen, R.E.; Hu, M.; Anderson, D.; Li, J.; Gopalakrishnan, M. Pharmacology of A7 Nicotinic Acetylcholine Receptor Mediated Extracellular Signal-Regulated Kinase Signalling in PC12 Cells. *Br. J. Pharmacol.* **2009**, *156*, 638–648. [CrossRef]
51. Kashiwagi, Y.; Yanagita, M.; Kojima, Y.; Shimabukuro, Y.; Murakami, S. Nicotine Up-Regulates IL-8 Expression in Human Gingival Epithelial Cells Following Stimulation with IL-1β or P. Gingivalis Lipopolysaccharide via Nicotinic Acetylcholine Receptor Signalling. *Arch. Oral Biol.* **2012**, *57*, 483–490. [CrossRef]
52. Chen, R.-J.; Ho, Y.-S.; Guo, H.-R.; Wang, Y.-J. Rapid Activation of Stat3 and ERK1/2 by Nicotine Modulates Cell Proliferation in Human Bladder Cancer Cells. *Toxicol. Sci.* **2008**, *104*, 283–293. [CrossRef] [PubMed]

53. De Jonge, W.J.; van der Zanden, E.P.; The, F.O.; Bijlsma, M.F.; van Westerloo, D.J.; Bennink, R.J.; Berthoud, H.-R.; Uematsu, S.; Akira, S.; van den Wijngaard, R.M.; et al. Stimulation of the Vagus Nerve Attenuates Macrophage Activation by Activating the Jak2-STAT3 Signaling Pathway. *Nat. Immunol.* **2005**, *6*, 844–851. [CrossRef] [PubMed]
54. Wickham, H. *Ggplot2*; Springer International Publishing: Cham, Switzerland, 2016.

marine drugs

MDPI

Commentary

Fucoidans of Brown Algae: Comparison of Sulfated Polysaccharides from *Fucus vesiculosus* and *Ascophyllum nodosum*

Anatolii I. Usov *, Maria I. Bilan, Nadezhda E. Ustyuzhanina and Nikolay E. Nifantiev *

The Laboratory of Glycoconjugate Chemistry, N.D. Zelinsky Institute of Organic Chemistry, Russian Academy of Sciences, Leninsky Prospect 47, 119991 Moscow, Russia
* Correspondence: usov@ioc.ac.ru (A.I.U.); nen@ioc.ac.ru (N.E.N.); Tel.: +7-495-135-8784 (N.E.N.)

Abstract: Preparations of sulfated polysaccharides obtained from brown algae are known as fucoidans. These biopolymers have attracted considerable attention due to many biological activities which may find practical applications. Two Atlantic representatives of Phaeophyceae, namely, *Fucus vesiculosus* and *Ascophyllum nodosum*, belonging to the same order Fucales, are popular sources of commercial fucoidans, which often regarded as very similar in chemical composition and biological actions. Nevertheless, these two fucoidan preparations are polysaccharide mixtures which differ considerably in amount and chemical nature of components, and hence, this circumstance should be taken into account in the investigation of their biological properties and structure–activity relationships. In spite of these differences, fractions with carefully characterized structures prepared from both fucoidans may have valuable applications in drug development.

Keywords: brown algae; *Fucus vesiculosus*; *Ascophyllum nodosum*; fucoidans

Citation: Usov, A.I.; Bilan, M.I.; Ustyuzhanina, N.E.; Nifantiev, N.E. Fucoidans of Brown Algae: Comparison of Sulfated Polysaccharides from *Fucus vesiculosus* and *Ascophyllum nodosum*. *Mar. Drugs* **2022**, *20*, 638. https://doi.org/10.3390/md20100638

Academic Editors: Vladimir I. Kalinin, Pavel S. Dmitrenok and Natalia V. Ivanchina

Received: 26 August 2022
Accepted: 11 October 2022
Published: 13 October 2022

1. Introduction

Sulfated polysaccharides containing L-fucose as the main monosaccharide component were discovered in several brown algae, including *Fucus vesiculosus* and *Ascophyllum nodosum*, more than a century ago [1]. Now, it is well known that preparations obtained by extraction of algae and designated by trivial name "fucoidans" usually represent complex mixtures of several chemically different polysaccharides [2–4], where fucan sulfate (FS, a polysaccharide built up of fucose and sulfate only) may (but not necessarily) be one of the main components. The procedure suitable for commercial production of fucoidans from brown algal biomass was suggested in 1952 [5], but a highly purified sample of more or less individual FS was obtained much later by numerous manipulations with crude fucoidan from *F. vesiculosus*, probably accompanied by considerable loss of the starting material [6]. The first data on the chemical structure of SF published in 1950 [7,8] were then revised several times. In this period, evidence appeared on the prospective biological activity of fucoidans, especially as anticoagulants [9,10].

The availability of commercial fucoidans resulted in the rapid appearance of a large amount of publications devoted mainly to the investigation of their biological properties [11]. At present, there are hundreds of such papers; for example, more than 400 references dedicated mainly to therapeutic applications of fucoidans were cited in three reviews by Fitton et al. [12–14]. Multiple biological actions of fucoidans depend primarily on their interaction with different proteins due to the presence of sulfate groups [15–19], but branching of molecules [20] and molecular weights [21,22] may also be very important factors. Fucoidans are traditionally regarded as promising anticoagulant [23–26], antitumor [27,28] and anti-inflammatory agents [29–33], but recently they acquired special importance as potential components of antiviral drugs [34–36], activators of hematopoiesis [37,38], and reagents for use in nanomedicine [39].

The elucidation of correlations between the biological properties of concrete samples of fucoidans and their chemical structures remains a very important task [23,40], but there are substantial difficulties connected with the conception of "fucoidan", which is not the term of strict carbohydrate nomenclature. It designates a preparation of water-soluble sulfated polysaccharides, obtained by extraction of brown algal biomass and separated (partially or completely) from other polysaccharide components of this biomass devoid of sulfate groups—alginates and laminarans. The class Phaeophyceae (brown algae) numbers more than 1000 species, which may differ considerably in their polysaccharide composition [41,42]. Thus, fucoidans isolated from different species may contain not only fucopyranose, but also fucofuranose units [43] together with other monosaccharides, such as galactose, xylose, mannose, and glucuronic acid, etc. [2–4,41].

Therefore, the algal species is the first factor determining the composition of extracted sulfated polysaccharides. Then it is necessary to bear in mind that representatives of the same species growing in different conditions should inevitably differ in chemical composition. Ecological factors influencing the chemical composition of biomass include age and physiological status of the alga [44], climate and season [45], water temperature and salinity, solar radiation, and the accessibility of biogenic elements. Additional influences on the quality of polysaccharide preparations should have the procedures of harvesting and storage of raw material, as well as conditions of biomass treatment, which should secure the completeness of polysaccharide extraction without their degradation and minimal dissolution of extraneous non-carbohydrate materials, such as proteins and polyphenols. Since crude preparations of sulfated polysaccharides, obtained by water extraction, are usually mixtures of biopolymers of different structures [46], the nature of products destined for structural analysis or the investigation of biological properties will be determined by the used methods of fractionation [47]. Taking into account all the factors listed above, it is not surprising that different groups of researchers, dealing with samples being commonly named by the same term "fucoidan", in fact are often working with very different polysaccharides [48].

Below we overview the published results of structural studies of polysaccharides obtained from two widely distributed brown algal species, namely, *F. vesiculosus* and *A. nodosum*. Both these species have been investigated for a long time and may be regarded as prospective sources for the large-scale preparation of sulfated polysaccharides suitable for diverse medical applications [49]. For example, based on such polysaccharides, biological vectors are being developed, which may be used for delivering drugs or diagnostic contrasting agents to tissues with increased P-selectin expression [50–54]. It should be emphasized that GMP-graded production of low-molecular-weight fucoidan from *A. nodosum* was recently reported [54], and it was suggested to apply this product as the biovector or contraster for the detection of P-selectin expression during cardiovascular diseases.

2. Sulfated Polysaccharides of *Fucus vesiculosus*

Representatives of the genus *Fucus* are widely distributed in the North Atlantic, as well in the Barents and the Whyte Seas. Being typical littoral species, they occupy spacious coastal plots uncovering at the low tide. Harvesting of this natural raw material is not a very difficult task.

F. vesiculosus was one of those several brown algal species wherein the presence of fucoidans was discovered [1]. Polysaccharide preparation from this species was used in the first attempt to elucidate its chemical structure [7,8]. The alga was heated with water in the boiling water bath, the extract obtained was treated with lead acetate to remove alginic acid and proteins, and the remaining soluble polysaccharides were reprecipitated several times and yielded a preparation, which was regarded as a fucan sulfate. According to the analytical data, an essentially linear structure was suggested for the backbone of this polysaccharide, built up of 1→2-linked α-L-fucopyranose residues, together with some unsubstituted or monosulfated fucose residues attached to position 3 of the backbone as single branches.

About 40 years later, a commercial fucoidan from *F. vesiculosus* was reinvestigated and the structure of hypothesized fucan sulfate was corrected [55]. It was shown that 1→3-linkage is the main type of linkage between the backbone residues. Possible positions of sulfate and branches are depicted in Figure 1. Similar figures are often used to illustrate structures of FS isolated from other brown algal species [56]. It should be noted that this formula does not belong to any concrete sample of FS, but shows only the set of different units, which may be found in different proportions in different samples of polysaccharides.

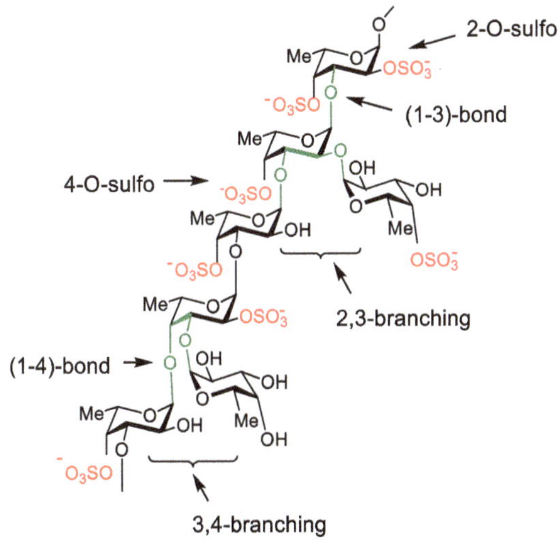

Figure 1. The current idealized formula of brown algal fucan sulfate (structural elements, which are possibly important for biological activity, are indicated).

The structure of the backbone was revised once more after the introduction of NMR spectroscopy in the practice of structural analysis of FS. Spectral data gave reliable confirmation on the presence of 1→3-linked backbone in FS isolated from *Chorda filum* [57] and several other algae belonging to the order Laminariales, but showed that fucoidans from *F. vesiculosus* and *A. nodosum* contain fractions with backbones built up of alternating (1→3)- and (1→4)-linked fucose residues [58,59]. Polysaccharides containing two alternating linkages in the backbones were also found in several other representatives of the order Fucales [60–62]. A hypothesis once appeared that (1→3)-linked FS backbones are characteristic for algae from Laminariales only, whereas backbones with alternating (1→3)- and (1→4)-linkages are typical for algae from Fucales, but the existence of such a firm correlation between the taxonomic position of the algae and the structure of their FS was not confirmed in recent investigations [48].

The formula of FS depicted in Figure 1 does not take into account the presence of several other monosaccharides, primarily galactose, xylose, mannose, and glucuronic acid, which usually may be found in fucoidan preparations. These monosaccharides may be components of other types of polysaccharides. According to contemporary evidence, several polysaccharides forming brown algal cell walls are linked with proteins and polyphenols in a complex [4], where the nature of linkages between components remains mostly unknown. To isolate polysaccharides, this complex should be destroyed by, for example, the action of dilute acids [5], although FS itself may be partially degraded under acid conditions. Crude fucoidans usually contain a wide set of molecules differing in composition and molecular weights and evidently need additional purification. Anion-exchange chromatography demonstrates the presence of continuous spectrum of molecules differing in charge and monosaccharide composition, from fractions with low sulfate and low fucose, containing other neutral monosaccharides and glucuronic acid, to highly sulfated fucans [63]. Simultaneous presence of several different sulfated polysaccharides was demonstrated for many brown algae [46,64]. The isolation of the components of these mixtures depends on their relative content and specific extraction and fractionation procedures. More detailed discussion on the problem is given below in the description of polysaccharides from *A. nodosum*.

One of the laboratory procedures of fucoidan isolation [60] recommends extraction of algal biomass with 2% aqueous calcium chloride at 85 °C. It results in dissolution of neutral laminaran and sulfated polysaccharides, whereas insoluble Ca-salts of alginic acids remain in the precipitate. Acid polysaccharides may be precipitated from extract by the addition of a cationic detergent, such as cetyltrimethylammonium bromide, transformed into the soluble sodium salts and then chromatographed on an anion-exchanger, such as DEAE-Sephacel, using stepwise elution with NaCl solutions of increasing concentrations. Neutral components are not absorbed on the column, while alginic acids are eluted with 0.5 M NaCl, and sulfated fractions are then eluted according to the increase of sulfate content, the most sulfated fractions appearing in the region of 2.0 M NaCl.

There are numerous modifications of isolation procedure aimed at acceleration of the process or increase of the yield of target polysaccharides. For these purposes it was suggested to isolate cell walls [64,65], to carry out extraction with solutions of acids, alkali, or detergents [64,66,67], to use autohydrolysis [68], microwave radiation [69], ultrasound [70], or treatment with enzymes capable of destroying the accompanying polysaccharides [71,72]. Fucoidans prepared by different procedures may differ considerably in composition and properties. Thus, seventeen fucoidans isolated by different authors from *F. vesiculosus* contain from 4% to 39% of sulfate and from 50% to 94% of fucose in carbohydrate moiety. These analytical characteristics were given in a recent review [48].

3. Sulfated Polysaccharides of *Ascophyllum nodosum*

This brown algal species belonging to the same family Fucaceae, as described above *F. vesiculosus*, practically coincides with it in geographical distribution, but is growing in sublittoral conditions. *A. nodosum* is used for industrial production of alginates and is available as feedstock in large amounts.

A peculiar sulfated heteropolysaccharide named "ascophyllan" was isolated from the mixture of water-soluble polysaccharides obtained by extraction of *A. nodosum* [73]. It contained approximately equimolar amounts of fucose, xylose, glucuronic acid, and sulfate, as well as a tightly bound polypeptide fragment. Partial acid hydrolysis of ascophyllan gave rise to 3-O-β-D-xylopyranosyl-L-fucose [74] and a non-dialysable polyuronide, indicating the presence of a backbone built up of glucuronic acid residues and side chains containing fucose and xylose. Using extractions under very mild conditions, it was possible to isolate a high-molecular complex, which was split under subsequent acid treatment into ascophyllan, alginate, and a fraction close to fucan sulfate in composition. Based on these data, it was supposed that a similar situation should also be found for polysaccharides of *F. vesiculosus*, but in this case the content of fucan sulfate predominates considerably over the content of hypothetical ascophyllan analogue [75]. Structural analysis of ascophyllan was the subject of a series of more recent publications [76–81]. Based on these data, a procedure was developed for isolation of a fraction of fucan sulfate from *A. nodosum*, having backbones of 1,3-linked fucose residues [82]. At the same time another fraction, containing backbones of alternating 1,3-1,4-linked fucose residues, was obtained from the same alga by another group of authors. The structure of the latter polysaccharide was carefully investigated using chemical methods together with NMR spectroscopy [58,59] and mass spectrometry [83]. Hence, fucan sulfate itself is heterogeneous and contains fractions having fundamental structural differences in their carbohydrate moieties.

Later on, a simplified procedure of biomass treatment was published, which gives the possibility to prepare ascophyllan and fucan sulfate separately [84] (Table 1). Polysaccharides isolated by this procedure were studied in biological tests, and it was found for the first time that ascophyllan can stimulate the growth of a culture of mammalian cells (under the same conditions fucan sulfate showed opposite action) [85]. Comparison of polysaccharide preparations, isolated from *A. nodosum* by usual extraction with dilute acid, and by several new procedures using microwave radiation [86], ultrasound, or enzymatic degradation of cell walls was described in a recent paper [87]. As expected, these preparations differ in yields, composition, and molecular weights, but have comparable prebiotic activity by stimulating, in vitro, the growth of lactic acid bacteria. Two polysaccharide samples differing in molecular weights and capable of inhibiting inflammation were isolated by treatment of *A. nodosum* with enzymes followed by anion-exchange chromatography [88]. Both preparations contained not only fucose, but also galactose and hence were fragments of a sulfated galactofucan. They exhibited different anti-inflammatory activities, indicating that molecular weight is an important factor for this type of biological action.

Table 1. Yields and composition (in %) of several polysaccharide preparations isolated from *A. nodosum* and *F. vesiculosus* *.

Preparation	Yield	Fuc	Xyl	Glc	Man	Gal	UA	SO_3Na
Ascophyllan [71]	1.9	15.5 (1.00)	13.4 (0.95)	0.3 (0.02)	3.4 (0.2)	0.6 (0.04)	21.4 (1.17)	9.6 (1.06)
Fucan sulfate from *A. nodosum* [71]	1.25	28.4 (1.00)	4.3 (0.16)	2.0 (0.06)	0.8 (0.03)	5.3 (0.17)	5.8 (0.17)	19.4 (1.17)
Fucan sulfate from *A. nodosum* [69]		(1.00)	(0.05)	-	-	-	-	(0.47)
Fucan sulfate from *F. vesiculosus* [5]		(1.00)	tr.	-	-	tr.	-	(0.47)
Commercial fucoidan from *F. vesiculosus* [71]		24.8 (1.00)	1.9 (0.09)	0.8 (0.03)	1.0 (0.04)	3.1 (0.11)	9.6 (0.33)	22.6 (1.56)

* Molar proportions relative to fucose content (Fuc = 1.00) are given in brackets.

4. Conclusions

Both brown algal species described in this review are convenient sources of so-called "fucoidans", which are crude mixtures of sulfated polysaccharides. Both "fucoidans" contain fucan sulfates as the main components, which are especially interesting as biologically active polysaccharides. Differences between "fucoidans" of these two algal species are connected mainly with the higher content of another main component, termed "ascophyllan", in *A. nodosum*. In fact, ascophyllan itself is a mixture of several heteropolysaccharides of moderate sulfation degree, containing, in addition to fucose, also xylose, glucuronic acid, and some other monosaccharides [73,77,81]. Ascophyllan has its own practically useful properties and may find application as a preparation with peculiar (distinct from fucan sulfates) biological activities [89]. The paper by W. Jin et al. [90] may be mentioned as a very impressive example of investigation devoted to correlating the biological activity and chemical structures of fucoidan components: the authors carefully analyzed the structural information of about sixty fucoidan samples isolated from different algae, used them in a cell surface tau-binding assay, and found that two different branched sulfated polysaccharide components of fucoidans, namely, a galactofucan and a fucoglucuronomannan, acted as effective inhibitors of tau spreading. Hence, these data may provide a basis for creation of drugs applicable for therapy in the earliest phase of Alzheimer's disease.

Resolution of "fucoidans" into individual polysaccharide components remains the most difficult problem. Since these polysaccharides are polyanions, anion-exchange chromatography is traditionally used for their separation, giving excellent results in some cases [61]. At the same time this procedure cannot resolve compounds differing in structure, but having similar charge densities. Similar limitations are typical for gel-permeation chromatography, where even structurally different polysaccharides cannot be resolved, if they have close molecular weights. Evidently, the resolution of fucoidans needs additional improvement. Specific enzymatic degradation of unnecessary components of the mixture [91] or affinity chromatography based on the property of sulfated polysaccharides to bind several proteins [92,93] may be mentioned as the possible approaches to new fractionation procedures.

Fucoidan preparations devoted to medical applications should satisfy the very strong demands concerning their reproducible composition and compatibility with manufacturing requirements under GMP-standards. A similar problem has been solved in the preparation of low-molecular-weight heparins [94]. The process of fucoidan isolation should be carefully controlled at several steps. One of the most important parameters is the standard quality of the raw material, which is highly varying and depending on the place and season of harvesting, as well on the procedure of its conservation and storage. Extraction may be carried out with dilute acids under moderate heating at the conditions, which are sufficient for the destruction of polysaccharide complexes of the cell walls without marked degradation of the target fucoidan. Mild oxidants may be used for bleaching, and high molecular mass may be diminished (if necessary) by careful partial depolymerization. The wanted fraction may be prepared using membrane filtration. Finally (this is especially important for algae similar in polysaccharide composition to *A. nodosum*) the most interesting biologically active highly sulfated fraction should be separated from lower-sulfated material (such as so-called ascophyllan) using chromatography on anion-exchange resins. The preparations obtained should be characterized by quantitative determination of monosaccharides (e.g., gas–liquid chromatography, spectrophotometry) sulfate (e.g., turbidimetry) and molecular-mass distribution (e.g., analytical gel-permeation chromatography). Similar standardization procedures are suggested for sulfated glycosaminoglycans, such as, for example, for chondroitin sulfates [95].

Regulatory requirements mentioned above shorten the fields of medical use of fucoidans by the case of superficial applications (ointments, gels, inhalable compositions, etc.), but injectable forms may be based on synthetic oligosaccharides related to fucoidan fragments. It should be emphasized that the preparation of such compounds is well elaborated to date [96–98]. In addition, one promising option for fucoidan standardization can be connected with enzymatic treatments. Recent studies discovered a series of fucoidan degrading and modifying enzymes [99–101], but up to now only a few agents of this type were obtained, while their applicability for industrial application was not studied yet and needs to be investigated.

Author Contributions: Conceptualization, A.I.U. and N.E.N.; resources, M.I.B. and N.E.U.; writing—original draft preparation, A.I.U.; writing—review and editing, N.E.N. and M.I.B.; supervision, N.E.N.; project administration, N.E.N.; funding acquisition, N.E.N. All authors have read and agreed to the published version of the manuscript.

Funding: This research was funded by the Ministry of Science and Higher Education of Russia (Grant Agreement No. 075-15-2020-792).

Data Availability Statement: Data are available from the corresponding author.

Conflicts of Interest: The authors declare no conflict of interest.

References

1. Kylin, H. Zur Biochemie der Meeresalgen. *Hoppe-Seyler's Z. Physiol. Chem.* **1913**, *83*, 171–197. [CrossRef]
2. Usov, A.I.; Bilan, M.I. Fucoidans—sulfated polysaccharides of brown algae. *Russ. Chem. Revs* **2009**, *78*, 846–862. [CrossRef]
3. Ale, M.T.; Meyer, A.S. Fucoidans from brown seaweeds: An update on structures, extraction techniques and use of enzymes as tools for structural elucidation. *RSC Advances* **2013**, *3*, 8131–8141. [CrossRef]
4. Deniaud-Bouët, E.; Hardouin, K.; Potin, P.; Kloareg, B.; Hervé, C. A review about brown algal cell walls and fucose-containing sulfated polysaccharides: Cell wall context, biomedical properties and key research challenges. *Carbohydr. Polym.* **2017**, *175*, 395–408. [CrossRef]
5. Black, W.A.P.; Dewar, E.T.; Woodward, F.N. Manufacture of algal chemicals. IV. Laboratory scale isolation of fucoidin from brown marine algae. *J. Sci. Food Agric.* **1952**, *3*, 122–129. [CrossRef]
6. Bernardi, G.; Springer, G.F. Properties of highly purified fucan. *J. Biol. Chem.* **1962**, *237*, 75–80. [CrossRef]
7. Percival, E.G.V.; Ross, A.G. Fucoidin. Part I. The isolation and purification of fucoidin from brown seaweeds. *J. Chem. Soc.* **1950**, 717–720. [CrossRef]

8. Conchie, J.; Percival, E.G.V. Fucoidin. Part II. The hydrolysis of a methylated fucoidin prepared from *Fucus vesiculosus*. *J. Chem. Soc.* **1950**, 827–832. [CrossRef]
9. Springer, G.F.; Wurzel, H.A.; McNeal, G.M., Jr.; Ansell, N.J.; Doughty, M.F. Isolation of anticoagulant fractions from crude fucoidin. *Proc. Soc. Exptl. Biol. Med.* **1957**, *94*, 404–409. [CrossRef]
10. Church, F.C.; Meade, J.B.; Treanor, R.E.; Whinna, H.C. Antithrombin activity of fucoidan. The interaction of fucoidan with heparin cofactor II, antithrombin III, and thrombin. *J. Biol. Chem.* **1989**, *264*, 3618–3623. [CrossRef]
11. Berteau, O.; Mulloy, B. Sulfated fucans, fresh perspectives: Structures, functions, and biological properties of sulfated fucans and an overview of enzymes active toward this class of polysaccharide. *Glycobiology* **2003**, *13*, 29R–40R. [CrossRef] [PubMed]
12. Fitton, J.H. Therapies from fucoidan; multifunctional marine polymers. *Mar. Drugs* **2011**, *9*, 1731–1760. [CrossRef] [PubMed]
13. Fitton, J.H.; Stringer, D.N.; Karpiniec, S.S. Therapies from fucoidan: An update. *Mar. Drugs* **2015**, *13*, 5920–5946. [CrossRef] [PubMed]
14. Fitton, J.H.; Stringer, D.N.; Park, A.Y.; Karpiniec, S.S. Therapies from fucoidan: New developments. *Mar. Drugs* **2019**, *17*, 571. [CrossRef]
15. Luthuli, S.; Wu, S.; Cheng, Y.; Zheng, X.; Wu, M.; Tong, H. Therapeutic effects of fucoidan: A review on recent studies. *Mar. Drugs* **2019**, *17*, 487. [CrossRef]
16. Kang, J.; Jia, X.; Wang, N.; Xiao, M.; Song, S.; Wu, S.; Li, Z.; Wang, S.; Cui, S.W.; Guo, Q. Insights into the-structure-bioactivity relationships of marine sulfated polysaccharides: A review. *Food Hydrocolloids* **2022**, *123*, 107049. [CrossRef]
17. Haroun-Bouhedja, F.; Ellouali, M.; Sinquin, C.; Boisson-Vidal, C. Relationship between sulfate groups and biological activities of fucans. *Thromb. Res.* **2000**, *100*, 453–459. [CrossRef]
18. Koyanagi, S.; Tanigawa, N.; Nakagawa, H.; Soeda, S.; Shimeno, H. Oversulfation of fucoidan enhances its anti-angiogenic and antitumor activities. *Biochem. Pharmacol.* **2003**, *65*, 173–179. [CrossRef]
19. Qiu, X.; Amarasekara, A.; Doctor, V. Effect of oversulfation on the chemical and biological properties of fucoidan. *Carbohydr. Polym.* **2006**, *63*, 224–228. [CrossRef]
20. Clément, M.-J.; Tissot, B.; Chevolot, L.; Adjadj, E. Du, Y.; Curmi, P.A.; Daniel, R. NMR characterization and molecular modeling of fucoidan showing the importance of oligosaccharide branching in its anticomplementary activity. *Glycobiology* **2010**, *20*, 883–894. [CrossRef]
21. Lahrsen, E.; Liewert, I.; Alban, S. Gradual degradation of fucoidan from *Fucus vesiculosus* and its effect on structure, antioxidant and antiproliferative activities. *Carbohydr. Polym.* **2018**, *192*, 208–216. [CrossRef] [PubMed]
22. Cabral, E.M.; Mondala, J.R.M.; Oliveira, M.; Przyborska, J.; Fitzpatrick, S.; Rai, D.K.; Sivagnanam, S.P.; Garcia-Vaquero, M.; O'Shea, D.; Devereux, M.; et al. Influence of molecular weight fractionation on the antimicrobial and anticancer properties of a fucoidan rich-extract from the macroalgae *Fucus vesiculosus*. *Int. J. Biol. Macromol.* **2021**, *186*, 994–1002. [CrossRef] [PubMed]
23. Cumashi, A.; Ushakova, N.A.; Preobrazhenskaya, M.E.; D'Incecco, A.; Piccoli, A.; Totani, L.; Tinari, N.; Morozevich, G.E.; Berman, A.E.; Bilan, M.I.; et al. A comparative study of the antiinflammatory, anticoagulant, antiangiogenic and antiadhesive activities of nine different fucoidans from brown seaweeds. *Glycobiology* **2007**, *17*, 541–552. [CrossRef] [PubMed]
24. Ustyuzhanina, N.E.; Ushakova, N.A.; Zyuzina, K.A.; Bilan, M.I.; Elizarova, A.L.; Somonova, O.V.; Madzhuga, A.V.; Krylov, V.B.; Preobrazhenskaya, M.E.; Usov, A.I.; et al. Influence of fucoidans on hemostatic system. *Mar. Drugs* **2013**, *11*, 2444–2458. [CrossRef] [PubMed]
25. Ustyuzhanina, N.E.; Ushakova, N.A.; Preobrazhenskaya, M.E.; Bilan, M.I.; Tsvetkova, E.A.; Krylov, V.B.; Anisimova, N.A.; Kiselevskiy, M.V.; Krukovskaya, N.V.; Li, C.; et al. Fucoidans as a platform for new anticoagulant drugs discovery. *Pure Appl. Chem.* **2014**, *86*, 1365–1375. [CrossRef]
26. Mourão, P.A.S. Perspective on the use of sulfated polysaccharides from marine organisms as a source of new antithrombotic drugs. *Mar. Drugs* **2015**, *13*, 2770–2784. [CrossRef]
27. Atashrazm, F.; Lowenthal, R.M.; Woods, G.M.; Holloway, A.F.; Dickinson, J.L. Fucoidan and cancer: A multifunctional molecule with anti-tumor potential. *Mar. Drugs* **2015**, *13*, 2327–2346. [CrossRef]
28. van Weelden, G.; Bobinski, M.; Jkla, K.; van Weelden, W.J.; Romano, A.; Pijnenborg, J.M.A. Fucoidan structure and activity in relation to anti-cancer mechanisms. *Mar. Drugs* **2019**, *17*, 32. [CrossRef]
29. Preobrazhenskaya, M.E.; Berman, A.E.; Mikhailov, V.I.; Ushakova, N.A.; Mazurov, A.V.; Semenov, A.V.; Usov, A.I.; Nifant'ev, N.E.; Bovin, N.V. Fucoidan inhibits leucocyte recruitment in a model peritoneal inflammation in rat and blocks interaction of P-selectin with its carbohydrate ligand. *Biochem. Mol. Biol. Int.* **1997**, *43*, 443–451. [CrossRef]
30. Phull, A.R.; Kim, S.J. Fucoidan as bio-functional molecule: Insight into the anti-inflammatory potential and associated molecular mechanisms. *J. Func. Food.* **2017**, *38*, 415–426. [CrossRef]
31. Apostolova, E.; Lukova, P.; Baldzhieva, A.; Katsarov, P.; Nikolova, M.; Iliev, I.; Peychev, L.; Trica, B.; Oancea, F.; Delattre, C.; et al. Immunomodulatory and anti-inflammatory effects of fucoidan: A review. *Polymers* **2020**, *123*, 107049. [CrossRef] [PubMed]
32. Fonseca, R.J.C.; Mourão, P.A.S. Pharmacological activities of sulfated fucose-rich polysaccharides after oral administration: Perspectives for the development of new carbohydrate-based drugs. *Mar. Drugs* **2021**, *19*, 425. [CrossRef] [PubMed]
33. Yang, J.-Y.; Lim, S.Y. Fucoidans and bowel health. *Mar. Drugs* **2021**, *19*, 436. [CrossRef]
34. Fitton, J.H.; Park, A.Y.; Karpiniec, S.S.; Stringer, D.N. Fucoidan and lung function: Value in viral infection. *Mar. Drugs* **2021**, *19*, 4. [CrossRef] [PubMed]

35. Lomartire, S.; Gonçalves, A.M.M. Antiviral activity and mechanisms of seaweeds bioactive compounds on enveloped viruses—A review. *Mar. Drugs* **2022**, *20*, 385. [CrossRef]
36. Pradhan, B.; Nayak, R.; Patra, S.; Bhuyan, P.P.; Behera, P.K.; Mandal, A.K.; Behera, C.; Ki, J.-S.; Adhikary, S.P.; MubarakAli, D.; et al. A state-of-the-art review on fucoidan as an antiviral agent to combat viral infections. *Carbohydr. Polym.* **2022**, *291*, 119551. [CrossRef]
37. Anisimova, N.; Ustyuzhanina, N.; Bilan, M.; Donenko, F.; Usov, A.; Kiselevskiy, M.; Nifantiev, N. Fucoidan and fucosylated chondroitin sulfate stimulate hematopoiesis in cyclophosphamide-induced mice. *Mar. Drugs* **2017**, *15*, 301. [CrossRef]
38. Anisimova, N.Y.; Ustyuzhanina, N.E.; Bilan, M.I.; Donenko, F.V.; Ushakova, N.A.; Usov, A.I.; Kiselevskiy, M.V.; Nifantiev, N.E. Influence of modified fucoidan and related sulfated oligosaccharides on hematopoiesis in cyclophosphamide-induced mice. *Mar. Drugs* **2018**, *16*, 333. [CrossRef]
39. Chollet, L.; Saboural, P.; Chauvierre, C.; Villemin, J.-N.; Letourneur, D.; Chaubet, F. Fucoidans in nanomedicine. *Mar. Drugs* **2016**, *14*, 145. [CrossRef]
40. Croci, D.O.; Cumashi, A.; Ushakova, N.A.; Preobrazhenskaya, M.E.; Piccoli, A.; Totani, L.; Ustyuzhanina, N.E.; Bilan, M.I.; Usov, A.I.; Grachev, A.A.; et al. Fucans, but not fucomannoglucuronans, determine the biological activities of sulfated polysaccharides from *Laminaria saccharina* brown seaweed. *PLoS ONE* **2011**, *6*, e17283. [CrossRef]
41. Ale, M.T.; Mikkelsen, J.D.; Meyer, A.S. Important determinants for fucoidan bioactivity: A critical review of structure-function relations and extraction methods for fucose-containing sulfated polysaccharides from brown seaweeds. *Mar. Drugs* **2011**, *9*, 2106–2130. [CrossRef] [PubMed]
42. Usov, A.I.; Smirnova, G.P.; Klochkova, N.G. Polysaccharides of algae. 55. Polysaccharide composition of several brown algae from Kamchatka. *Russ. J. Bioorg. Chem.* **2001**, *27*, 395–399. [CrossRef] [PubMed]
43. Bilan, M.I.; Vinogradova, E.V.; Tsvetkova, E.A.; Grachev, A.A.; Shashkov, A.S.; Nifantiev, N.E.; Usov, A.I. A sulfated glucuronofucan containing both fucofuranose and fucopyranose residues from the brown alga *Chordaria flagelliformis*. *Carbohydr. Res.* **2008**, *343*, 2605–2612. [CrossRef] [PubMed]
44. Skriptsova, A.V.; Shevchenko, N.M.; Tarbeeva, D.V.; Zvyagintseva, T.N. Comparative study of polysaccharides from reproductive and sterile tissues of five brown seaweeds. *Mar. Biotechnol.* **2012**, *14*, 304–311. [CrossRef]
45. Skriptsova, A.V.; Shevchenko, N.M.; Zvyagintseva, T.N.; Imbs, T.I. Monthly changes in the content and monosaccharide composition of fucoidan from *Undaria pinnatifida* (Laminariales, Phaeophyta). *J. Appl. Phycol.* **2010**, *22*, 79–86. [CrossRef]
46. Bilan, M.I.; Grachev, A.A.; Shashkov, A.S.; Kelly, M.; Sanderson, C.J.; Nifantiev, N.E.; Usov, A.I. Further studies on the composition and structure of a fucoidan preparation from the brown alga *Saccharina latissima*. *Carbohydr. Res.* **2010**, *345*, 2038–2047. [CrossRef]
47. Sichert, A.; Le Gall, S.; Klau, L.J.; Laillet, B.; Rogniaux, H.; Aachmann, F.L.; Hehemann, J.-H. Ion-exchange purification and structural characterization of five sulfated fucoidans from brown algae. *Glycobiology* **2021**, *31*, 352–357. [CrossRef]
48. Ponce, N.M.A.; Stortz, C.A. A comprehensive and comparative analysis of the fucoidan compositional data across the Phaeophyceae. *Front. Plant Sci.* **2020**, *11*, 556312. [CrossRef]
49. Keleszade, E.; Patterson, M.; Trangmar, S.; Guinan, K.J.; Costabile, A. Clinical efficacy of brown seaweeds *Ascophyllum nodosum* and *Fucus vesiculosus* in the prevention or delay progression of the metabolic syndrome: A review of clinical trials. *Molecules* **2021**, *26*, 714. [CrossRef]
50. Dellaquila, A.; Le Bao, C.; Letourneur, D.; Simon-Yarza, T. In vitro strategies to vascularize 3D physiologically relevant models. *Adv. Sci.* **2021**, *8*, 2100798. [CrossRef]
51. Forero Ramirez, L.M.; Gobin, E.; Aid-Launais, R.; Journe, C.; Moraes, F.C.; Picton, L.; Le Cerf, D.; Letourneur, D.; Chauvierre, C.; Chaubet, F. Gd(DOTA)-grafted submicronic polysaccharide-based particles functionalized with fucoidan as potential MR contrast agent able to target human activated platelets. *Carbohydr. Polym.* **2020**, *245*, 116457. [CrossRef] [PubMed]
52. Li, X.; Bauer, W.; Israel, I.; Kreissl, M.C.; Weirather, J.; Richter, D.; Bauer, E.; Herold, V.; Jakob, P.; Buck, A.; et al. Targeting P-selectin by gallium-68-labeled fucoidan positron emission tomography for noninvasive characterization of vulnerable plaques: Correlation with in vivo 17.6T MRI. *Arterioscler. Thromb. Vasc. Biol.* **2014**, *34*, 1661–1667. [CrossRef] [PubMed]
53. Abdollah, M.R.A.; Carter, T.J.; Jones, C.; Kalber, T.L.; Rajkumar, V.; Tolner, B.; Gruettner, C.; Zaw-Thin, M.; Torres, J.B.; Ellis, M.; et al. Fucoidan prolongs the circulation time of dextran-coated iron oxide nanoparticles. *ACS Nano* **2018**, *12*, 1156. [CrossRef] [PubMed]
54. Chauvierre, C.; Aid-Launais, R.; Aerts, J.; Chaubet, F.; Maire, M.; Chollet, L.; Rolland, L.; Bonafé, R.; Rossi, S.; Bussi, S.; et al. Pharmaceutical development and safety evaluation of a GMP-grade fucoidan for molecular diagnosis of cardiovascular diseases. *Mar. Drugs* **2019**, *17*, 699. [CrossRef] [PubMed]
55. Patankar, M.S.; Oehninger, S.; Barnett, T.; Williams, R.L.; Clark, G.F. A revised structure for fucoidan may explain some of its biological activities. *J. Biol. Chem.* **1993**, *268*, 21770–21776. [CrossRef]
56. Usov, A.I.; Smirnova, G.P.; Bilan, M.I.; Shashkov, A.S. Polysaccharides of algae. 53. Brown alga *Laminaria saccharina* (L.) Lam. As a source of fucoidan. *Russ. J. Bioorg. Chem.* **1998**, *24*, 382–389.
57. Chizhov, A.O.; Dell, A.; Morris, H.R.; Haslam, S.M.; McDowell, R.A.; Shashkov, A.S.; Nifant'ev, N.E.; Khatuntseva, E.A.; Usov, A.I. A study of fucoidan from the brown seaweed *Chorda filum*. *Carbohydr. Res.* **1999**, *320*, 108–119. [CrossRef]
58. Chevolot, L.; Mulloy, B.; Ratiscol, J.; Foucault, A.; Colliec-Jouault, S. A disaccharide repeat unit is the major structure in fucoidans from two species of brown algae. *Carbohydr. Res.* **2001**, *330*, 529–535. [CrossRef]

59. Chevolot, L.; Foucault, A.; Chaubet, F.; Kervarec, N.; Sinquin, C.; Fischer, A.-M.; Boisson-Vidal, C. Further data on the structure of brown seaweed fucans: Relationships with anticoagulant activity. *Carbohydr. Res.* **1999**, *319*, 154–165. [CrossRef]
60. Bilan, M.I.; Grachev, A.A.; Ustuzhanina, N.E.; Shashkov, A.S.; Nifantiev, N.E.; Usov, A.I. Structure of a fucoidan from the brown seaweed *Fucus evanescens* C.Ag. *Carbohydr. Res.* **2002**, *337*, 719–730. [CrossRef]
61. Bilan, M.I.; Grachev, A.A.; Ustuzhanina, N.E.; Shashkov, A.S.; Nifantiev, N.E.; Usov, A.I. A highly regular fraction of a fucoidan from the brown seaweed *Fucus distichus* L. *Carbohydr. Res.* **2004**, *339*, 511–517. [CrossRef] [PubMed]
62. Descamps, V.; Colin, S.; Lahaye, M.; Jam, M.; Richard, C.; Potin, P.; Barbeyron, T.; Yvin, J.-C.; Kloareg, B. Isolation and culture of a marine bacterium degrading the sulfated fucans from marine brown algae. *Mar. Biotechnol.* **2006**, *8*, 27–39. [CrossRef] [PubMed]
63. Nishino, T.; Nishioka, C.; Ura, H.; Nagumo, T. Isolation and partial characterization of a novel amino sugar-containing fucan sulfate from commercial *Fucus vesiculosus* fucoidan. *Carbohydr. Res.* **1994**, *255*, 213–224. [CrossRef]
64. Mabeau, S.; Kloareg, B.; Joseleau, J.-P. Fractionation and analysis of fucans from brown algae. *Phytochem.* **1990**, *29*, 2441–2445. [CrossRef]
65. Mabeau, S.; Kloareg, B. Isolation and analysis of the cell walls of brown algae: *Fucus spiralis, F. ceranoides, F. vesiculosus, F. serratus, Bifurcaria bifurcata* and *Laminaria digitata*. *J. Exp. Bot.* **1987**, *38*, 1573–1580. [CrossRef]
66. Bourne, E.J.; Brush, P.; Percival, E. The active carbohydrate metabolites of the brown seaweed, *Fucus vesiculosus*. *Carbohydr. Res.* **1969**, *9*, 415–422. [CrossRef]
67. Ruperez, P.; Ahrazem, O.; Leal, J.A. Potential antioxidant capacity of sulfated polysaccharides from the edible marine brown seaweed *Fucus vesiculosus*. *J. Agric. Food Chem.* **2002**, *50*, 840–845. [CrossRef]
68. Rodríguez-Jasso, R.M.; Mussatto, S.I.; Pastrana, L.; Aguilar, C.N.; Teixeira, J.A. Extraction of sulfated polysaccharides by autohydrolysis of brown seaweed *Fucus vesiculosus*. *J. Appl. Phycol.* **2013**, *25*, 31–39. [CrossRef]
69. Rodríguez-Jasso, R.M.; Mussatto, S.I.; Pastrana, L.; Aguilar, C.N.; Teixeira, J.A. Microwave-assisted extraction of sulfated polysaccharides (fucoidan) from brown seaweed. *Carbohydr. Polym.* **2011**, *86*, 1137–1144. [CrossRef]
70. Obluchinskaya, E.D.; Makarova, M.N.; Pozharitskaya, O.N.; Shikov, A.N. Effect of ultrasound treatment on the chemical composition and anticoagulant properties of dry *Fucus* extract. *Pharm. Chem. J.* **2015**, *49*, 183–186. [CrossRef]
71. Hahn, T.; Lang, S.; Ulber, R.; Muffler, K. Novel procedures for the extraction of fucoidan from brown algae. *Process Biochem.* **2012**, *47*, 1691–1698. [CrossRef]
72. Flórez-Fernández, N.; Torres, M.D.; Gonsález-Muñoz, M.J.; Domínguez, H. Potential of intensification for the extraction and depolymerization of fucoidan. *Algal Res.* **2018**, *30*, 128–148. [CrossRef]
73. Larsen, B.; Haug, A.; Painter, T.J. Sulphated polysaccharides in brown algae. *I. Isolation and preliminary characterisation of three sulphated polysaccharides from Ascophyllum nodosum. Acta Chem. Scand.* **1966**, *20*, 219–230. [CrossRef]
74. Larsen, B. Sulphated polysaccharides in brown algae. II. Isolation of 3-O-β-D-xylopyranosyl-L-fucose from ascophyllan. *Acta Chem. Scand.* **1967**, *21*, 1395–1396. [CrossRef]
75. Larsen, B.; Haug, A.; Painter, T.J. Sulphated polysaccharides in brown algae. III. The native state of fucoidan in *Ascophyllum nodosum* and *Fucus vesiculosus*. *Acta Chem. Scand.* **1970**, *24*, 3339–3352. [CrossRef]
76. Percival, E. Glucuronoxylofucan, a cell wall polysaccharide of the Phaeophyceae. *Chem. Ind.* **1967**, 511.
77. Percival, E. Glucuronoxylofucan, a cell wall component of *Ascophyllum nodosum*. Part I. *Carbohydr. Res.* **1968**, *7*, 272–283. [CrossRef]
78. Percival, E. Glucuronoxylofucan, a cell wall component of *Ascophyllum nodosum*. Part II. Methylation. *Carbohydr. Res.* **1971**, *17*, 121–126. [CrossRef]
79. Medcalf, D.G.; Larsen, B. Fucose-containing polysaccharides in the brown algae *Ascophyllum nodosum* and *Fucus vesiculosus*. *Carbohydr. Res.* **1977**, *59*, 531–537. [CrossRef]
80. Medcalf, D.G.; Larsen, B. Structural studies on ascophyllan and the fucose-containing complexes from the brown alga *Ascophyllum nodosum*. *Carbohydr. Res.* **1977**, *59*, 539–546. [CrossRef]
81. Medcalf, D.G.; Schneider, T.L.; Barnett, R.W. Structural features of a novel glucuronogalactofucan from *Ascophyllum nodosum*. *Carbohydr. Res.* **1978**, *66*, 167–171. [CrossRef]
82. Marais, M.-F.; Joseleau, J.-P. A fucoidan fraction from *Ascophyllum nodosum*. *Carbohydr. Res.* **2001**, *336*, 155–159. [CrossRef]
83. Daniel, R.; Chevolot, L.; Carrascal, M.; Tissot, B.; Mourão, P.A.S.; Abian, J. Electrospray ionization mass spectrometry of oligosaccharides derived from fucoidan of *Ascophyllum nodosum*. *Carbohydr. Res.* **2007**, *342*, 826–834. [CrossRef]
84. Nakayasu, S.; Soegima, R.; Yamaguchi, K.; Oda, T. Biological activities of fucose-containing polysaccharide ascophyllan isolated from the brown alga *Ascophyllum nodosum*. *Biosci. Biotech. Biochem.* **2009**, *73*, 961–964. [CrossRef]
85. Jiang, Z.; Okimura, T.; Yokose, T.; Yamasaki, Y.; Yamaguchi, K.; Oda, T. Effects of sulfated fucan, ascophyllan, from the brown alga *Ascophyllum nodosum* on various cell lines: A comparative study on ascophyllan and fucoidan. *J. Biosci. Bioeng.* **2010**, *110*, 113–117. [CrossRef] [PubMed]
86. Yuan, Y.; Macquarrie, D. Microwave assisted extraction of sulfated polysaccharides (fucoidan) from *Ascophyllum nodosum* and its antioxidant activity. *Carbohydr. Polym.* **2015**, *129*, 101–107. [CrossRef] [PubMed]
87. Okolie, C.L.; Mason, B.; Mohan, A.; Pitts, N.; Udenigwe, C.C. The comparative influence of novel extraction technologies on in vitro prebiotic-inducing chemical properties of fucoidan extracts from *Ascophyllum nodosum*. *Food Hydrocolloids* **2019**, *90*, 462–471. [CrossRef]
88. Wang, L.; Wang, L.; Yan, C.; Ai, C.; Wen, C.; Guo, X.; Song, S. Two *Ascophyllum nodosum* fucoidans with different molecular weights inhibit inflammation via blocking of TLR/NF-κB signaling pathway discriminately. *Foods* **2022**, *11*, 2381. [CrossRef]

89. Liang, Y.; Zha, S.; Tentaku, M.; Okimura, T.; Jiang, Z.; Ueno, M.; Hirasaka, K.; Yamaguchi, K.; Oda, T. Supressive effect of sulfated polysaccharide ascophyllan isolated from *Ascophyllum nodosum* on the production of NO and ROS in LPS-stimulated RAW264.7 cells. *Biosci. Biotech. Biochem.* **2021**, *85*, 882–889. [CrossRef]
90. Jin, W.; Lu, C.; Zhu, Y.; Zhao, J.; Zhang, W.; Wang, L.; Linchardt, R.J.; Wang, C.; Zhang, F. Fucoidans inhibited tau interaction and cellular uptake. *Carbohydr. Polym.* **2022**, *299*, 120176. [CrossRef]
91. Sakai, T.; Kimura, H.; Kojima, K.; Shimanaka, K.; Irai, K.; Kato, I. Marine bacterial sulfated fucoglucuronomannan (SFGM) lyase digests brown algal SFGM into trisaccharides. *Mar. Biotechnol.* **2003**, *5*, 70–78. [CrossRef] [PubMed]
92. Sanjeewa, K.K.A.; Herath, K.H.I.N.M.; Yang, H.-W.; Choi, C.S.; Jeon, Y.-J. Anti-inflammatory mechanisms of fucoidans to treat inflammatory diseases: A review. *Mar. Drugs* **2021**, *19*, 678. [CrossRef] [PubMed]
93. Ho, C.-H.; Chen, M.-L.; Huang, H.-L.; Lai, C.-J.; Liu, C.-H.; Chuu, C.-P.; Lin, Y.-H. Active targeting of P-selectin by fucoidan modulates the molecular profiling of metastasis in docetaxel-resistant prostate cancer. *Mar. Drugs* **2022**, *20*, 542. [CrossRef] [PubMed]
94. Harenberg, J.; Walenga, J.; Torri, G.; Dahl, O.E.; Drouet, L.; Fareed, J. Update of the recommendations on biosimilar low-molecular-weight heparins from the Scientific Subcommittee on Control of Anticoagulation of the International Society on Thrombosis and Hemostasis. *J. Thromb. Haemost.* **2013**, *11*, 1421–1425. [CrossRef]
95. Restaino, O.F.; Schiraldi, C. Chondroitin sulfate: Are the purity and the structural features well assessed? A review on the analytical challenges. *Carbohydr. Polym.* **2022**, *292*, 119690. [CrossRef]
96. Ustuzhanina, N.E.; Krylov, V.B.; Grachev, A.A.; Gerbst, A.G.; Nifantiev, N.E. Synthesis, NMR and conformational studies of fucoidan fragments. 8. Convergent block-wise synthesis of long chain linear and 2,3-branched oligosaccharides. *Synthesis* **2006**, *23*, 4017–4031. [CrossRef]
97. Krylov, V.B.; Kaskova, Z.M.; Vinnitskiy, D.Z.; Ustyuzhanina, N.E.; Grachev, A.A.; Chizhov, A.O.; Nifantiev, N.E. Acid-promoted synthesis of per-O-sulfated fucooligosaccharides related to fucoidan fragments. *Carbohydr. Res.* **2011**, *346*, 540–550. [CrossRef]
98. Vinnitskiy, D.Z.; Krylov, V.B.; Ustyuzhanina, N.E.; Dmitrenok, A.S.; Nifantiev, N.E. The synthesis of heterosaccharides related to the fucoidan from *Chordaria flagelliformis* bearing α-L-fucofuranosyl unit. *Org. Biomol. Chem.* **2016**, *14*, 598–611. [CrossRef]
99. Silchenko, A.S.; Ustyuzhanina, N.E.; Kusaykin, M.I.; Krylov, V.B.; Shashkov, A.S.; Dmitrenok, A.S.; Usoltseva (Menshova), R.V.; Zueva, A.O.; Nifantiev, N.E.; Zvyagintseva, T.N. Expression and biochemical characterization and substrate specificity of the fucoidanase from *Formosa algae*. *Glycobiology* **2017**, *27*, 254–263. [CrossRef]
100. Shvetsova, S.V.; Shabalin, K.A.; Bobrov, K.S.; Ivanen, D.R.; Ustyuzhanina, N.E.; Krylov, V.B.; Nifantiev, N.E.; Naryzhny, S.N.; Zgoda, V.G.; Eneyskaya, E.V.; et al. Characterization of a new α-L-fucosidase isolated from Fusarium proliferatum LE1 that is regioselective to α-(1/4)-L-fucosidic linkage in the hydrolysis of α-L-fucobiosides. *Biochimie* **2017**, *132*, 54–65. [CrossRef]
101. Silchenko, A.S.; Rubtsov, N.K.; Zueva, A.O.; Kusaykin, M.I.; Rasin, A.B.; Ermakova, S.P. Fucoidan-active α-L-fucosidases of the GH29 and GH95 families from a fucoidan degrading cluster of the marine bacterium *Wenyingzhuangia fucanilytica*. *Arch. Biochem. Biophys.* **2022**, *728*, 109373. [CrossRef] [PubMed]

marine drugs

Article

Physicochemical Properties and Antiherpetic Activity of κ-Carrageenan Complex with Chitosan

Viktoriya N. Davydova [1,*], Natalya V. Krylova [2], Olga V. Iunikhina [2], Aleksandra V. Volod'ko [1], Evgeniya A. Pimenova [3], Mikhail Y. Shchelkanov [2] and Irina M. Yermak [1]

1 G.B. Elyakov Pacific Institute of Bioorganic Chemistry, Far-Eastern Branch of the Russian Academy of Sciences, 690022 Vladivostok, Russia; morskaia@list.ru (A.V.V.); imyer@mail.ru (I.M.Y.)
2 G.P. Somov Institute of Epidemiology and Microbiology, Rospotrebnadzor, 690087 Vladivostok, Russia; krylovanatalya@gmail.com (N.V.K.)
3 A.V. Zhirmunsky National Scientific Center of Marine Biology, Far Eastern Branch, Russian Academy of Sciences, Palchevskogo 17, 690041 Vladivostok, Russia
* Correspondence: vikdavidova@yandex.ru

Abstract: Nanoparticles formation is one of the ways to modulate the physicochemical properties and enhance the activity of original polysaccharides. For this purpose, based on the polysaccharide of red algae, κ-carrageenan (κ-CRG), it polyelectrolyte complex (PEC), with chitosan, were obtained. The complex formation was confirmed by ultracentrifugation in a Percoll gradient, with dynamic light scattering. According to electron microscopy and DLS, PEC is dense spherical particles with sizes in the range of 150–250 nm. A decrease in the polydispersity of the initial CRG was detected after the PEC formation. Simultaneous exposure of Vero cells with the studied compounds and herpes simplex virus type 1 (HSV-1) showed that the PEC exhibited significant antiviral activity, effectively inhibiting the early stages of virus–cell interaction. A two-fold increase in the antiherpetic activity (selective index) of PEC compared to κ-CRG was shown, which may be due to a change in the physicochemical characteristics of κ-CRG in PEC.

Keywords: carrageenan; chitosan; polyelectrolyte complex; antiherpetic activity; selective index

Citation: Davydova, V.N.; Krylova, N.V.; Iunikhina, O.V.; Volod'ko, A.V.; Pimenova, E.A.; Shchelkanov, M.Y.; Yermak, I.M. Physicochemical Properties and Antiherpetic Activity of κ-Carrageenan Complex with Chitosan. *Mar. Drugs* **2023**, *21*, 238. https://doi.org/10.3390/md21040238

Academic Editor: Hitoshi Sashiwa

Received: 21 March 2023
Revised: 6 April 2023
Accepted: 12 April 2023
Published: 13 April 2023

1. Introduction

Polysaccharides of marine origin are macromolecules that have engaged researchers for many decades. Such attractiveness of these compounds is not accidental, since they combine biocompatibility and low toxicity with unique biological properties and low cost. In recent years, studies related to the development drugs for pharmaceuticals and biomedicine based on marine biopolymers have been carried out most intensively [1–3]. This is due to the activities found for them: antitumor [4], antibacterial [5], antiviral [6], anticoagulant [7] and a number of others [8].

Some of the well known polysaccharides of marine origin are the polysaccharides of red algae—carrageenans (CRGs). They are a family of sulfated copolymers of 3-linked β-D-galactopyranose and 4-linked α-D-galactopyranose (D-units) or 4-linked 3,6-anhydro-α-D-galactopyranose [9]. Depending on the number and position of sulfate groups, as well as the presence or absence of 3,6-anhydrogalactose in the molecule, CRGs are classified into various structural types, denoted by Greek letters. The most famous of which are κ-, λ- and ι-CRGs [10]. Carrageenans are the most studied among algae polysaccharides in terms of toxicity, pyrogenicity, and allergenicity [11]. Their safety of use for food and medical purposes has been confirmed by various studies of different scientific groups in experiments in vivo [12–14].

It is known that CRGs, mimicking heparan sulfate, are potential antivirals that can interfere with the early stages of viral replication, including virus entry, by masking the positive charge of the virus surface receptors to prevent them from binding to the heparan sulfate proteoglycans in the host cell surface [15]. CRGs can be included in pharmaceutical compositions for the prevention or treatment of viral infections with no side effects [16]. Earlier, we tested the effect of CRGs, isolated from a different family of red algae of the Pacific coast, at the stage of viral infection, and we showed the dependence of antiviral activity on the structural features of these polysaccharides [17]. The results of pre-treatment of herpes simplex virus type 1 (HSV-1) with CRGs and pre-treatment of cells with CRGs revealed that these polysaccharides affect the earliest stages of the viral life cycle, which are the attachment and the penetration ones. CRGs, containing 3,6-anhydrogalactose units, such as κ- and κ/β-CRGs, significantly reduced the attachment of HSV-1 to cells. However, they showed lower virucidal activity compared to other CRGs.

One of the ways to increase the bioavailability of polymers can be the production of nanoparticles (NPs). They have a unique property of combining a high charge density per unit surface and small sizes [18]. The most promising approach to obtain NPs is the utilization of green chemistry methods—to use aqua solution and biomolecules without the application of toxic solvents and synthetic cross-linking agents.

CRGs find use in the production of nanoparticles and microparticles. The presence of sulfate groups allows the formation of polyelectrolyte complexes (PECs) with oppositely charged polymers, which has proven to be very useful in the development of new drugs of various forms [19]. The advantage of PEC-based NPs is not only their high stability at different pH values, but also their increased mechanical strength compared to that of NPs obtained by other methods, such as the ionic gelation method [20]. Chitosan (CH) is most often used to form such complexes. CH is a copolymer of D-glucosamine and N-acetylglucosamine linked by β-(1→4) bonds obtained from the alkaline deacetylation of chitin, the main component of crustacean shells [21]. It exhibits low toxicity, biodegradability, good cost performance, and strong mucoadhesive properties, and it is a useful material for PEC preparation. Rodrigues et al. [22] produced complexes by mixing CH and κ-CRG, at room temperature, and they showed that the physical state of PEC formed depended on experimental conditions. Depending on the ratios, the PEC was presented as a solution or precipitate. If the CH charges are neutralized, the electrostatic repulsions are reduced or eliminated, and, in this case, the precipitation can take place. Another possibility is to use polyelectrolytes at nonstoichiometric charge ratios, as a result of which the PECs formed will remain in solution in the form of NPs [19,22].

Previously, we showed that the interaction of CRGs with the CH leads to the formation of soluble polyelectrolyte complexes (PECs) of the nanometer range [23,24]. It is known that obtaining NPs can be a way to increase the activity of the initial components. In this regard, as well as the high antiherpetic activity of κ-CRG, it was of interest to evaluate the antiviral effect of the nanosized form of PEC κ-CRG in comparison with the effect of the initial polysaccharide.

2. Results and Discussion

2.1. Characterization of the Original Polysaccharides

Chitosan was isolated from chitin of king crab shell by alkaline deacetylation. The structure of the obtained polysaccharide was confirmed by H^1 NMR spectroscopy (Figure S1). The degree of N-deacetylation (DD) of polysaccharide was calculated based on the values of the integral intensity, which was calculated by using integrals of the peak of proton H1 of deacetylated monomer (H1) at 5.0 md and of the peak of the three protons of acetyl group (H-Ac) at 2.19 md, and it was equal to 93% [25].

Carrageenan was isolated from the red alga *Chondrus armatus* by water extraction followed by fractionation with 4% KCl solution. The structure of KCl-insoluble fraction was studied by ^{13}C-NMR and IR-Fourier spectroscopy, and the obtained spectra have been compared with the spectra of polysaccharides isolated by us earlier from this specie of algae [26].

In the FTIR spectra (Figure S2), the KCl-insoluble fraction of polysaccharides from *C. armatus* showed a poorly resolved absorption band with a maximum at about 1263 and

1212 cm^{-1}, indicating the presence of sulfate groups (asymmetric stretching vibrations, including two and three S=O bonds of the SO_3 group). This spectrum showed absorption bands at 934 cm^{-1} for 3,6-anhydrogalactose (C(3′)–O–C(6′) stretching vibration) and the stretching vibration S–O bonds of 847 cm^{-1} of the axial sulfate group at C-4 of the 3-linked β-D-galactose. The two signals at 103.1 ppm and 95.9 ppm in the anomeric carbon resonance area of the ^{13}C-NMR spectra (Figure S3) of KCl-insoluble polysaccharide from *C. armatus* were assigned to C-1 of the 3-linked β-D-galactose residue (G4S) and C-1 of the 4-linked 3,6-anhydro-α-D-galactose (DA) of κ-CRG. Therefore, spectroscopy data suggest that gelling polysaccharide from *C. armatus* was represented by κ-CRG. The sulfate content was 22%.

As shown earlier [12,27,28], the biological activity of polysaccharides may change significantly due to the presence of low molecular weight fractions in the sample. In this regard, the CH and κ-CRG were passed through ultrafiltration membranes to remove low molecular weight polysaccharides. Molecular weights of the polysaccharides were obtained by the viscosimetric method and were 520 ± 30 and 560 ± 45 kDa, respectively, for CH and κ-CRG.

2.2. *Preparation and Characterization of κ-CRG-CH PEC*

A number of methods are used to obtain NPs. These are ionotropic gelation, microemulsion, diffusion of emulsifying solvent, formation of PEC, and the reverse micellar method [29,30]. Of these, the method of PEC formation seems to be the most attractive. It is quite simple, not expensive, and does not require the use of organic solvents.

Soluble PECs in NPs form can be obtained in non-stoichiometric charge ratios [31]. This can be achieved, either with a significant difference in the molecular weights of the polymers, or with an excess amount of one of the components. Based on previous studies of the process of PEC formation [24], a 1:10 weight ratio (1:5 charge ratio) of CH-CRG was chosen to obtain the soluble form of PEC. The resulting complexes did not precipitate after centrifugation at 13,000 g/10 min.

2.2.1. Ultracentrifugation in a Percoll Gradient

Earlier, we applied FTIR spectroscopy to prove the fact the formation of a complex between κ-CRGand CH, based on the decomposition of the spectrum into its individual components. A detailed analysis of the IR spectra is described in our previous work [32].

The use of gradient centrifugation makes it possible to separate macromolecules that differ little in size or density. This method is one of the few applications for direct registration of complex formation and has been successfully used by a number of researchers [33,34]. Here, the ultracentrifugation in a Percoll gradient was used to confirm a complex formation. The distribution curves of the initial CH and κ-CRG, as well as their mixture, are shown in Figure 1. According to the presented results, the profiles of the initial polysaccharides (Figure 1a) differ from their profiles in the mixture (Figure 1b). The coincidence of the maxima of the distribution profiles of CH and κ-CRG in the mixture proves the formation of PEC.

2.2.2. Dynamic Light Scattering (DLS)

DLS is a very powerful tool for studying the diffusion behavior of macromolecules in solution. The diffusion coefficient, and the hydrodynamic radii calculated from it, depend on the size and shape of the macromolecules. Modern data analysis, using various approaches, makes it possible to estimate the size and homogeneity of macromolecules. DLS is employed to study different types of interaction in solutions [35]. DLS data confirmed the existence of PEC of CH-CRG in these weight ratios. In the case of the mixtures of κ-CRG and CH (10:1 *w/w*), a significant decrease in polydispersity in comparison with initial polysaccharides and monomodal distribution of the particles were registered (Figure 2). The complex represents particles with an average hydrodynamic radius of 138 nm. A partial neutralization of κ-CRG surface charge from −61.9 mV to −38.4 mV also indicates the PEC formation. A sufficiently high negative charge of PEC indicates the electrostatic stability of the obtained particles, and this also suggests the surface localization of sulfate groups.

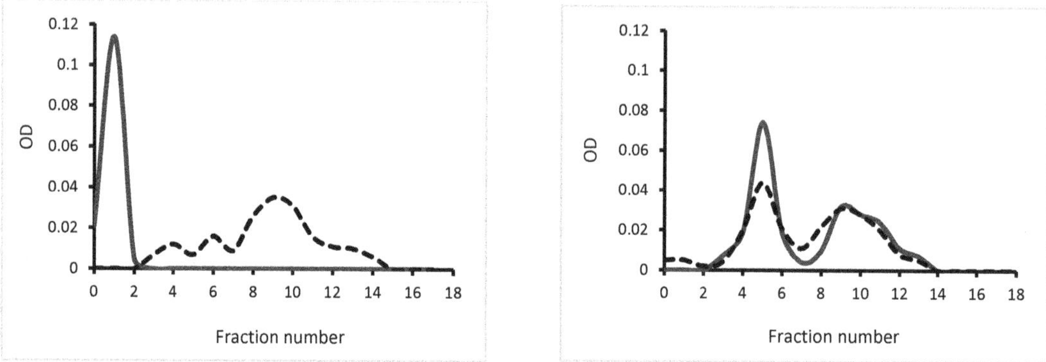

(**a**) (**b**)

Figure 1. Ultracentrifugation in a Percoll gradient: (**a**) initial CH and κ-CRG and (**b**) their mixture in a ratio CH:CRG of 1:10 *w/w*. Contents of NH_2-groups in CH—gray solid line; contents of SO_4^{2-}—groups in CRG—black dash line.

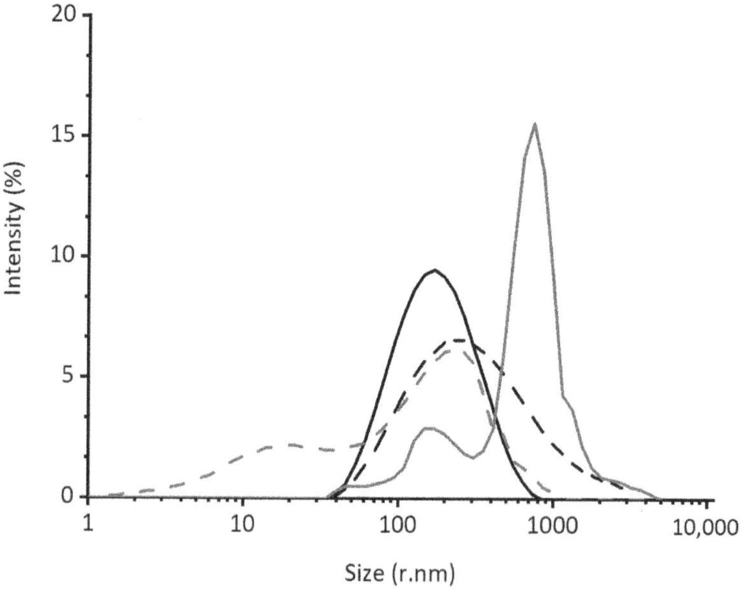

Sample	PdI	Hydrodynamic radius (Z-average), nm	ζ-potential, mV
κ-CRG	0.842±0.029	801.5±45.8	−61.9±2.5
CH	0.710±0.040	359.7±38.7	+54.9±0.8
PEC	0.294±0.007	138.3±6.9	−38.4±1.8
PEC-lyoph	0.416±0.006	232.4±18.8	−35.4±1.3

Figure 2. Distribution by intensity of CH-CRG 1:10 *w/w* particles: solid black line—freshly prepared particles; dot black line—re-dissolved particles; solid grey line—κ-CRG; dot grey line—CH.

According to European standards, NPs are materials with a specific surface area above 60 m^2/cm^3 [36]. With a material density of 1, these are particles with a diameter of 100 nm. However, world practice does not adhere to such strict standards, and diameters up to 500 nm are allowed in the pharmaceutical literature [37]. Thus, the obtained PEC of κ-CRG-CH can be considered as an NP.

One of the widely used ways to store and transport NPs is lyophilization. However, they do not always retain their properties after dissolution. To control the parameters of the re-dissolved particles, freshly obtained PECs were lyophilized and, after storage for six months, they were re-dissolved, and their characteristics were determined by the DLS. As can be seen from Figure 2, the re-dissolved PEC particles have parameters quite close to those of a freshly prepared sample. Although, their polydispersity and Z-average increase slightly.

2.2.3. Electron Microscopy

Scanning electron microscopy (SEM). Imaging techniques using microscopy make it possible to observe various surface characteristics of polymers, such as surface texture, shape, and thickness. Among other things, researchers prefer scanning electron microscopy (SEM) and transmission electron microscopy as reliable methods for studying the surface morphology of polymers. A comparative study of the morphology of the initial polysaccharides and their complexes can explain the difference in the macromolecular structure of the studied compounds and, as a result, their biological activity.

Microstructure and surface morphology of κ-CRG and PEC were characterized by SEM. κ-CRG (Figure 3a,b) is represented by extended, curved worm-like structures of more than 1 μm, the surface of which is rough and folded. PEC (Figure 3c,d), in contrast to κ-CRG, represents compact corpuscular structures of spherical or oblong shapes of about 200–300 nm. At higher magnification (100 K), surface roughness, and wrinkling, as in the case of κ-CRG, can be seen.

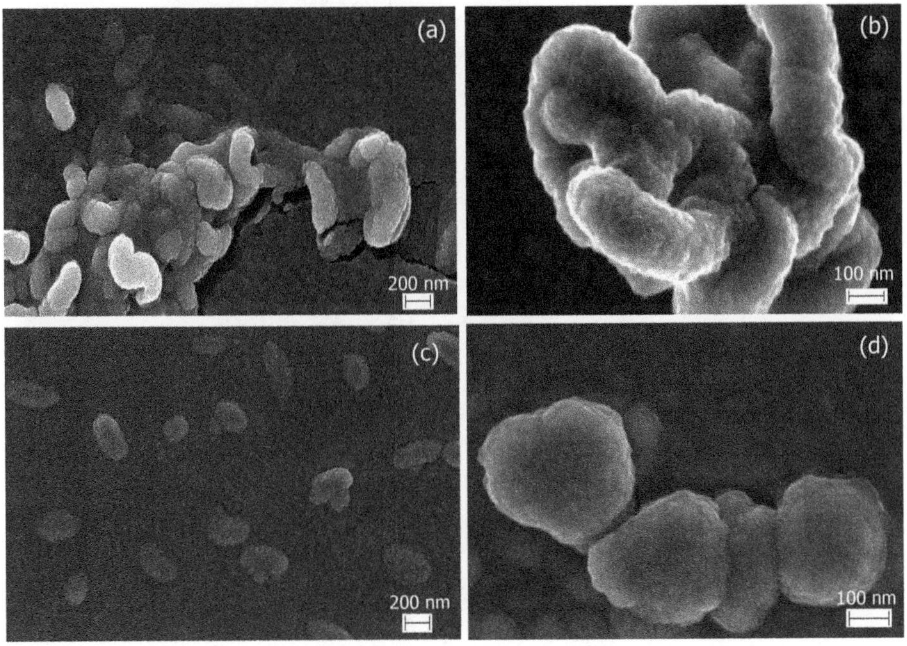

Figure 3. Scanning electron microscopy images of κ-CRG (**a**,**b**), C = 0.1 mg × mL^{-1}, and PEC κ-CRG-CH 10:1 *w*/*w* (**c**,**d**).

Figure 3 clearly shows the changes in κ-CRG morphology after the PEC formation: polydisperse aggregated structures are transformed into more compact formations with smaller sizes. The SEM data are in good agreement with the DLS results. The size of the structures, on SEM images, is close to the particles size determined by DLS.

Transmission electron microscopy (TEM). TEM can provide more details at the atomic scale, such as crystal structure, which is more powerful and competitive than SEM.

The TEM images of κ-CRG presented in Figure 4a show the fibrous entangled structures typical of this type of polysaccharides [38,39].

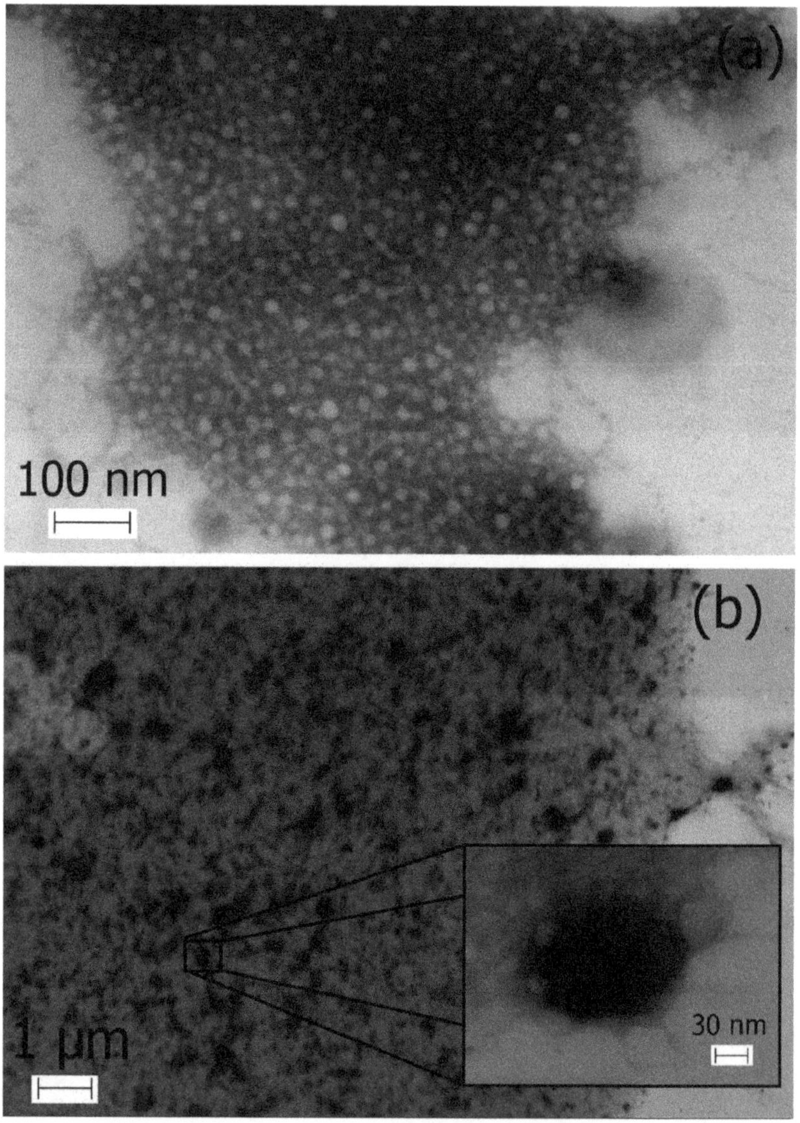

Figure 4. Transmission electron microscopy images of κ-CRG (**a**), C = 0.5 mg × mL^{-1}, and PEC κ-CRG-CH 10:1 *w*/*w* (**b**).

TEM of the PEC (Figure 4b) revealed the presence of a large number of spherical particles of different sizes. The features of the preparation of preparations for TEM suggest the aggregation of the PEC particles during its drying on a substrate. More detailed analysis at higher magnification of single particles evidences a solid and compact structure, with an average diameter of about 150–200 nm. The PEC particle size, according to TEM data, is slightly smaller than that determined by DLS. The results were in agreement with other reports [40,41], showing that the mean size of NPs, measured by TEM, is subtle and less than that determined by DLS. This slight difference is widely considered to be influenced by the difference methods between TEM and DLS.

The centrifugation in a Percoll gradient (Figure 1) and DLS (Figure 2) indicate that κ-CRG forms stable complex with CH. The complex formation significantly reduces the polydispersity of the initial polysaccharides. According to the data of DLS and electron microscopy, the formed complex has nanometer sizes. The complex remains stable after lyophilization, followed by re-dissolution. The retention of the negative surface charge of the particles and the excess of κ-CRG in the PEC composition allows us to assume the surface localization of sulfate groups in the obtained NPs, which are known to provide antiviral activity of natural polysaccharides. Therefore, it can be supposed that the antiviral properties of CRG, which we previously established in relation to the herpes simplex virus [17], will also be characteristic of its PEC with CH.

2.3. Antiherpetic Activity of CH, CRG and Their PEC

Chitosan-based NPs and their biological properties are widely described [18,20,31,41,42]. NPs are used as a way to increase the antibacterial activity of the initial polysaccharide. This may be due to the small size and positive surface charge of NPs, which may improve their stability in the presence of biological cations and increase their antibacterial activity through interaction with negatively charged biological membranes and site-specific targeting in vivo [43].

CRG-based NPs are less studied. More often, this polysaccharide is used in the gel form as a matrix for incorporating different types of NPs for drug delivery or tissue engineering [44], as well as for coating for stabilization of already formed protein [45] and metal-containing NPs [46]. In this regard, it seemed interesting to us to study the antiviral properties of κ-CRG-based NPs.

The study of antiviral activity was carried out on the Vero cell line against the herpes simplex virus type 1 (HSV-1), strain L2.

The evaluation of the cytotoxicity of the polysaccharides and their PEC showed that the studied compounds had low cytotoxic activity against Vero cells. Their 50% cytotoxic concentrations (CC_{50}) exceeded 2000 μg/mL (Table 1). For further study of antiviral activity, concentrations of compounds below 500 μg/mL were used.

Table 1. Antiherpetic activity of CH, CRG and their PEC 1:10 *w/w*.

Sample	CC_{50}, μg/mL	IC_{50}, μg/mL	SI
κ-CRG	>2000	56 ± 8 *	36 ± 5 *
CH	>2000	374 ± 41 *	5.3 ± 0.6 *
PEC	>2000	28 ± 4	70 ± 9

* Significance of the differences between the parameters of PEC compared to other compounds (CH, κ-CRG) ($p \leq 0.05$).

The antiviral effect of the studied compounds was assessed by the inhibition rate (IR, %) of the cytopathogenic effect (CPE) of the HSV-1 in Vero cells using the MTT test (Figure 5). Simultaneous treatment of Vero cells with the studied compounds and HSV-1 virus showed that CH at a concentration of 250 μg/mL inhibits the cytopathogenic effect of the virus by 35 ± 4%. At the same time, κ-CRG and PEC at this concentration almost completely suppressed the CPE of the virus: their inhibition rates did not differ significantly ($p > 0.05$) and amounted to 86 ± 9% and 96 ± 10%, respectively. However, when significantly lower concentrations of compounds (50 μg/mL) were used, PEC suppressed the CPE of the virus more effectively than κ-CRG: 89 ± 9% and 66 ± 7%, respectively ($p \leq 0.05$) (Figure 5). At low concentrations of the tested compounds (10 µg/mL), the inhibition of viral replication using PEC was higher than that of κ-CRG—43 ± 5% and 30 ± 3%, respectively ($p \leq 0.05$). The CH at the indicated concentrations was ineffective (Figure 5).

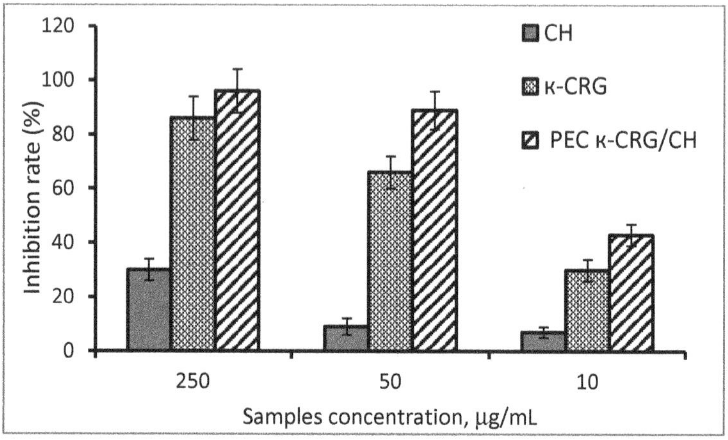

Figure 5. Inhibitory activity of CH, κ-CRG, and their PEC during simultaneous treatment of Vero cells with HSV-1 virus and compounds.

The low activity of CH is probably associated with a somewhat different mechanism of its antiviral activity. Although, despite these extensive published reports, the exact antiviral mechanism of chitosan and its derivatives is not fully understood [47], it is known that the CH ability to inactivate the virus by binding to its membrane was observed only at pH above 6.0 and only for certain viruses [48]. Inhibiting viral adsorption and subsequent host cell invasion was noted only for sulfated chitosan derivatives [49]. Most researchers tend to believe that the antiviral activity of CH is associated with its ability to activate immune responses. Thus, intranasal administration of CH caused activation of mucosal immune responses due to a significant increase in leukocyte infiltration and an increase in the level of pro-inflammatory cytokines in tissues [50], and it also contributed to the protection of mice from lethal doses of the influenza virus [51].

The 50% inhibitory concentrations (IC_{50}) and selective indexes (SI) of the test compounds calculated by regression analysis are presented in Table 1. According to the presented data, the IC_{50} of the PEC was two times lower, and SI were two times higher than those of κ-CRG. At the same time, PEC was 13 times more effective than CH.

So, simultaneous treatment of Vero cells with the studied compounds and HSV-1 showed that the κ-CRG/CH PEC exhibits significant antiviral activity, effectively inhibiting the early stages of virus–cell interaction. Since the ratio of the initial components of κ-CRG and CH in the complex is 10:1 *w*/*w*, it is likely that the inhibitory activity of the complex is largely due to the action of κ-CRG. In turn, an increase in the selective index of PEC by two times was observed, compared to κ-CRG, and it was 13 times greated compared to CH, which indicates a greater safety and efficacy of PEC compared to its constituent components.

3. Materials and Methods

3.1. Polysacharides

A CH was obtained by alkaline treatment of crab chitin according to the published protocol [52], followed by deacetylation with the mixture of 40% aqueous solution of NaOH with isopropyl alcohol (1:16 *v*:*v*) under heating for 7 h at 100 °C. The pellet was filtered and dissolved in water acidified with hydrochloric acid (pH 3.5), dialyzed against water, and lyophilized. The DD of the CH sample was calculated according to ^{1}H NMR spectroscopic data, according to the formula [25].

$$DD = \frac{I(H1)}{\left(I(H1) + \frac{I(HAc)}{3}\right)} \times 100\% \quad (1)$$

where *I*(*H*1) and *I*(*HAc*) represent integral intensity of the signal of *H*1 at 5.0 md and signal of protons of the acetate group at 2.19 md, respectively. The κ-CRG was isolated with hot water extraction from the red algae *Chondrus armatus* (Gigartinaceae), according to an earlier published protocol [26]. Briefly, the algae were collected at Peter the Great Bay (Sea of Japan), and they were washed with tap water in order to remove an excess of salt. Bleaching of the seaweed was achieved by maintaining the specimen in pure acetone for three days prior to being dried in the air. Dried and milled algae (50 g) were suspended in hot water (1.5 L), and the polysaccharides were extracted at 90 °C for 2 h in a water bath. The polysaccharides were fractionated into gelling KCl-insoluble and non-gelling KCl-soluble fractions. The structure of polysaccharides was established on the basis of NMR and FTIR spectroscopy data according to a previously published protocol [53]. The sulfate content was determined, as described previously [54].

3.2. Molecular Weight Determination

The molecular masses of polysaccharides were calculated by the Mark-Houwink-Kuhn-Sakurada equation: $[\eta] = K \times M^{\alpha}$, where [η] is the intrinsic viscosity, and K and α are empirical constants for κ-CRG [55] and CH [56]. For this, the viscosity of κ-CRG and CH solutions was measured (0.1–1.0 mg/mL in 0.1 M NaCl and 2.0–10.0 mg/mL in 0.2 M NaCl/0.2 M AcOH, respectively) on a modified Ubbelohde viscometer (OKB Pushchino, Pushchino, Russia) with a capillary diameter of 0.3 mm at 25 °C, and the timing accuracy is ±0.1 s. For each concentration, five measurements were made. The intrinsic viscosity of the samples was calculated by extrapolating the $\ln(\eta) \times C^{-1}$ dependence to infinite dilution using the least squares method, and viscosimetric molecular weights were calculated.

3.3. Complexes CRG:CH

The initial polysaccharides were dissolved in deionized water with stirring for 3 h and filtered through 0.45 μm. The CH-chloride solution (0.1 mg/mL) was added to equal volume of the κ-CRG solution (1 mg/mL) and stirred for 30 min at 25 °C and lyophilized.

3.4. Centrifugation in a Percoll Gradient

Percoll (Sigma, St. Louis, MO, USA, 26 mL, 30%) in NaCl solution (0.15 M) was placed into a 28 mL centrifuge tube. A sample of κ-CRG, CH, or their mixture (2 mL) was layered on the Percoll and centrifuged in an angled rotor Heraeus Biofuge Stratos (Hanau, Germany, Germany) at 20,000× *g* for 60 min. After centrifuging, the tube contents were removed through the top using a peristaltic pump. Fractions (1.5 mL) were collected. The density of the Percoll solution in each fraction was calculated using the refractive index determined on a RF-4 refractometer (LOMO, St. Peterburg, Russia). The presence of CH and κ-CRG in the fractions was determined by the reaction of polysaccharide amino group with 2,4,6-trinitrobenzenesulfonic acid [57] and by the reaction of sulfate group with Taylor's blue (1,9-dimethylmethylene blue) [58], respectively. The optical density at 410 nm 535 nm, respectively, for CH and κ-CRG, was measured with a μ-Quant spectrophotometer (Bio-Tek Instruments Inc., Winuskey, VT, USA).

3.5. Dynamic Light Scattering (DLS) and Electrophoretic Properties of the CRG:CH Complexes

The hydrodynamic radius and ζ-potentials of the initial polysaccharides and their PEC in solution were determined using a ZetaSizer NanoZS system (Malvern, UK) operating at 633 nm. The measurements were performed at 25 °C. The parameters were automatically calculated with the instrument's software, based on analysis of the autocorrelation function.

3.6. Microscopy Study

3.6.1. Transmission Electron Microscopy

TEM images of samples were examined using a LIBRA-120 electron microscope (Karl Zeiss Group, Jena, Germany). The solutions of κ-CRG (0.5 mg/mL) and PEC (0.5 mg/mL κ-CRG and 0.05 mg/mL CH) in deionized water were obtained. The drops of the solutions were placed on grids coated with formvar film stabilized with carbon, desiccated using a piece of filter paper, and negatively stained for 3 min with 2% uranyl acetate. Then, the samples were carefully desiccated and examined.

3.6.2. Scanning Electron Microscopy

For the scanning electron microscopy studies, solution samples in deionized water were placed onto the surface of Thermanox® plastic coverslips ("ThermoFisher Scientific", Waltham, MA, USA), fixed with 2.5% glutaraldehyde, incubated for 1 h for sedimentation, and dehydrated in alcohols of increasing concentrations and acetone. Thereafter, the samples were completely dried in carbon dioxide, according to the critical-point drying method, using a BALTEC 030 (Amstetten, Germany). The samples were then placed on the surfaces of aluminium substrates and coated with chromium. The samples were analysed using a Zeiss Sigma 300 VP scanning electron microscope (Carl Zeiss Ltd., Cambridge, UK).

3.7. Antiherpetic Activity

3.7.1. Virus and Cell Culture

In this study, HSV-1 strain L2 was obtained from N.F. Gamaleya Federal Research Centre for Epidemiology and Microbiology, Moscow, Russia. Determination of the cytotoxicity and antiviral activity of the compounds was carried out on a Vero cell culture (kidney epithelial cells of the African green monkey *Chlorocebus* sp.), obtained from N.F. Gamaleya Federal Research Centre for Epidemiology and Microbiology, Moscow, Russia. HSV-1 was grown in Vero cells using Dulbecco's Modified Eagle's Medium (DMEM, Biolot, St. Petersburg, Russia), which was supplemented with 10% fetal bovine serum (FBS) (Biolot, St. Petersburg, Russia) and 100 U/mL of gentamycin (Dalkhimpharm, Khabarovsk, Russia) at 37 °C in a CO_2 incubator. In the maintenance medium, the FBS concentration decreased to 1%. The cell concentration in all experiments was 10^4 cells/mL.

3.7.2. Cytotoxicity of the Tested Compounds

The cytotoxicity evaluation of the studied samples was performed using the MTT assay, as previously described [17]. In brief, confluent Vero cells (1×10^4 cells/well) in 96-well microplates were incubated with various concentrations of the tested compounds (1–2000 µg/mL) at 37 °C for 72 h (5% CO_2); untreated cells were used as controls. MTT solution (methylthiazolyltetrazolium bromide, Sigma, St. Louis, MO, USA) was added to cells at a concentration of 5 mg/mL, following incubation for 2 h at 37 °C. Then, the MTT solution was removed, and isopropanol was added to dissolve the insoluble formazan crystals. The optical density was read at 540 nm (Labsystems Multiskan RC, Vantaa, Finland). Cytotoxicity was expressed as 50% cytotoxic concentration (CC_{50}) of the tested compound, which reduced the viability of treated cells by 50%, which was compared to control cells [59]. Experiments were performed in triplicate and repeated three times.

3.7.3. Anti-HSV-1 Activity of the CH, κ-CRG and Their PEC

The inhibitory effects of the CH, κ-CRG, and their PEC on HSV-1-infection in Vero cells were evaluated by the cytopathic effect (CPE) inhibition assay. Test compound concentrations ranged from 1 to 250 µg/mL; the infectious dose of the HSV-1 virus was 100 TCID50/mL (50% tissue cytopathic infectious dose). Simultaneous treatment of the cells with the tested compounds and the virus was used. The virus and samples (1:1 *v*/*v*) were applied to Vero cell monolayers, which were grown in 96-well plates, simultaneously, and incubated for one hour at 37 °C. After virus absorption, the virus–compound mixture was removed; the cells were washed, and maintenance medium was added. The plates were incubated at 37 °C for 72 h (5% CO_2) until 90% CPE was observed in virus control compared to cell control. Antiviral activity of compounds was assessed by MTT assay, and the viral inhibition rate (IR, %) was calculated according to the formula [60]:

$$IR = (OD_{tv} - OD_{cv})/(OD_{cd} - OD_{cv}) \times 100, \quad (2)$$

where OD_{tv} represents the OD of cells infected with virus and treated with the tested compounds; OD_{cv} corresponds to the OD of the untreated virus-infected cells; and, OD_{cd} is the OD of the control (untreated and noninfected) cells. The 50% inhibitory concentration (IC_{50}) of each compound was determined as the compound concentration that reduced virus-mediated CPE by 50%, and it was calculated using a regression analysis of the dose–response curve [61]. The selectivity index (SI) was calculated as the ratio of CC_{50} to IC_{50} for each compound.

4. Conclusions

The problem of searching for compounds that can effectively act on a wide range of viral infections without harming the host organism still remains relevant. Natural sulfated polysaccharides, in this regard, seem to be very attractive. The study showed that the preparation of κ-CRG-based NPs, in the form of PEC with CH, makes it possible to increase the antiviral activity of the initial κ-CRG. This may be due to a change in the physicochemical characteristics of κ-CRG in PEC. The polysaccharide in the form of NPs has a low polydispersity, a significantly smaller size, a compact molecular shape, and, at the same time, retains a negative charge. This, in turn, can facilitate its interaction with both viral particles and the cell surface, and this thereby contributes to the manifestation of a higher antiherpetic activity.

Supplementary Materials: The following supporting information can be downloaded at: https://www.mdpi.com/article/10.3390/md21040238/s1, Figure S1: ^{1}H NMR spectrum of hydrochloric salt of CH; Figure S2: FTIR-spectrum of κ-CRG; Figure S3: ^{13}C NMR spectrum of κ-CRG.

Author Contributions: V.N.D.—conceptualization, methodology, validation, writing—original draft preparation, writing—review and editing, visualization; N.V.K.—methodology, validation, data curation; A.V.V.—resources, investigation, visualization; E.A.P.—investigation, visualization; O.V.I.—investigation; M.Y.S.—data curation; I.M.Y.—conceptualization, methodology, data curation, writing—review and editing, supervision, funding acquisition. All authors have read and agreed to the published version of the manuscript.

Funding: This research was funded by the Russian Science Foundation, grant number 21-74-20019.

Institutional Review Board Statement: Not applicable.

Data Availability Statement: The original data are available from the correspondent author on request.

Acknowledgments: The study was carried out on the equipment of the Collective Facilities Center, The Far Eastern Center for Structural Molecular Research (NMR/MS) PIBOC FEB RAS.

Conflicts of Interest: The authors declare no conflict of interest.

References

1. Aziz, T.; Ullah, A.; Ali, A.; Shabeer, M.; Shah, M.N.; Haq, F.; Iqbal, M.; Ullah, R.; Khan, F.U. Manufactures of bio-degradable and bio-based polymers for bio-materials in the pharmaceutical field. *J. Appl. Polym. Sci.* **2022**, *139*, e52624. [CrossRef]
2. Zhang, H.; Wu, X.; Quan, L.; Ao, Q. Characteristics of Marine Biomaterials and Their Applications in Biomedicine. *Mar. Drugs* **2022**, *20*, 372. [CrossRef]
3. Bhovi, V.K.; Melinmath, S.P.; Gowda, R. Biodegradable Polymers and their Applications: A Review. *Mini Rev. Med. Chem.* **2022**, *22*, 2081–2101. [CrossRef]
4. Jha, A.; Kumar, M.; Mishra, B. Marine Biopolymer-Based Anticancer Drug Delivery Systems. In *Marine Biomaterials*; Springer: Singapore, 2022; pp. 351–401. [CrossRef]
5. Das, A.; Ringu, T.; Ghosh, S.; Pramanik, N. A comprehensive review on recent advances in preparation, physicochemical characterization, and bioengineering applications of biopolymers. *Polym. Bull.* **2022**, 1–66. [CrossRef]

6. Balasubramaniam, B.; Prateek; Ranjan, S.; Saraf, M.; Kar, P.; Singh, S.P.; Thakur, V.K.; Singh, A.; Gupta, R.K. Antibacterial and Antiviral Functional Materials: Chemistry and Biological Activity toward Tackling COVID-19-like Pandemics. *ACS Pharmacol. Transl. Sci.* **2021**, *4*, 8–54. [CrossRef] [PubMed]
7. Chandika, P.; Tennakoon, P.; Kim, T.H.; Kim, S.C.; Je, J.Y.; Kim, J., II; Lee, B.; Ryu, B.M.; Kang, H.W.; Kim, H.W.; et al. Marine Biological Macromolecules and Chemically Modified Macromolecules; Potential Anticoagulants. *Mar. Drugs* **2022**, *20*, 654. [CrossRef]
8. Carvalho, D.N.; Reis, R.L.; Silva, T.H. Marine origin materials on biomaterials and advanced therapies to cartilage tissue engineering and regenerative medicine. *Biomater. Sci.* **2021**, *9*, 6718–6736. [CrossRef]
9. Rees, D.A. The carrageenan system of polysaccharides. Part I. the relation between the κ- and λ-components. *J. Chem. Soc.* **1963**, 1814–1821. [CrossRef]
10. Knutsen, S.H.; Myslabodski, D.E.; Larsen, B.; Usov, A.I. A Modified System of Nomenclature for Red Algal Galactans. *Bot. Mar.* **1994**, *37*, 163–170. [CrossRef]
11. Qureshi, D.; Nayak, S.K.; Maji, S.; Kim, D.; Banerjee, I.; Pal, K. Carrageenan: A Wonder Polymer from Marine Algae for Potential Drug Delivery Applications. *Curr. Pharm. Des.* **2019**, *25*, 1172–1186. [CrossRef]
12. McKim, J.M.; Willoughby, J.A.; Blakemore, W.R.; Weiner, M.L. Clarifying the confusion between poligeenan, degraded carrageenan, and carrageenan: A review of the chemistry, nomenclature, and in vivo toxicology by the oral route. *Crit. Rev. Food Sci. Nutr.* **2019**, *59*, 3054–3073. [CrossRef] [PubMed]
13. Lee, C. Carrageenans as Broad-Spectrum Microbicides: Current Status and Challenges. *Mar. Drugs* **2020**, *18*, 435. [CrossRef] [PubMed]
14. Weiner, M.L. Food additive carrageenan: Part II: A critical review of carrageenan in vivo safety studies. *Crit. Rev. Toxicol.* **2014**, *44*, 244–269. [CrossRef] [PubMed]
15. Álvarez-Viñas, M.; Souto, S.; Flórez-Fernández, N.; Torres, M.D.; Bandín, I.; Domínguez, H. Antiviral activity of carrageenans and processing implications. *Mar. Drugs* **2021**, *19*, 437. [CrossRef]
16. Park, J.H.; Kim, Y.H.; Kwon, T.K.; Park, J.H.; Woo, J.S. Antiviral composition comprising a carrageenan as an active ingredient. Korea Patent KR20160003964A, 12 January 2016.
17. Krylova, N.V.; Kravchenko, A.O.; Iunikhina, O.V.; Pott, A.B.; Likhatskaya, G.N.; Volod'ko, A.V.; Zaporozhets, T.S.; Shchelkanov, M.Y.; Yermak, I.M. Influence of the Structural Features of Carrageenans from Red Algae of the Far Eastern Seas on Their Antiviral Properties. *Mar. Drugs* **2022**, *20*, 60. [CrossRef]
18. Yue, Z.G.; Wei, W.; Lv, P.P.; Yue, H.; Wang, L.Y.; Su, Z.G.; Ma, G.H. Surface charge affects cellular uptake and intracellular trafficking of chitosan-based nanoparticles. *Biomacromolecules* **2011**, *12*, 2440–2446. [CrossRef]
19. Grenha, A.; Gomes, M.E.; Rodrigues, M.; Santo, V.E.; Mano, J.F.; Neves, N.M.; Reis, R.L. Development of new chitosan/carrageenan nanoparticles for drug delivery applications. *J. Biomed. Mater. Res. Part A* **2010**, *92*, 1265–1272. [CrossRef]
20. Valente, J.F.A.; Gaspar, V.M.; Antunes, B.P.; Countinho, P.; Correia, I.J. Microencapsulated chitosan-dextran sulfate nanoparticles for controled delivery of bioactive molecules and cells in bone regeneration. *Polymer* **2013**, *54*, 5–15. [CrossRef]
21. Kou, S.G.; Peters, L.M.; Mucalo, M.R. Chitosan: A review of sources and preparation methods. *Int. J. Biol. Macromol.* **2021**, *169*, 85–94. [CrossRef]
22. Rodrigues, S.; Da Costa, A.M.R.; Grenha, A. Chitosan/carrageenan nanoparticles: Effect of cross-linking with tripolyphosphate and charge ratios. *Carbohydr. Polym.* **2012**, *89*, 282–289. [CrossRef]
23. Davydova, V.N.; Volod'ko, A.V.; Mishchenko, N.P.; Yermak, I.M. Chitosan-Based Mucoadhesive Systems for the Inclusion of the Echinochrome Active Substance. *Appl. Biochem. Microbiol.* **2018**, *54*, 478–483. [CrossRef]
24. Volodko, A.V.; Davydova, V.N.; Barabanova, A.O.; Soloveva, T.F.; Ermak, I.M. Formation of soluble chitosan-carrageenan polyelectrolyte complexes. *Chem. Nat. Compd.* **2012**, *48*, 353–357. [CrossRef]
25. Lavertu, M.; Xia, Z.; Serreqi, A.N.; Berrada, M.; Rodrigues, A.; Wang, D.; Buschmann, M.D.; Gupta, A. A validated 1H NMR method for the determination of the degree of deacetylation of chitosan. *J. Pharm. Biomed. Anal.* **2003**, *32*, 1149–1158. [CrossRef] [PubMed]
26. Yermak, I.M.; Kim, Y.H.; Titlynov, E.A.; Isakov, V.V.; Solov'eva, T.F. Chemical structure and gel properties of carrageenans from algae belonging to the Gigartinaceae and Tichocarpaceae, collected from the Russian Pacific Coast. *J. Appl. Phycol.* **1999**, *11*, 41–48. [CrossRef]
27. Lyu, F.; Xu, X.; Zhang, L. Natural polysaccharides with different conformations: Extraction, structure and anti-tumor activity. *J. Mater. Chem. B* **2020**, *8*, 9652–9667. [CrossRef]
28. Kulikov, S.N.; Chirkov, S.N.; Il'ina, A.V.; Lopatin, S.A.; Varlamov, V.P. Effect of the molecular weight of chitosan on its antiviral activity in plants. *Prikl. Biokhimiia i Mikrobiol.* **2006**, *42*, 224–228. [CrossRef]
29. Tiyaboonchai, W. Chitosan Nanoparticles: A Promising System for Drug Delivery. *Naresuan Univ. J.* **2013**, *11*, 51–66.
30. Sailaja, K.A.; Amareshwar, P.; Chakravarty, P. Different techniques used for the preparation of nanoparticles using natural polymers and their application. *Int. J. Pharm. Pharm. Sci.* **2011**, *3*, 45–50.
31. Kulkarni, A.D.; Vanjari, Y.H.; Sancheti, K.H.; Patel, H.M.; Belgamwar, V.S.; Surana, S.J.; Pardeshi, C.V. Polyelectrolyte complexes: Mechanisms, critical experimental aspects, and applications. *Artif. Cells Nanomed. Biotechnol.* **2016**, *44*, 1615–1625. [CrossRef]
32. Volod'ko, A.V.; Davydova, V.N.; Glazunov, V.P.; Likhatskaya, G.N.; Yermak, I.M. Influence of structural features of carrageenan on the formation of polyelectrolyte complexes with chitosan. *Int. J. Biol. Macromol.* **2016**, *84*, 434–441. [CrossRef]

33. De Jonge, M.I.; Pehau-Arnaudet, G.; Fretz, M.M.; Romain, F.; Bottai, D.; Brodin, P.; Honoré, N.; Marchal, G.; Jiskoot, W.; England, P.; et al. ESAT-6 from Mycobacterium tuberculosis dissociates from its putative chaperone CFP-10 under acidic conditions and exhibits membrane-lysing activity. *J. Bacteriol.* **2007**, *189*, 6028–6034. [CrossRef] [PubMed]
34. Pisarev, A.V.; Hellen, C.U.T.; Pestova, T.V. Recycling of Eukaryotic Posttermination Ribosomal Complexes. *Cell* **2007**, *131*, 286–299. [CrossRef] [PubMed]
35. Stetefeld, J.; McKenna, S.A.; Patel, T.R. Dynamic light scattering: A practical guide and applications in biomedical sciences. *Biophys. Rev.* **2016**, *8*, 409–427. [CrossRef] [PubMed]
36. Bleeker, E.A.J.; de Jong, W.H.; Geertsma, R.E.; Groenewold, M.; Heugens, E.H.W.; Koers-Jacquemijns, M.; van de Meent, D.; Popma, J.R.; Rietveld, A.G.; Wijnhoven, S.W.P.; et al. Considerations on the EU definition of a nanomaterial: Science to support policy making. *Regul. Toxicol. Pharmacol.* **2013**, *65*, 119–125. [CrossRef]
37. Malhaire, H.; Gimel, J.C.; Roger, E.; Benoît, J.P.; Lagarce, F. How to design the surface of peptide-loaded nanoparticles for efficient oral bioavailability? *Adv. Drug Deliv. Rev.* **2016**, *106*, 320–336. [CrossRef]
38. Yermak, I.M.; Davydova, V.N.; Kravchenko, A.O.; Chistyulin, D.A.; Pimenova, E.A.; Glazunov, V.P. Mucoadhesive properties of sulphated polysaccharides carrageenans from red seaweed families Gigartinaceae and Tichocarpaceae. *Int. J. Biol. Macromol.* **2020**, *142*, 634–642. [CrossRef]
39. Borgström, J.; Piculell, L.; Viebke, C.; Talmon, Y. On the structure of aggregated kappa-carrageenan helices. A study by cryo-TEM, optical rotation and viscometry. *Int. J. Biol. Macromol.* **1996**, *18*, 223–229. [CrossRef]
40. Wasupalli, G.K.; Verma, D. Molecular interactions in self-assembled nano-structures of chitosan-sodium alginate based polyelectrolyte complexes. *Int. J. Biol. Macromol.* **2018**, *114*, 10–17. [CrossRef]
41. Chen, T.; Li, S.; Zhu, W.; Liang, Z.; Zeng, Q. Self-assembly pH-sensitive chitosan/alginate coated polyelectrolyte complexes for oral delivery of insulin. *J. Microencapsul.* **2019**, *36*, 96–107. [CrossRef]
42. Divya, K.; Jisha, M.S. Chitosan nanoparticles preparation and applications. *Environ. Chem. Lett.* **2018**, *16*, 101–112. [CrossRef]
43. Qi, L.; Xu, Z.; Jiang, X.; Hu, C.; Zou, X. Preparation and antibacterial activity of chitosan nanoparticles. *Carbohydr. Res.* **2004**, *339*, 2693–2700. [CrossRef] [PubMed]
44. Mokhtari, H.; Tavakoli, S.; Safarpour, F.; Kharaziha, M.; Bakhsheshi-Rad, H.R.; Ramakrishna, S.; Berto, F. Recent advances in chemically-modified and hybrid carrageenan-based platforms for drug delivery, wound healing, and tissue engineering. *Polymers* **2021**, *13*, 1744. [CrossRef] [PubMed]
45. Cheng, C.J.; Jones, O.G. Stabilizing zein nanoparticle dispersions with ι-carrageenan. *Food Hydrocoll.* **2017**, *69*, 28–35. [CrossRef]
46. Das, N.; Kumar, A.; Rayavarapu, R.G. The role of deep eutectic solvents and carrageenan in synthesizing biocompatible anisotropic metal nanoparticles. *Beilstein J. Nanotechnol.* **2021**, *12*, 924–938. [CrossRef] [PubMed]
47. Matica, M.A.; Aachmann, F.L.; Tøndervik, A.; Sletta, H.; Ostafe, V. Chitosan as a wound dressing starting material: Antimicrobial properties and mode of action. *Int. J. Mol. Sci.* **2019**, *20*, 5889. [CrossRef]
48. Chirkov, S.N. The Antiviral Activity of Chitosan. *Appl. Biochem. Microbiol. Vol.* **2000**, *97*, 131–141.
49. Jaber, N.; Al-Remawi, M.; Al-Akayleh, F.; Al-Muhtaseb, N.; Al-Adham, I.S.I.; Collier, P.J. A review of the antiviral activity of Chitosan, including patented applications and its potential use against COVID-19. *J. Appl. Microbiol.* **2022**, *132*, 41–58. [CrossRef]
50. Zheng, M.; Qu, D.; Wang, H.; Sun, Z.; Liu, X.; Chen, J.; Li, C.; Li, X.; Chen, Z. Intranasal Administration of Chitosan Against Influenza A (H7N9) Virus Infection in a Mouse Model. *Sci. Rep.* **2016**, *6*, 28729. [CrossRef]
51. Sui, Z.; Chen, Q.; Wu, R.; Zhang, H.; Zheng, M.; Wang, H.; Chen, Z. Cross-protection against influenza virus infection by intranasal administration of M2-based vaccine with chitosan as an adjuvant. *Arch. Virol.* **2010**, *155*, 535–544. [CrossRef]
52. Naberezhnykh, G.A.; Gorbach, V.I.; Likhatskaya, G.N.; Davidova, V.N.; Solov'eva, T.F. Interaction of chitosans and their N-acylated derivatives with lipopolysaccharide of gram-negative bacteria. *Biochemistry* **2008**, *73*, 432–441. [CrossRef]
53. Yermak, I.M.; Kravchenko, A.O.; Khasina, E.I.; Menchinskaya, E.S.; Pislyagin, E.A.; Sokolova, E.V.; Likhatskaya, G.N.; Aminin, D.L. The Anti-Inflammatory Effect of Carrageenan/Echinochrom Complex at Experimental Endotoxemia. *Int. J. Mol. Sci.* **2022**, *23*, 11702. [CrossRef] [PubMed]
54. Dodgson, K.S.; Price, R.G. A note on the determination of the ester sulphate content of sulphated polysaccharides. *Biochem. J.* **1962**, *84*, 106–110. [CrossRef] [PubMed]
55. Rochas, C.; Rinaudo, M.; Landry, S. Role of the molecular weight on the mechanical properties of kappa carrageenan gels. *Carbohydr. Polym.* **1990**, *12*, 255–266. [CrossRef]
56. Harding, S.E. The intrinsic viscosity of biological macromolecules. Progress in measurement, interpretation and application to structure in dilute solution. *Prog. Biophys. Mol. Biol.* **1997**, *67*, 207–262. [CrossRef] [PubMed]
57. Snyder, S.L.; Sobocinski, P.Z. An improved 2,4,6-trinitrobenzenesulfonic acid method for the determination of amines. *Anal. Biochem.* **1975**, *64*, 284–288. [CrossRef]
58. Keler, T.; Nowotny, A. Metachromatic assay for the quantitative determination of bacterial endotoxins. *Anal. Biochem.* **1986**, *156*, 189–193. [CrossRef] [PubMed]
59. Mosmann, T. Rapid colorimetric assay for cellular growth and survival: Application to proliferation and cytotoxicity assays. *J. Immunol. Methods* **1983**, *65*, 55–63. [CrossRef]

60. 60Ngan, L.T.M.; Jang, M.J.; Kwon, M.J.; Ahn, Y.J. Antiviral activity and possible mechanism of action of constituents identified in Paeonia lactiflora root toward human rhinoviruses. *PLoS ONE* **2015**, *10*, e0121629. [CrossRef]
61. Weislow, O.S.; Kiser, R.; Fine, D.L.; Bader, J.; Shoemaker, R.H.; Boyd, M.R. New soluble-formazan assay for HIV-1 cytopathic effects: Application to high-flux screening of synthetic and natural products for AIDS-antiviral activity. *J. Natl. Cancer Inst.* **1989**, *81*, 577–586. [CrossRef]

marine drugs

MDPI

Review

Host Defense Proteins and Peptides with Lipopolysaccharide-Binding Activity from Marine Invertebrates and Their Therapeutic Potential in Gram-Negative Sepsis

Tamara Fedorovna Solov'eva, Svetlana Ivanovna Bakholdina * and Gennadii Alexandrovich Naberezhnykh

G.B. Elyakov Pacific Institute of Bioorganic Chemistry, Far-Eastern Branch of the Russian Academy of Science, Vladivostok 690022, Russia; soltaf@mail.ru (T.F.S.); naber1953@mail.ru (G.A.N.)

* Correspondence: sibakh@mail.ru

Abstract: Sepsis is a life-threatening complication of an infectious process that results from the excessive and uncontrolled activation of the host's pro-inflammatory immune response to a pathogen. Lipopolysaccharide (LPS), also known as endotoxin, which is a major component of Gram-negative bacteria's outer membrane, plays a key role in the development of Gram-negative sepsis and septic shock in humans. To date, no specific and effective drug against sepsis has been developed. This review summarizes data on LPS-binding proteins from marine invertebrates (ILBPs) that inhibit LPS toxic effects and are of interest as potential drugs for sepsis treatment. The structure, physicochemical properties, antimicrobial, and LPS-binding/neutralizing activity of these proteins and their synthetic analogs are considered in detail. Problems that arise during clinical trials of potential anti-endotoxic drugs are discussed.

Keywords: lipopolysaccharide (LPS, endotoxin); LPS-binding proteins/peptides; host defense proteins/peptids; innate immune system; marine invertebrates; Gram-negative sepsis; endotoxic shock

Citation: Solov'eva, T.F.; Bakholdina, S.I.; Naberezhnykh, G.A. Host Defense Proteins and Peptides with Lipopolysaccharide-Binding Activity from Marine Invertebrates and Their Therapeutic Potential in Gram-Negative Sepsis. *Mar. Drugs* **2023**, *21*, 581. https://doi.org/10.3390/md21110581

Academic Editor: Alba Silipo

Received: 21 September 2023
Revised: 16 October 2023
Accepted: 30 October 2023
Published: 7 November 2023

1. Introduction

Lipopolysaccharide (LPS, endotoxin), the major structural component of the Gram-negative bacteria's outer membrane, serves as a physical barrier that protects bacteria from the external environment and can be released into the surrounding medium during cell division or death [1]. These molecules are the potent stimulators of the innate immune system, playing an important role in the pathogenesis of Gram-negative infections in animals. When it enters the body of a warm-blooded host, LPS binds and activates the cellular Toll-like receptor 4 (TLR4), which leads to the development of an inflammatory reaction and, ultimately, as a result, to the death and elimination of the pathogen. However, the accumulation of endotoxin in the bloodstream in large quantities (with Gram-negative infection, invasive procedures, etc.) can cause the excessive activation of immunocompetent cells, inducing the overproduction of pro-inflammatory cytokines and a systemic inflammatory response, which leads to sepsis and septic shock within a few days (Figure 1). Lipid A, the predominantly lipophilic and most conserved fragment of the LPS molecule, directly interacts with TLR4 and is responsible for most of the immunobiological and toxic properties of endotoxin [1].

Gram-negative sepsis remains a serious unresolved problem in clinical medicine. This type (according to the etiology of pathogens) of sepsis is clinically the most severe, often accompanied by septic shock, and has a significantly higher mortality rate than other types. The increasing use of new technologies in medical practice—cytostatic and immunosuppressive therapy and transplantation and prosthetics—as well as the HIV and COVID-19 pandemics and the increasing resistance of pathogens to antibiotics contribute to the growth of septic complications in patients. The aging population, accompanied by an increase in chronic diseases, also leads to an increase in the incidence of sepsis and septic shock. As a result, despite the great advances in antimicrobial chemotherapy, mortality rates from septic shock remain the highest and do not decrease worldwide. The development and introduction of new antibiotics into medical practice does not fundamentally solve the problem; moreover, their use in some cases can lead to a deterioration in the condition of a septic patient.

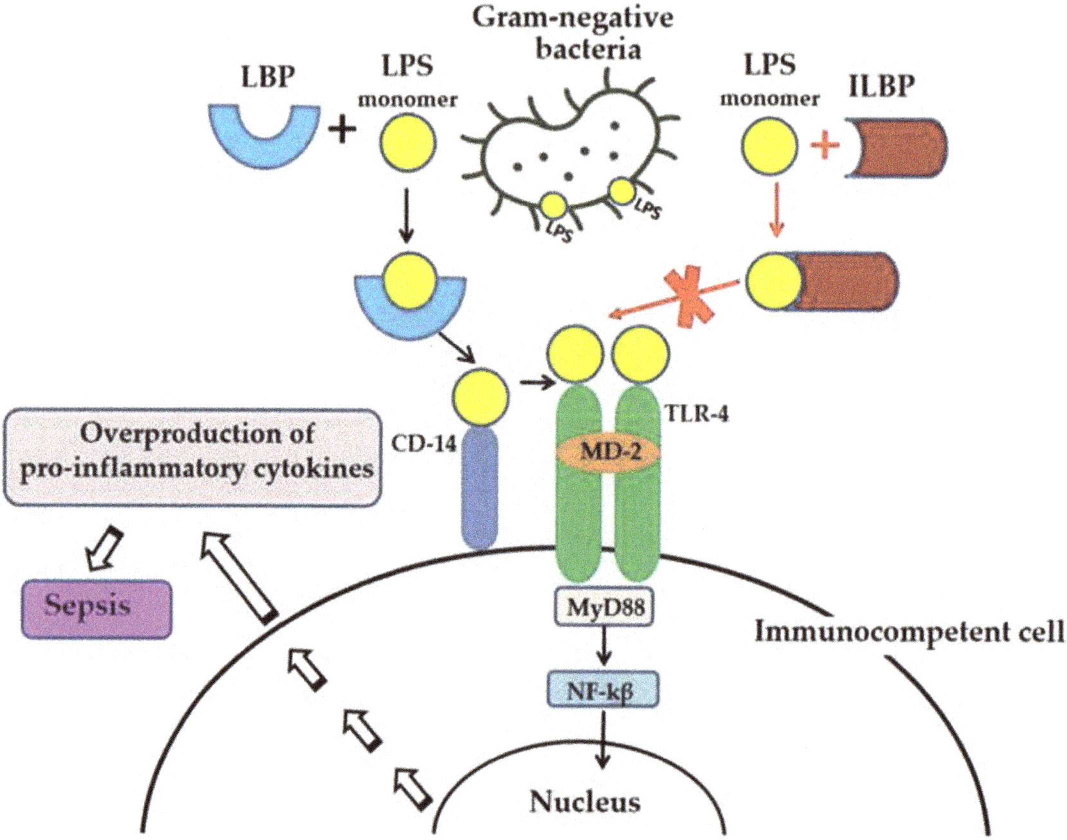

Figure 1. LPS-induced inflammatory response of the innate immune system and the anti-inflammatory effect of LPS-binding peptides/proteins from marine invertebrates. Serum protein LBP (LPS-binding protein) binds the monomer of LPS and delivers it to a CD14 molecule. CD14 transfers LPS to the ectodomain of the TLR4/MD-2 receptor complex, which leads to homodimerization of TLR4. This change in TLR4 conformation provides a binding site for adaptor molecule MyD88 (myeloid differentiation primary-response protein 88). The MyD88-dependent signaling pathway leads to the activation of nuclear factor-κβ (NF-κβ), which regulates the expression of target genes encoding pro-inflammatory mediators. The overproduction of pro-inflammatory cytokines may lead to an uncontrolled inflammatory reaction and eventually to sepsis. The binding of ILBP to LPS blocks CD14–LPS interaction and prevents the transfer of LPS to the TLR4/MD-2 complex, thus interfering with TLR4 dimerization and downstream inflammatory responses.

Modern medicine does not have specific and effective anti-sepsis drugs whose molecular target is LPS. Molecules that can selectively block TLR4, preventing endotoxin from binding to receptor and the development of a systemic inflammatory response, may have therapeutic potential for the treatment of sepsis. In particular, TLR-4 receptor antagonists are structural analogs of lipid A with low toxicity, including native lipid A from a number of marine proteobacteria (Proteobacteria) [2,3]. Another approach to the endotoxin-neutralizing drug design is based on the use of synthetic or natural substances that can suppress the biological activity of LPS due to the formation of strong complexes with it (Figure 1) [4]. Searching for such compounds among the host defense proteins of marine invertebrates—which represent one of the most extensive and diverse groups of animals, numbering 153,434 species—seems very promising (UN data for 2019).

Marine invertebrates only have an innate immune system, including a huge set of defense proteins, which was formed during a long evolution and allowed them to survive in the environment enriched with pathogenic microorganisms [5]. Host defense proteins are traditionally referred to as antimicrobial peptides, although they have been shown to be polyfunctional compounds. These proteins are constitutively expressed and rapidly induced in various cells and tissues, interact directly with infectious agents, and/or activate immune responses to eliminate pathogens. Defense proteins usually recognize and bind to the surface of the pathogen the most conservative biopolymers that are common and vital for a large group of microorganisms but not present in the host. These molecules, known as pathogen-associated molecular patterns (PAMPs), trigger innate immune responses in the host [6]. In Gram-negative bacteria, this PAMP is LPS.

Host defense proteins and peptides with lipopolysaccharide-binding capacity from marine invertebrates (ILBPs, invertebrate lipopolysaccharide-binding proteins) may possess different biological properties depending on the structure and nature of the interaction with endotoxins. With a high affinity for LPS, these proteins may have antimicrobial or endotoxin-neutralizing activity, or both [4]. ILBPs that combine both of these properties are considered today as the most effective potential drugs for the treatment of human sepsis.

This review summarizes data on host defense proteins (antimicrobial peptides) of marine invertebrates that can recognize and bind LPS. Particular attention is focused on the group of ILBPs and their synthetic analogs, which neutralizes endotoxin. The structure, physicochemical properties, LPS-binding/neutralizing activities, and antimicrobial activities of these proteins are considered in detail as far as the sources for their isolation among marine invertebrates. The structural basis of the endotoxin-neutralizing action of ILBPs is discussed.

2. Anti-Lipopolysaccharide Factor (ALF)

A cationic protein that inhibited the LPS-induced activation of the crab hemolymph coagulation system has been found in hemocyte lysates of Japanese (*Tachypleus tridentatus*) and American (*Limulus polyphemus*) horseshoe crabs [7–9]. This protein, called the anti-lipopolysaccharide factor (ALF), was able to bind LPS, neutralize its biological activity (in vitro and in vivo), and inhibit the growth of R-type Gram-negative bacteria. The natural and recombinant anti-LPS factor Limulus (ALF-L) also suppressed endotoxin-mediated activation of cultured endothelial and B cells, reduced the concentration of endotoxin and TNF-α in the blood serum of experimental animals, and protected them from death in the late stages of endotoxemia and sepsis [10–14].

Numerous ALF homologues have been identified and characterized in different types of crustaceans: shrimp, lobster, crabs, and crayfish [15–22]. The most extensively studied ALFs are isolated from shrimp of the Penaeidae family, which includes many economically important species that are of interest as objects of industrial fishing and breeding [23].

ALFs show a broad spectrum of antimicrobial activity against Gram-negative and Gram-positive bacteria, fungi, human enveloped viruses (herpesvirus type 1, adenovirus), and white spot syndrome virus (WSSV), which is widely distributed throughout the world and considered as one of the most destructive and pathogenic viruses in shrimp farms [24–26].

In crustaceans, one organism usually contains several isoforms of ALF, which are either encoded by different genes or formed as a result of alternative mRNA splicing [27]. Thus, six isoforms were identified in the tiger shrimp *Penaeus monodon*, and seven isoforms were found in the Chinese shrimp *Fenneropenaeus chinensis* and the South Korean blue crab *Portunus trituberculatus* [15,28–32]. The isoforms differed in tissue distribution and antimicrobial properties. The wide diversity of ALF sequences within a species may provide a synergistic enhancement of their protective action against bacterial infection.

ALFs are a group of small single-domain antimicrobial proteins consisting of 97–124 amino acid residues with a relatively short 16–28 residue signal sequence. The molecular weight of the mature protein is about 11 kDa: ALFs from *L. polyphemus* and shrimp of various species have masses of 11.8 and 10.74 to 12.23 kDa, respectively [9,33]. According to the values of the isoelectric points (pI), ALFs were classified as cationic peptides, but more and more data are emerging on the existence of anionic proteins among them [33–35]. The theoretical pI values of mature shrimp ALFs range from 5.02 to 10.29 [33]. Typically, ALF molecules have a highly hydrophobic N-terminal region and conserved cluster of positively charged and hydrophobic amino acid residues within a loop fixed by a disulfide bond between two conserved cysteine residues, which is commonly referred to as the LPS-binding domain [36]. This amphipathic loop is an important functional molecule moiety, which is responsible for the biological activity of ALFs. Indeed, synthetic peptides corresponding to this fragment from various ALFs have shown antimicrobial activity, the ability to inhibit WSSV virus replication, and a protective effect in sepsis [20,37–39].

Despite the large number of registered ALFs (more than 300 proteins of this class from crustaceans were isolated and characterized until 2021), only one crystal structure of them has been established to date. The X-ray structure of recombinant ALF-L consists of three α-helices (one at the N-terminus and two at the C-terminus) packed against a four-stranded β-sheet, giving rise to a wedge-shaped molecule [36]. The potential LPS-binding domain includes an amphipathic β-hairpin formed by the longest β-strands S2 and S3 of the β-sheet and stabilized by the single disulfide bond (Cys31–Cys52). The positively charged residues within the β-hairpin of ALF-L are supposed to interact with the negatively charged phosphate groups of lipid A. However, the lipid A binding site on ALF remains poorly understood to date. Later, the spatial ALF structure from the shrimp *P. monodon*, expressed in yeast cells, rALFPm3, was determined using NMR (Figure 2) [40]. The structure of rALF-Pm3, like ALF-L, is composed of three α-helices, a four-stranded β-sheet, and contains a β-hairpin formed by S2 and S3 β-strands closely linked via a Cys34-Cys55 disulfide bond. A Comparison of the 3D structures of these proteins revealed highly similar clusters of positively charged and hydrophobic residues on the β-sheet surface. This suggests that ALF-L and ALFPm3 have a similar LPS binding site, which is located on the β-sheet and mainly consists of 5–6 positively charged and several hydrophobic residues capable of binding lipid A through electrostatic and hydrophobic interactions.

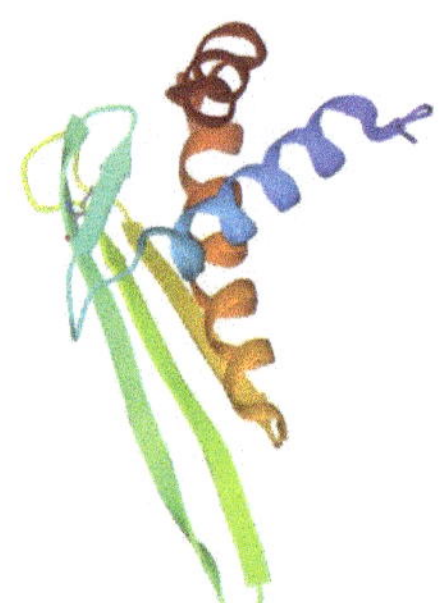

Figure 2. Structure of the anti-lipopolysaccharide factor from *Penaeus monodon* (pdb, 2job).

A series of peptides of various lengths, including cyclic ones, derived from the ALF-L sequence, were synthesized [41]. These peptides demonstrated high endotoxin-binding and neutralizing activities, comparable with those of the parent recombinant protein, and were non-toxic for erythrocytes or cultured human monocytes. A new class of peptides based on the LPS-binding domain of ALF-L or part of it with significant changes in length and primary sequence calculated for optimal lipid A binding was also designed [42]. A preclinical study revealed that these peptides have high selectivity for LPS, as well as high LPS-neutralizing activity in vitro and the ability to protect against sepsis in vivo. An analysis of the obtained data showed that the endotoxin-neutralizing activity of the peptides is closely related to their affinity for LPS and the ability to incorporate into LPS aggregates with changes in their structure. The authors highly appreciate the potential of synthetic peptides as drugs for the treatment of endotoxemia and sepsis.

3. β-Hairpin Peptides

3.1. *Arenicins*

Several antimicrobial peptides were isolated from the coelomocytes of marine polychaeta lugworm *Arenicola marina* called arenicins-1, -2, and -3 [43,44]. Arenicin molecules consist of 21 amino acid residues and have the amphipathic β-hairpin structure, formed by the two-stranded antiparallel β-sheet stabilized by one (arenicin-1,-2) or two disulfide (arenicin-3) bridges (Figure 3a).

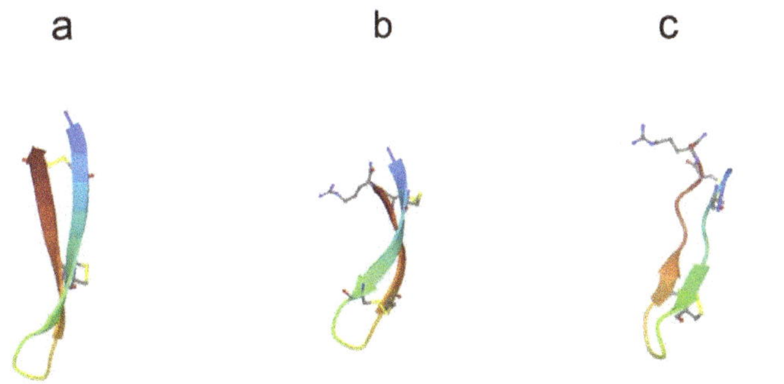

Figure 3. Structure of β-hairpin peptides. (**a**) Arenicin-3 (pdb, 5v0y), (**b**) tachyplesin-1 (pdb,1wo0), and (**c**) polyphemusin-1 (pdb, 1rkk).

Conformational analysis via NMR spectroscopy revealed that the β-sheet in arenicins had a marked right-handed twist in an aqueous solution. This distortion effectively shields the hydrophobic side of the β-sheet from contacts with polar solvent, thus reducing the peptide surface amphipathicity [45,46]. When interacting with membranes or in membrane-mimetic environments, arenicins, form dimers stabilized by hydrogen bonds between parallel N-terminal β-strands in two neighboring molecules. Dimerization induces a substantial conformational change so that the molecules adopt almost planar amphipathic β-sheet structures [47,48]. A significant decrease in the twist of the β-hairpin as a result of arenicin dimerization leads to an increase in its amphiphilicity and stability. Natural arenicins exhibit pronounced antimicrobial activity against a broad spectrum of Gram-positive and Gram-negative bacteria, pathogenic fungi, and yeasts even under high-ionic-strength conditions [43,49]. Mode-of-action studies strongly suggest that the antibacterial activity of arenicins is consistent with their ability to disrupt the integrity of bacterial membranes.

More recently, arenicin-3, a member of the arenicin family, was discovered by the pharmaceutical company Adenium Biotech ApS, which develops novel antibiotics for the treatment of Gram-negative bacterial infections [44]. This peptide is very attractive due to its potent broad-spectrum antibacterial activity, even against multidrug-resistant clinical isolates, and its ability to bind LPS. However, arenicin-3 is toxic to mammalian cells and causes hemolysis of human erythrocytes. To solve the problem of toxicity, the structural analogs of arenicin-3 were designed by changing the number of disulfide bonds, hydrophobicity, or charge of the molecule [50,51]. As a result, peptides were synthesized with low cytotoxicity while maintaining antibacterial properties. The optimized synthetic arenicin-3 derivatives, which retained the β-hairpin structure stabilized by one or two disulfide bonds, were found to be the most active against Gram-negative bacteria. These peptides have demonstrated the ability to bind LPS with higher affinity than polymyxin B and neutralize its toxic effects. Also, these synthetic analogs increased the survival of mice during LPS-induced peritonitis and sepsis, protected from lethal LPS challenge in vivo, endotoxin-induced lung injury, and death caused by bacterial infection (*E. coli* and *S. enteritidis*) and also inhibited the production of proinflammatory cytokines. At the same time, they had low hemolytic activity and cytotoxicity, as well as higher antimicrobial activity than the natural peptide. According to the researchers, these optimized arenicin-3 analogs may be potential candidates for the creation of dual-acting drugs with antibacterial and anti-endotoxin activities. Currently, one of the obtained variants is undergoing preclinical trials [52].

3.2. Tachyplesins and Polyphemusins

In horseshoe crab hemocytes, in addition to ALF, another group of antimicrobial LPS-binding peptides was found. These relatively short and structurally closely related peptides, known as polyphemusins I and II and tachyplesins I–III, were isolated from *L. polyphemus* and *T. tridentatus*, *Tachypleus gigas*, and *Carcinoscorpius rotundicauda*, respectively [53–55].

Their concentration in hemocytes is extremely high, up to 10 mg in the total hemolymph of an individual horseshoe crab [56]. These peptides are 17–18 amino acid residues in length and have an amidated C-terminal arginine residue and a net positive charge. The spatial structure of the peptides in aqueous solutions is an amphiphilic, antiparallel β-hairpin connected by a β-turn and stabilized by two disulfide bonds (Figure 3b,c) [57–60]. The peptide structure is highly stable and is preserved when samples are heated to 100 °C in neutral pH buffers and kept at low pH [55]. This stability seems to be due to the rigid structure imposed by the two disulfide linkages. In the presence of dodecylphosphocholine micelles, conformational changes in tachyplesin structure occur, which are accompanied by an increase in the amphiphilicity of the molecule and the formation of a contiguous well-defined hydrophobic surface [58]. The ability of peptides to adopt distinct conformations in solution and upon membrane association appears to be partly responsible for their wide range of biological activities, including antimicrobial, antitumor, and anti-inflammatory. So far, for these peptides, the relationships between their structures and functions are not well understood.

These peptides recognize and bind LPS and quickly integrate into LPS monolayers. They are able to displace divalent cations from their binding sites with LPS and penetrate into the outer membrane of bacteria and also demonstrate resistance to inhibition from these cations of incorporation into LPS monolayers [61–63]. The recognition site for the peptide on the LPS molecule is the lipid A moiety. The peptides were found to show higher (280-fold for tachyplesin I) affinity for LPS compared with acidic phospholipids. When tachyplesin interacts with LPS, slight changes in its secondary structure are observed: the β-sheet is elongated and twisted, and the whole structure is stabilized. A twisting β-sheet structure may be important for tachyplesin to recognize LPS. According to the proposed model of the complex of tachyplesin I with LPS, the peptide lies across two D-glucosamine residues of lipid A, and its cationic and aromatic residues interact with phosphate groups and acyl chains of lipid A moiety, respectively [62].

The binding of peptides to LPS is accompanied by the neutralization of its toxic effect on the macroorganism. Polyphemusins inhibit the production of pro-inflammatory cytokines TNF-α and IL-6 through LPS-stimulated macrophages, protect mice from endotoxemia, and block the development of endotoxin shock in an animal model. The structural analog study of polyphemusins and tachyplesins made it possible to establish that the antiendotoxin activity of peptides increases with an increase in their binding affinity to LPS and the amphiphilicity of the molecule [63].

Tachyplesins and polyphemusins show pronounced activity against a wide range of microorganisms and, along with arenicins, are considered the most active antimicrobial peptides of animal origin. They inhibit the growth of both Gram-positive and Gram-negative bacteria, as well as some fungi at sub-micromolar and micromolar concentrations [53,63–65].

Despite the commonality of the beta-hairpin fold stabilized by disulfide bonds and a wide spectrum of activity against both Gram-negative and Gram-positive bacteria, tachyplesins/polyphemusins and arenicins have a rather low degree of amino acid sequence similarity (up to 35%) [66] and differ in the mechanism of antimicrobial action. Tachyplesins and polyphemusins have been shown to translocate across membranes without significant disruption of lipid bilayers [67], while arenicins disrupt the cell membrane through the formation of higher oligomeric states [46].

4. Big Defensins

Big defensins were first discovered in horseshoe crabs. A novel defensin-like protein was isolated from *T. tridentatus* hemocytes, which contained 79 amino acid residues and was named "big defensin" (BigDef). This protein had a pronounced ability to bind LPS, as well as to inhibit the growth of Gram-positive and Gram-negative bacteria and fungi [68]. The BigDef molecule consists of a highly hydrophobic N-terminal domain and a cationic C-terminal domain containing six cysteine residues involved in three internal disulfide bridges. These two different domains are connected via a flexible linker. The spatial structure of the C-terminal domain is a twisted three-stranded antiparallel β-sheet, stabilized by three disulfide bonds, and the N-terminal domain adopts a conformation formed by parallel β-sheet and two α-helices, which in the lipid environment are transformed into an elongated single α-helix (Figure 4a) [69,70].

The C-terminal domain is structurally similar to human β-defensins, HβD-2 and HβD-3 [71] and differs from invertebrate defensins. Trypsin cleaves BigDef at the Arg-37 residue to form two peptide fragments that have diverse activities. The N-terminal hydrophobic peptide acts predominantly against Gram-positive bacteria, while the C-terminal cationic peptide is more active against Gram-negative bacteria. Both generated peptides showed weak LPS-binding activity, whereas the activity of intact full-length BigDef was significant compared with that of anti-LPS factor ALF peptide from *T. tridentatus* hemocytes. Thus, binding to LPS requires the native conformation of the entire molecule.

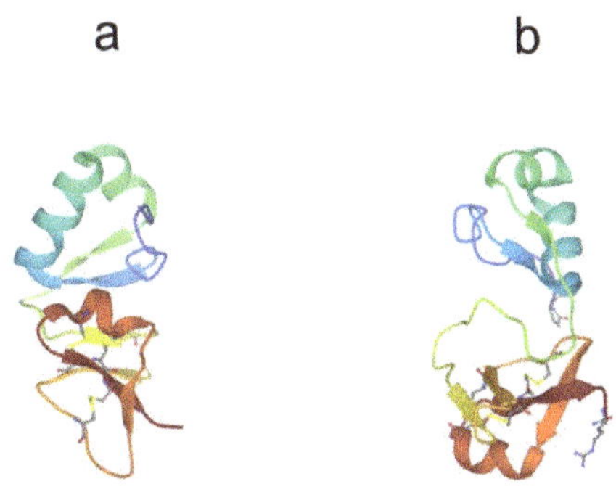

Figure 4. Structures of big defensins isolated from (**a**) horseshoe crab *Tachypleus tridentatus* (pdb, 2rng) and (**b**) oyster *Crassostrea gigas* (pdb, 6qbl).

Phylogenetic analysis of all currently known sequences of the BigDef genes showed that these proteins form a group predominantly represented in marine invertebrates, mainly in mollusks [72–74] and, to a much lesser extent, in horseshoe crabs and lancelets [75]. The spatial structure of the big defensin from the Pacific oyster *Crassostrea gigas* (Cg-BigDef1) was recently determined (Figure 4b) [76]. This is the second currently known structure for a protein from the BigDef family. NMR spectroscopy revealed that oyster Cg-BigDef1, like horseshoe crab Tt-BigDef, possesses two structural domains. Cg-BigDef differs from Tt-BigDef in the orientation of the N- and C-terminal domains, the length of the linker sequence, and, as a result, surface properties. The big defensin overall structure from the oyster is mainly hydrophobic, while that from the horseshoe crab is amphiphilic. This suggests that Cg-BigDef1 binding to bacterial membranes occurs through hydrophobic rather than electrostatic interactions and is not impaired at high salt concentrations. Cg-BigDef1 exhibited salt-stable activity against both Gram-positive and Gram-negative bacteria and fungi. Genes encoding homologous proteins from two other mollusk species, *Venerupis philippinarum* and *Argopecten irradians*, have been cloned and expressed, and the obtained recombinant proteins have been characterized [77,78]. These molecules are cationic peptides with a molecular weight of 8–11 kDa and an isoelectric point of 8.6–9.2. They exhibit antibacterial and antifungal activities. The ability of these proteins to bind LPS has not been studied.

5. Factor C

Factor C is a unique LPS-binding protein found in horseshoe crabs. It is a trypsin-like serine protease zymogen. The zymogen is activated by picogram amounts of LPS and is the initiator of the coagulation cascade, which the horseshoe crab uses as one of the defense mechanisms against pathogens [79]. It can specifically bind LPS on the surface of hemocytes and directly recognize Gram-negative bacteria in an LPS-dependent manner, acting as a pattern-recognizing receptor [80]. Recombinant Factor C (rFC) interacts at extremely high affinity with LPS and lipid A: the dissociation constant (K_D) of its complex with lipid A is 7.6×10^{-10} M [81]. Due to high specificity and sensitivity to LPS, it is widely used by pharmaceutical companies to detect endotoxin contamination of parenteral drug products and medical devices (Limulus Amoebocyte Lysate (LAL) test and rFC test) [82]. Recombinant factor C has been shown to effectively inhibit LPS-induced production of TNF-α and IL-8 via human macrophages and is therefore a potential LPS-neutralizing agent [83]. Furthermore, rFC is non-toxic to human monocytes and *HeLa* cells. In horseshoe crab amoebocyte lysate, Factor C was found in two molecular species: in a single-chain form and a two-chain form, which consists of heavy and light chains linked by (a) disulfide bond(s) [84]. Both molecules have the same molecular mass. A single-chain form is transformed into a two-chain form as a result of the LPS-mediated activation of Factor C [85,86]. It is believed that molecular reorganization may occur during the isolation and purification of Factor C due to the presence of LPS trace impurities.

Factor C is a glycoprotein and, depending on the source of isolation, *L. polyphemus* and *T. tridentatus hemocytes* or *Carcinoscorpius rotundicauda* hemocytes, has a molecular weight of 120 or 132 kDa with heavy and light chain sizes of 80 and 43 or 80 and 52 kDa, respectively [86,87]. The Factor C molecule consists of domains that are structurally related to proteins of the mammalian complement system [88]. Along with the typical serine protease domain at the C-terminus, Factor C also includes a cysteine-rich (Cys) region; a domain homologous to epidermal growth factor (EGF); five complement control protein (CCP) modules, also known as Sushi domains; an LCCL segment (a fragment common to a number of proteins such as Coch-5b2 and Lg11 [89]); and a domain similar to a C-type lectin (Figure 5).

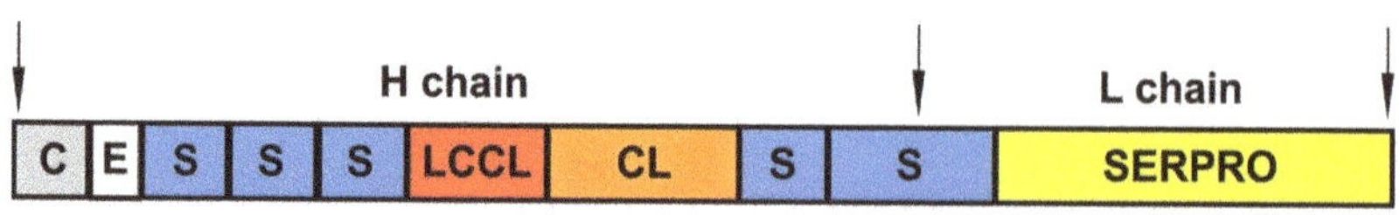

Figure 5. The modular structure of mature Factor C. C—cysteine-rich region (Cys); E—epidermal growth factor (EGF)-like domain; S—complement control protein (CCP) module, also known as Sushi domain; LCCL—domain-type identified in cochlear protein Coch-5b2 and late gestation lung protein Lgl1; CL—C-type lectin-like domain; SERPRO—serine protease domain. The positions of the heavy and light chains are indicated.

The CCP module consists of approximately 60 amino acid residues, including four cysteine residues, forming two internal conserved disulfide bonds, and has a β-sandwich spatial structure with a compact hydrophobic core. One face of the β-sandwich is made up of three β-strands linked by hydrogen bonds, and the other face formed by two separate β-strands [90]. The regions of the polypeptide chain between the β-strands are composed of both well-defined turns and less well-defined loops. An analysis of CCP sequence alignments reveals a high degree of conservation among residues of obvious structural importance, while almost all insertions, deletions, or substitutions are found in the region of the loops. This suggests that the structure of the 16th CCP module from human complement factor H (Figure 6a) gives sufficient understanding of the structure of these modules in the Factor C molecule.

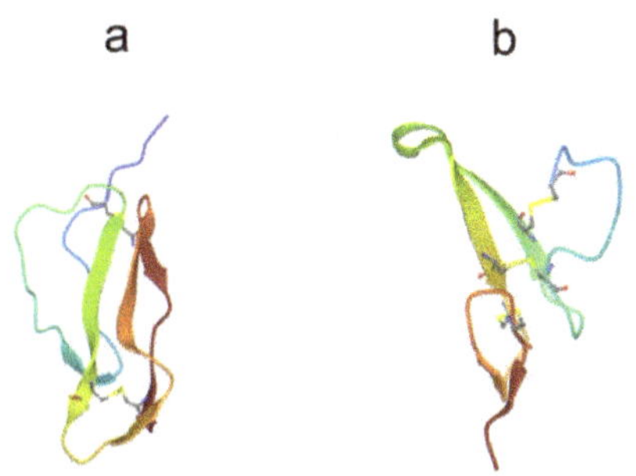

Figure 6. Structures of (**a**) complement control protein module 16 from human complement factor H (pdb, 1hcc) and (**b**) the human epidermal growth factor (pdb, 1jl9).

The EGF-like domains are about 50 amino acid residues in length and contain six cysteine residues that form disulfide bonds, resulting in a very compact configuration [91,92]. The secondary structure of these polypeptides in water contains two antiparallel β-sheets and several β-turns. Figure 6b shows the crystal structure of human EGF (hEGF), which consists of an N region and a C region [93]. The N region has an irregular N-terminal peptide segment and an anti-parallel β-sheet. The C region contains a short anti-parallel β-sheet and a C-terminal segment, which are probably disordered in isolation. There are two hEGF molecules in the asymmetric unit of the crystals, which form a potential dimer.

The N-terminal fragment of Factor C is completely responsible for binding to LPS [94,95]. This fragment, which includes a Cys-rich region, an EGF-like domain, and three CCP (Sushi) modules, has several LPS (lipid A) binding sites and demonstrates strong positive cooperativity of binding to the ligand [96]. This, apparently, determines its significantly higher ability to neutralize endotoxin, in comparison with polymyxin B, as well as the high sensitivity of Factor C to LPS. At low concentrations, the Factor C fragment completely inhibits LPS-induced production of TNF-α and IL-8 by human monocyte cells THP-1 and peripheral blood mononuclear cells. N-terminal fragment, which has low cytotoxicity, protects mice from LPS-induced lethality.

Structural and functional analysis of the N-terminal region of factor C from the horseshoe crab *C. rotundicauda* showed that LPS-recognizing regions are localized in the CCP 1 (Sushi 1) and CCP 3 (Sushi 3) modules [97]. These modules have high-affinity LPS binding sites with K_D from 10^{-9} to 10^{-10} M, which are located in two 34-mer peptides, S1 and S3. Both S1 and S3 can inhibit the LAL reaction with endotoxin and the LPS-induced production of TNF-α by human macrophages with different efficiencies. The analysis of the obtained data showed that at least two S1 peptides cooperatively bind to one LPS molecule with a Hill coefficient of 2.42. In contrast, the binding of LPS to S3 is non-cooperative. Synthetic peptides developed on the basis of CCP modules inhibited the LPS-induced secretion of TNF-α by human THP-1 cells and protected D-galactosamine-sensitized mice from a lethal dose of *E. coli* LPS [96,97], demonstrating the promise of their use for the immunotherapy of Gram-negative sepsis.

In the case of the Factor C orthologue from *T. tridentatus*, the LPS-binding region of the molecule was found not inside the tandem CCP (Sushi) modules but in the N-terminal cysteine-rich Cys region and the EGF-like domain [80]. The Cys-rich region specifically binds LPS not only in free form but also on the bacterial surface. The LPS binding site in the Cys region contains a conserved tripeptide sequence (Arg36-Trp37-Arg38) consisting of an aromatic residue flanked by two basic residues, which is also found in other LPS-recognizing proteins [98]. Mutations in this tripeptide prevent its binding to both LPS and Gram-negative bacteria, which determines the key role of this conserved motif in interaction with LPS. It is assumed that the binding of this peptide to LPS occurs according to the mechanism previously proposed for the ALF peptide: the basic residues interact with D-glucosamine-1-phosphate of lipid A, and the aromatic residue associates with its hydrophobic part.

Full-length Factor C binds and neutralizes LPS more effectively than individual LPS-binding peptides derived from it [83,97]. This fact indicates that interdomain interactions in the molecule enhance the overall interaction between factor C and LPS. The tandem arrangement of Sushi's LPS-binding domains in Factor C has been reported to be responsible for its high affinity for LPS. In addition, the lectin-like and CCP 4 domains have been shown to contribute to the binding of factor C to LPS [99]. These modules can either influence the conformation of LPS-binding domains or directly participate in LPS binding.

6. Bactericidal/Permeability-Increasing Proteins

The bactericidal/permeability-increasing protein (BPI) is a member of the LBP/BPI family of LPS-binding proteins and a component of the innate immune system that acts selectively against Gram-negative bacteria [100]. These proteins have a direct cytotoxic and opsonizing effect on bacteria and also bind LPS in the lipid A region and neutralize its biological activities [101–103]. These properties of BPI have been used therapeutically for endotoxin-related complications of various diseases [104]. In animal models of sepsis, pneumonia, and endotoxemia, as well as in preclinical and clinical trials, recombinant BPI peptides have been shown to neutralize many of the biological effects of LPS.

The first BPIs were isolated from rabbit and human polymorphonuclear leukocytes [105,106]. Human BPI (hBPI), one of the most well-studied representatives of the LBP/BPI family, is a cationic protein with a molecular weight of 55 kDa, whose three-dimensional structure was determined by X-ray diffraction (Figure 7) [107,108]. At the same time, orthologues of this protein have been found in several non-mammalian vertebrate species and various invertebrates [109].

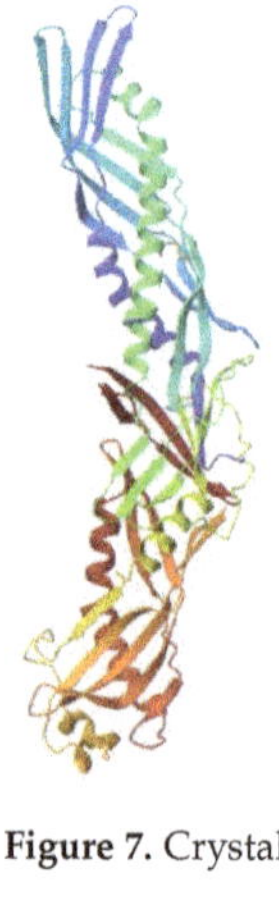

Figure 7. Crystal structure of human bactericidal/permeability-increasing protein (pdb, 1bp1).

The hBPI homologue was revealed in the oyster *C. gigas* [110]. This was the first time that BPI was identified in an invertebrate. The protein, named Cg-BPI, was obtained in recombinant form and characterized. Mature Cg-BPI is a cationic protein (50.1 kDa) with a calculated pI of 9.3, close to that of hBPI, pI 9.4. Amino acid sequence analysis, as well as structure modeling and electrostatic surface potential prediction, showed that Cg-BPI has a high degree of structural similarity to hBPI. It contains two conserved domains, N-terminal and C-terminal, which are separated by a proline-rich region. Although these domains have a low degree of amino acid sequence similarity, they exhibit the same spatial structure. The Cg-BPI, like the hBPI, has a boomerang shape and consists of two identical barrels formed by a beta-sheet and two alpha-helices, which are connected via a central β-sheet (Figure 7). The N-terminal domain of Cg-BPI contains functional regions previously characterized in hBPI as responsible for LPS binding and neutralization, as is bactericidal activity [111,112]. This domain has three conserved lysine residues that can bind to negatively charged LPS groups through electrostatic interactions [113], as well as two cysteines characteristic of mammalian BPI and forming a disulfide bond, which has been shown to be important for the rhBPI function [114]. Both domains contain an apolar pocket that serves as a binding site for lipids and probably lipid A.

Recombinant Cg-BPI bound LPS and lipid A with high affinity (K_D 3.1 $\times$ 10^{-8} M for *E. coli* LPS) [110]. It had a strong bactericidal effect on Gram-negative bacteria and disrupted their cytoplasmic membranes. Thus, the BPI protein from *C. gigas* combines LPS-binding activity with antibacterial and membrane-permeabilizing properties. It is noteworthy that Cg-BPI was highly active against *E. coli* SBS363, which contains LPS with short O-polysaccharide chains but was 30 times less active against *E. coli* ML35 with long-chain LPS. A similar result was obtained for hBPI and was explained by the greater accessibility of anionic and hydrophobic sites in and near the lipid A region of the LPS molecule in *E. coli* SBS363 due to a decrease in the shielding effect of O-polysaccharide chains [115,116].

In further research, a second BPI, Cg-BPI2, was found in *C. gigas*, which showed the highest sequence identity with the already-known Cg-BPI [117]. According to the results of molecular modeling, Cg-BPI2, like hBPI and Cg-BPI, has a structure consisting of an N- and C-terminal barrels and a central β-sheet. At the same time, a comparison of the electrostatic surface potentials revealed that Cg-BPI2 has a higher surface charge than hBPI and Cg-BPI. The recombinant N-terminal domain of Cg-BPI2 exhibited a high affinity for LPS and was effective against Gram-negative bacteria. Thus, the antibacterial activity of *C. gigas* BPIs, as well as human BPI, is determined by their N-terminal domain.

Recently, BPIs EsBPI2 and EsBPI4 from the squid *Euprymna scolopes* have been characterized [118]. Based on amino acid sequence analysis and comparative modeling data, EsBPI2/4 were predicted to have molecular characteristics typical of hBPI. These two proteins have a two-domain "boomerang-like" structure. They share with other BPIs the predicted LPS-binding regions in their N-terminal domains and conserved cysteines, which are involved in the formation of disulfide bonds crucial for the functional activity of this family of proteins. Both proteins isolated from squid tissue extract showed potent bactericidal activity against Gram-negative bacteria *Vibrio fischeri*. Host exposure to LPS derivatives (lipid A) led to increased EsBPI2 gene expression.

Using genomic technologies, proteins of the LBP/BPI family have been identified in a number of marine invertebrates, such as marine annelids [119], sea urchins [120], and mollusks [121]. The LBP/BPI gene expression in the invertebrates has been shown to occur after challenge with LPS. These results suggest that BPIs contribute to the elimination of Gram-negative bacteria through interaction with LPS.

7. Lipopolysaccharide-Binding Lectins

Lectins are non-immunoglobulin-type proteins or glycoproteins that selectively recognize and reversibly bind to specific carbohydrates and carbohydrate moieties without changing the structure of glycan. Marine animals, including invertebrates, have a large and complex set of lectins that vary considerably in their structure and carbohydrate specificity. Invertebrate lectins are potential molecules involved in the immune recognition and phagocytosis of microorganisms through opsonization. They are able to interact with LPS on the surface of bacterial cells. However, this fact only applies to those LPS that have carbohydrate motifs recognized by the lectins. Carbohydrate ligands in the LPS molecule are localized mainly in O-specific polysaccharide chains (hypervariable structural element) and rather less frequently in the core oligosaccharide (relatively conserved structure) and in lipid A (very conserved part of the molecule) [122].

In marine invertebrates, horseshoe crab LPS-binding lectins are the best studied. Five types of such lectins have been isolated from the hemolymph of *T. tridentatus*, of which four, called tachylectins (TL-1 to TL-4), are from hemocytes and one, TPL2 (Tachypleus plasma lectin 2), is from plasma. Unlike tachylectin proteins-1-3, TPL-2 is a glycoprotein. A study of recombinant TPL2 with a mutation in the glycosylation site shows that glycosylation does not appear to be important for LPS binding [123].The hemolymph lectins differ in their carbohydrate specificity. Indeed, TL-1 (L6) interacts with the core oligosaccharide of the LPS molecule, probably through the 2-keto-3-deoxyoctonic acid (KDO) residue [124]; TL-2 (L10) exhibits specific activity for D-GlcNAc ($K_D = 5.13 \times 10^{-5}$ M and 1.54×10^{-8} M for free and immobilized (GlcNAc–BSA) monosaccharide, respectively) and D-GalNAc [125,126]; TL-3 specifically binds to S-type LPS from several Gram-negative bacteria through a specific structural fragment of O-polysaccharide, similar to that of the blood group A antigen [127]; and TL-4, like TL-3, specifically recognizes S-type LPS, but not R-LPS lacking O-polysaccharide. The most likely specific ligand for TL-4 is colitose (3-deoxy-L-fucose), a monosaccharide that is structurally similar to L-fucose, to which the lectin is also able to bind, but with lower affinity [128]. The D-isomer of colitose, abequose, is also a candidate for another ligand. TPL2 shows an 80% sequence identity with TL-3 and, like TL-3, specifically interacts with the O-polysaccharide fragment of LPS [123,129,130]. This lectin binds to *E. coli* LPS with a K_D of 1.03×10^{-6} M [131]. The specific ligand for TPL2 is L-rhamnose.

A structural feature of tachylectins is the presence of tandem repeats in their amino acid sequence, which are at least 30 residues long and encode the secondary and tertiary structure of the protein. The TL-1 (27 kDa), TL-2 (27 kDa), and TL-3 (14 kDa) sequences include six, five, and two repeats, respectively [124,125]. TL-3 (14 kDa) is present as a dimer (29 kDa) in solution, while TL-4 (30 kDa) exists under physiological conditions as a high molecular weight oligomer (470 kDa) consisting of 30 kDa subunits [127,128]. Wild-type TPL-2 (18 kDa) exists in solution as a covalent dimer (36 kDa), and the cleavage of the intermolecular disulfide bond results in monomer formation and loss of LPS-binding activity. LPS induces TPL2 oligomerization, in which tetramers and hexamers are formed. In hemolymph plasma, TPL-2 is predominantly present as oligomers with a molecular weight above 60 kDa. Carbohydrate chains of TPL-2 glycoprotein have been suggested to be responsible for the formation of the oligomers' stable cluster through protein–carbohydrate interactions. Unlike most other LPS-binding proteins, TPL-2 has a near-neutral pI of 7.65. However, there are three clusters of basic amino acids in the TPL-2 sequence that may be critical for its binding to LPS. TPL2 inhibited the growth of Gram-negative *E. coli* but was almost unable to detect Gram-positive bacteria.

The X-ray structures of tachylectin-2 and its complex with N-acetyl-D-glucosamine were solved with a resolution of 2.0 Å [132]. The lectin has a five-bladed β-propeller structure: five four-stranded antiparallel interconnected beta-sheets of W-shaped topology are located around a central water-filled tunnel, with the water molecules arranged as a pentagonal dodecahedron (Figure 8a).

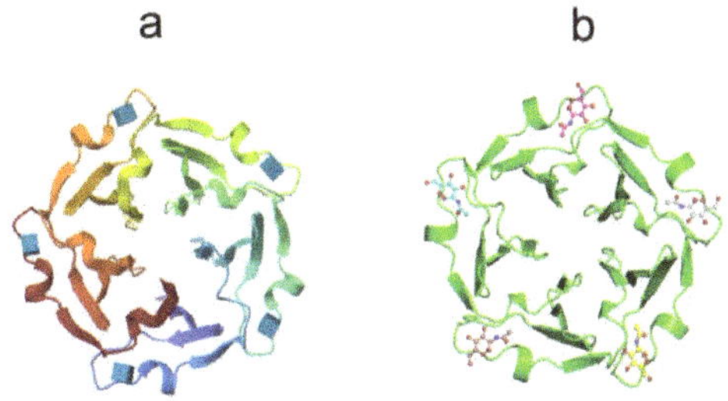

Figure 8. Crystal structures of (**a**) tachylectin-2 (pdb, 1tl2) and (**b**) complex of tachylectin-2 with N-acetyl-D-glucosamine (pdbe, 1tl2). Bound GlcNAc is shown as a ball-and-stick model.

The TL-2 molecule has five equivalent carbohydrate-binding sites located between adjacent β-sheets. The binding sites are formed by a large loop between the outermost strands of β-sheets and the connecting segment of the previous β-sheet (Figure 8b). According to crystal structure analysis, TL-1 is the protein of the six-bladed β-propeller structure [133]. The non-covalently bound TL-3 dimer is expected to have a four-blade β-propeller structure. A large number of binding sites in one polypeptide chain convincingly indicates the ability of the lectin to recognize carbohydrate surface structures of pathogens with a sufficiently high density of ligands.

The high affinity and specificity of horseshoe crab lectins binding with a propeller-like fold or oligomeric organization to a ligand is achieved due to their multivalence, short distances between individual binding sites (for example, 25 Å and 40 Å for TL-2), and low structural flexibility: upon ligand binding, the conformation of the main or side chains of tachylectins does not change at all. These observations suggest that these lectins can recognize parameters characterizing the distribution of glycan ligands on the cell surface, such as density, mobility and spatial arrangement, and this enables them to distinguish between simple ligands (monosaccharides and oligosaccharides) expressed on both the pathogen and the host and thus to discriminate between self and nonself.

The LPS-binding lectin, which is structurally related to tachylectins, has been found in the marine sponge *Suberites domuncula* [134]. This lectin (27 kDa), like TL-1, has six tandem repeats of 30–38 amino acid residues in sequence and exhibits high activity against Gram-negative bacteria, which is inhibited by D-GlcNAc, but not by D-GlcN. A number of proteins of the LPS- and β-1,3-glucan-binding proteins (LGBP) family can be assigned to lectins interacting with LPS. Many of them show a high degree of homology with invertebrate 1,3-β-glucanases. Probably, during the evolution, one of the gene copies of this enzyme evolved towards the specialization of the encoded protein exclusively for the function of binding to LPS and 1,3-β-glucans as a trigger mechanism for the body's defense response to the introduction of pathogenic microorganisms. At the same time, the protein lost its enzymatic activity, but its catalytic domain became a binding site not only for β-1,3-glucan but also for LPS [135]. Proteins of this family have been most extensively studied in crustaceans [136–138]. Recombinant LGBP (PmLGBP) from the *P. monodon* shrimp with a calculated molecular weight and pI of 39.8 kDa and 4.28, respectively, binds LPS with the apparent dissociation constant of 3.55×10^{-7} M [139]. Strong binding to LPS, as well as agglutinating activity against Gram-negative and Gram-positive bacteria, was shown by rLGBP from the scallop *Chlamys farreri* [140].

8. Conclusions

The present literature review showed that the LPS-binding proteins from marine invertebrates are understudied, which is confirmed by the data summarized in Table 1. These proteins have been characterized in a small number of invertebrate species that mainly inhabit the tropical seas of Southeast Asia and are objects of mariculture in the countries of the region. In addition, the study of ILBPs from new species is often limited by the search for homologues of already-known proteins. Basically, the field of researchers' interest is focused on the antimicrobial properties of these proteins, while their potential LPS-binding and LPS-neutralizing activities remain unidentified. Currently, it is impossible to exclude the discovery of new ILBP structural types under their targeted search involving new species of marine invertebrates. So a recent screening of marine invertebrates from the Sea of Okhotsk belonging to different taxonomic groups revealed a large number of species with LPS-binding activity that are of interest as new sources of ILBPs [141].

It is noteworthy that most of the well-studied ILBPs were originally isolated from the hemolymph of horseshoe crabs (lat. Xiphosura). This is for several reasons. The horseshoe crabs have the best-characterized immune system of any long-lived invertebrate. The study of immunity in the horseshoe crab has been facilitated by the availability and ease of collecting large volumes of blood. In addition, these marine animals have existed on Earth for about half a billion years and are often referred to as living fossils. The horseshoe crab habitat is rich in pathogenic microorganisms and contains a vast amount of endotoxin, since most aquatic bacteria are of the Gram-negative variety. During their long evolution, horseshoe crabs have formed a unique and very efficient host defense system, which includes a large set of proteins and peptides with high antimicrobial and endotoxin neutralizing activity. These defense molecules are attracting much attention from researchers as potential therapeutic agents.

Endotoxin-neutralizing ILBPs exhibit some structural features that provide optimal parameters for their interaction with LPS and inhibition of LPS toxicity. These proteins/peptides are mostly cationic amphiphilic molecules that have clusters of hydrophobic and hydrophilic amino acid residues on their surface [41,142]. Positively charged residues play a key role in binding ILBP to LPS and neutralizing endotoxin. Anionic ALF peptides that have lost most of these residues in the LPS-binding domains are unable to interact with LPS and exhibit low antimicrobial activity [143]. A high positive charge allows ILBPs to replace divalent cations, approach LPS molecules through strong electrostatic interactions, and neutralize and even overcompensate their negative charge. It could be argued, however, that the proper positioning of the basic amino acids in the three-dimensional structure of the protein is more important than the overall basic (cationic) nature of the protein for binding to the negatively charged LPS. An important factor is the distance between the positively charged residues in the ILBPs. The charged amino groups of Arg and Lys in ILBPs bound to LPS micelles show a typical distance range of 12–15 Å, which is in good accordance with an average distance between phosphate groups in lipid A [144,145]. This fact may mean that the positively charged amino acids residues in ILBPs mainly interact with the lipid A phosphate groups.

The presence of hydrophobic residues in the molecule, along with cationic ones, allows ILBPs to penetrate deeply into LPS micelles and bilayers and interact with lipid A acyl chains. A strong positive correlation is observed between hydrophobicity and LPS-binding activity of ILBPs [146]. An increase in the ratio between hydrophobicity (after reaching a certain threshold) and the net molecule positive charge increases the ability of ILBPs to neutralize LPS. At high hydrophobicity (outside the range), the activity drops, probably due to the strong self-association of ILBPs.

Table 1. LPS-binding peptides and proteins from marine invertebrates and their characteristics.

Peptides/Protein	Size, kDa	Structural Characteristics	Biological Activity	Source (Phylum, Species)	Reference
Anti-LPS factor (ALF)	11–12	α-helices, four-stranded β-sheet, disulfide bond	LPS-binding and -neutralization, antibacterial (G−, G+), antiviral, antifungal	Arthropoda—horseshoe crabs *Limulus polyphemus*, *Tachypleus tridentatus*, shrimps *Penaeus monodon*, *Fenneropenaeus chinensis*, *Litopenaeus vannamei*, crabs *Portunus trituberculatus*, *Scylla serrata*, *Eriocheir sinensis*, *Scylla paramamosai*, lobster *Homarus americanus*	[7–9,15–22,35]
Arenicins	2.62	antiparallel β-hairpin, disulfide bonds	anti-endotoxin (optimized arenicin derivatives), antibacterial (G−, G+), antifungal	Annelida—lugworm *Arenicola marina*	[43,47,48]
Tachyplesins, polyphemusins	2.27–2.46	antiparallel β-hairpin, amidated C-terminal arginine residue, disulfide bonds	LPS-binding and -neutralization, antibacterial (G−, G+), antifungal	Arthropoda—horseshoe crabs *L. polyphemus*, *T. tridentatus*, *Tachypleus gigas*, *Carcinoscorpius rotundicauda*	[53–63]
Big defensins	8–11	α-helices, β-sheets (antiparallel and parallel); two domains, disulfide bonds	LPS- binding; antibacterial (G−, G+), antifungal	Arthropoda—horseshoe crab *T. tridentatus*; Mollusca—oyster *Crassostrea gigas*, clam *Venerupis philippinarum*, scallops *Argopecten irradiant*, *Chlamys nobilis*; Chordata—amphioxus *Branchiostoma japonicum*	[69–78]
Factor C	120 and 132	β-sheets, disordered segments (loops), multidomain structure, disulfide bonds, tandem modules	LPS-binding (binding sites with K_D from 10^{-9} to 10^{-10} M) and -neutralization	Arthropoda—horseshoe crabs *L. polyphemus*, *T. tridentatus*, *Carcinoscorpius rotundicauda*	[80,81,86–88,96,97]
Bactericidal/permeability-increasing proteins (BPI)	50.1	α-helix, β-sheet; two-domain "boomerang-like" structure, disulfide bonds	LPS-binding and -neutralization; antibacterial (G−)	Mollusca—oyster *Crassostrea gigas*, squid *Euprymna scolopes*; Annelida—worm *Platynereis dumerilii*; Echinodermata—urchin *Sterechinus neumayeri*	[110,118–121]
Tachylectins (TL-1 to TL-4 and TPL2)	27, 27, 14, 30 (TL-1,-2,-3,-4); 18 (TPL-2)	β-sheets (four-stranded antiparallel interconnected), propeller-like fold or oligomeric organization, tandem repeats in sequence	LPS-binding (K_D 1.03 × 10^{-6} M for *E. coli* LPS); antibacterial activity (G-)	Arthropoda—horseshoe crab *T. tridentatus*; Porifera—sponge *Suberites domuncula*	[125,126,134]
LPS- and β-1,3-glucan-binding proteins (LGBP)	40-60		LPS-binding (K_D 3.55 × 10^{-7} M for *E. coli* LPS); antibacterial (G−, G+)	Arthropoda—shrimps *Penaeus monodon*, *Fenneropenaeus merguiensis*, crab *Eriocheir sinensis*; Mollusca—scallop *Chlamys farreri*	[136,137,139,140]

G−, Gram-negative bacteria, G+, Gram-positive bacteria.

ILBP incorporation into LPS aggregates leads to a change in the endotoxin supramolecular structure [42,147]. The protein-induced conversion of the unilamellar, cubic, or mixed unilamellar/cubic aggregate structures of LPS and lipid A into a multilamellar form is considered as a necessary condition for LPS inactivation. The degree of LPS multilamellarization can directly correlate with the endotoxin-neutralizing activity of ILBPs.

The specific arrangement of amino acid residues in the ILBP molecules is important for the expression of their anti-LPS activity. As illustrated by synthetic peptides (based on the

Limulus anti-LPS factor), it was shown that, while they have the same number of cationic and hydrophobic residues at similar sequence positions, they differ from each other through their LPS-neutralizing activity [42]. Moreover, these peptides differ in their ability to neutralize LPS in isolated forms and as constituents of Gram-negative bacteria. Apparently, this fact is explained by the fact that LPS molecules in the aggregate with a cubic structure and in the outer leaflet of the outer membrane of bacteria have different conformations, and peptides with different spatial structures are required for their neutralization. Thus, the geometric correspondence between the ILBP and LPS conformations, which allows positively charged protein residues to bind with high efficiency to the phosphate groups of lipid A, and hydrophobic residues to incorporate into its lipophilic part, can determine the endotoxin-neutralizing activity of the protein. In this regard, a well-organized, stable spatial structure of ILBPs, providing this structural compatibility, may be an important condition for binding lipid A with high affinity, which leads to LPS neutralization. The packaging of aromatic amino acid side chains, perhaps in part because they play an important role in stabilizing the compact structure of ILBPs, has a remarkable impact on the LPS-binding affinity [148]. A significant contribution to the stability of the ILBP structure is made by disulfide bonds. The substitution of cysteine residues in ALF, accompanied by the removal of a disulfide bond, can lead to a loss of endotoxin-neutralizing activity [40]. At the same time, the fully unfolded analog of tachyplesin-1, which has lost disulfide bonds, acquires a well-ordered structure upon binding to the LPS bilayer [144]. Moreover, ILBPs, which are mainly unstructured in solution, can gain an ordered conformation upon interaction with LPS micelles.

The enhancement of LPS-binding and -neutralizing activities of ILBP can be achieved by creating tandem repeats of the LPS-binding units in its molecule or by forming oligomeric forms of the protein. The effectiveness of this multivalent strategy for an improvement in the activity is demonstrated by ILBPs such as Factor C and tachyplesins [97,123].

Although ILBPs remain one of the most promising molecules for the development of endotoxin-neutralizing drugs, there are serious limitations to their introduction into medical practice. They can be unstable under physiological conditions (in particular, they are attacked by proteolytic enzymes) and toxic to mammalian cells and are quickly eliminated from the body [149]. The effective neutralization of LPS requires high therapeutic concentrations of ILBP, which causes serious side effects. To address these shortcomings, the structure of ILBPs is modified, or their synthetic derivatives are obtained. Thus, the stability of LBPs and their resistance to enzymatic degradation can be increased by the cyclization of the peptide (linking the C- and N-terminus), the introduction of D-isomers or unnatural amino acids into the peptide sequence, and their association with nanoparticles [150]. Covalent binding to polyethylene glycol increases the bioavailability of LBPs due to a decrease in the rate of renal clearance [151]. High manufacturing costs represent another major challenge for therapeutic applications of ILBPs. The development of recombinant DNA technologies and the solid-phase peptide synthesis method (SPPS) will contribute to solving this problem.

Synthetic peptides based on the structure of LPS-binding domains of known natural endotoxin-neutralizing ILBPs are of great interest as potential drugs for the treatment of sepsis [42]. Such peptides differ from their natural counterparts in size, amino acid substitutions, and other structural modifications introduced to increase their potential pharmacological efficacy and safety and considering the relationship between structure and biological activity established in the study of native ILBPs. The designed peptides demonstrate significant protective effect against septic shock in animal sepsis models even at a low peptide dose, a rather long half-life, low cytotoxic, and hemolytic activity.

However, ILBP molecules that show high efficacy in the treatment of sepsis in laboratory animals have been unsuccessful in human trials. This can partly be explained by the incomplete adequacy of mouse models of sepsis and septic shock [152]. Another reason may be the variability in the pathogenesis of various septic complications and the heterogeneity of patients, which must be taken into account when developing a strategy for clinical trials. For a correct assessment of the effectiveness of potential anti-endotoxic drugs, certain conditions must be met during trials: clinically proven selection of a suitable type of patients; the early recognition of sepsis and well-timed initiation of goal-directed therapy, which can interrupt the inflammatory cascade, preventing the progression to septic shock with multiple organ dysfunction; and the optimal duration of therapy.

Despite all the difficulties and disappointments, ILBPs remain one of the most promising molecules that can effectively neutralize bacterial endotoxins and inhibit the development of a systemic inflammatory response with cytokine overproduction (a cytokine storm). These molecules are often multifunctional [153]. In addition to the antiendotoxic activity, many peptides may exhibit antimicrobial and multifaceted immunomodulatory properties, possibly resulting in their wider therapeutic possibility. However, further research is required to evaluate the potential of LPS-neutralizing molecules with additional beneficial properties in the treatment of sepsis.

Author Contributions: Writing original draft, conceptualization, T.F.S.; Writing—Review and editing, T.F.S., S.I.B. and G.A.N. All authors have read and agreed to the published version of the manuscript.

Funding: This research received no external funding.

Institutional Review Board Statement: Not applicable.

Informed Consent Statement: Not applicable.

Data Availability Statement: Protein database (PDB) was used to obtain structural information for the proteins presented in the review.

Conflicts of Interest: The authors declare no conflict of interest.

Abbreviations

LPS	lipopolysaccharide
ILBPs	invertebrate lipopolysaccharide-binding proteins.

References

1. Alexander, C.; Rietschel, E.T. Bacterial lipopolysaccharides and innate immunity. *J. Endotoxin Res.* **2001**, *7*, 167–202. [CrossRef] [PubMed]
2. Lynn, M.; Rossignol, D.P.; Wheeler, J.L.; Kao, R.J.; Perdomo, C.A.; Noveck, R.; Vargas, R.; D'Angelo, T.; Gotzkowsky, S.; McMahon, F.G. Blocking of responses to endotoxin by E5564 in healthy volunteers with experimental endotoxemia. *J. Infect. Dis.* **2003**, *187*, 631–639. [CrossRef] [PubMed]
3. Solov'eva, T.; Davydova, V.; Krasikova, I.; Yermak, I. Marine compounds with therapeutic potential in Gram-negative sepsis. *Mar. Drugs* **2013**, *11*, 2216–2229. [CrossRef]
4. Jerala, R.; Porro, M. Endotoxin neutralizing peptides. *Curr. Top. Med. Chem.* **2004**, *4*, 1173–1184. [CrossRef]
5. Iwanaga, S.; Lee, B.L. Recent advances in the innate immunity of invertebrate animals. *J. Biochem. Mol. Biol.* **2005**, *38*, 128–150. [CrossRef] [PubMed]
6. Medzhitov, R.; Janeway, C., Jr. Innate immune recognition: Mechanisms and pathways. *Immunol. Rev.* **2000**, *173*, 89–97. [CrossRef]
7. Tanaka, S.; Nakamura, T.; Morita, T.; Iwanaga, S. Limulus anti-LPS factor: An anticoagulant which inhibits the endotoxin-mediated activation of Limulus coagulation system. *Biochem. Biophys. Res. Commun.* **1982**, *105*, 717–723. [CrossRef]
8. Morita, T.; Ohtsubo, S.; Nakamura, T.; Tanaka, S.; Iwanaga, S.; Ohashi, K.; Niwa, M. Isolation and biological activities of limulus anticoagulant (anti-LPS factor) which interacts with lipopolysaccharide (LPS). *J. Biochem.* **1985**, *97*, 1611–1620. [CrossRef]
9. Muta, T.; Miyata, T.; Tokunaga, F.; Nakamura, T.; Iwanaga, S. Primary structure of anti-lipopolysaccharide factor from American horseshoe crab, *Limulus polyphemus*. *J. Biochem.* **1987**, *101*, 1321–1330. [CrossRef]
10. Roth, R.I.; Su, D.; Child, A.H.; Wainwright, N.R.; Levin, J. Limulus anti-lipopolysaccharide factor prevents mortality late in the course of endotoxemia. *J. Infect. Dis.* **1998**, *177*, 388–394. [CrossRef]
11. Warren, H.S.; Glennon, M.L.; Wainwright, N.; Amato, S.F.; Black, K.M.; Kirsch, S.J.; Riveau, G.R.; Whyte, R.I.; Zapol, W.M.; Novitsky, T.J. Binding and neutralization of endotoxin by limulus anti-lipopolysaccharide factor. *Infect. Immun.* **1992**, *60*, 2506–2513. [CrossRef] [PubMed]
12. Desch, C.; O'Hara, P.; Harlan, J. Anti-lipopolysaccharide factor from horseshoe crab, *Tachypleus tridentatus*, inhibits lipopolysaccharide activation of cultured human endothelial cells. *Infect. Immun.* **1989**, *57*, 1612–1614. [CrossRef] [PubMed]

13. Alpert, G.; Baldwin, G.; Thompson, C.; Wainwright, N.; Novitsky, T.J.; Gillis, Z.; Parsonnet, J.; Fleisher, G.R.; Siber, G.R. Limulus antilipopolysaccharide factor protects rabbits from meningococcal endotoxin shock. *J. Infect. Dis.* **1992**, *165*, 494–500. [CrossRef] [PubMed]
14. Saladino, C.T.; Garcia, C.M.; Thompson, B.K.; Hammer, J.; Parsonnet, T.J.; Novitsky, G.R.; Siber, G.R.; Fleisher, G. Efficacy of a recombinant endotoxin neutralizing protein in rabbits with *Escherichia coli* sepsis. *Circ. Shock* **1994**, *42*, 104–110. [PubMed]
15. Supungul, P.; Klinbunga, S.; Pichyangkura, R.; Hirono, I.; Aoki, T.; Tassanakajon, A. Antimicrobial peptides discovered in the black tiger shrimp *Penaeus monodon* using the EST approach. *Dis. Aquat. Organ.* **2004**, *61*, 123–135. [CrossRef]
16. Liu, F.; Liu, Y.; Li, F.; Dong, B.; Xiang, J. Molecular cloning and expression profile of putative antilipopolysaccharide factor in chinese shrimp (*Fenneropenaeus chinensis*). *Mar. Biotechnol.* **2005**, *7*, 600–608. [CrossRef] [PubMed]
17. de la Vega, E.; O'Leary, N.A.; Shockey, J.E.; Robalino, J.; Payne, C.; Browdy, C.L.; Warr, G.W.; Gross, P.S. Anti-lipopolysaccharide factor in *Litopenaeus vannamei* (LvALF): A broad spectrum antimicrobial peptide essential for shrimp immunity against bacterial and fungal infection. *Mol. Immunol.* **2008**, *45*, 1916–1925. [CrossRef] [PubMed]
18. Beale, K.M.; Towle, D.W.; Jayasundara, N.; Smith, C.M.; Shields, J.D.; Small, H.J.; Greenwood, S.J. Anti-lipopolysaccharide factors in the american lobster *Homarus americanus*: Molecular characterization and transcriptional response to *Vibrio fluvialis* challenge. *Comp. Biochem. Physiol. Part D Genom. Proteom.* **2008**, *3*, 263–269. [CrossRef]
19. Zhang, Y.; Wang, L.; Wang, L.; Yang, J.; Gai, Y.; Qiu, L.; Song, L. The second anti-lipopolysaccharide factor (EsALF-2) with antimicrobial activity from *Eriocheir sinensis*. *Dev. Comp. Immunol.* **2010**, *34*, 945–952. [CrossRef]
20. Imjongjirak, C.; Amparyup, P.; Tassanakajon, A.; Sittipraneed, S. Antilipopolysaccharide factor (ALF) of mud crab *Scylla paramamosain*: Molecular cloning, genomic organization and the antimicrobial activity of its synthetic LPS binding domain. *Mol. Immunol.* **2007**, *44*, 3195–3203. [CrossRef]
21. Afsal, V.V.; Antony, S.P.; Sathyan, N.; Philip, R. Molecular characterization and phylogenetic analysis of two antimicrobial peptides: Anti-lipopolysaccharide factor and crustin from the brown mud crab, *Scylla serrata*. *Results Immunol.* **2011**, *1*, 6–10. [CrossRef]
22. Afsal, V.V.; Antony, S.P.; Sanjeevan, V.N.; Kumar, P.A.; Singh, I.B.; Philip, R. A new isoform of anti-lipopolysaccharide factor identified from the blue swimmer crab, *Portunus pelagicus*: Molecular characteristics and phylogeny. *Aquaculture* **2012**, *356*, 119–122. [CrossRef]
23. Tassanakajon, A.; Amparyup, P.; Somboonwiwat, K.; Supungul, P. Cationic antimicrobial peptides in Penaeid shrimp. *Mar. Biotechnol.* **2011**, *13*, 639–657. [CrossRef]
24. Somboonwiwat, K.; Marcos, M.; Tassanakajon, A.; Klinbunga, S.; Aumelas, A.; Romestand, B.; Gueguen, Y.; Boze, H.; Moulin, G.; Bachère, E. Recombinant expression and anti-microbial activity of anti-lipopolysaccharide factor (ALF) from the black tiger shrimp *Penaeus monodon*. *Dev. Comp. Immunol.* **2005**, *29*, 841–851. [CrossRef]
25. Yedery, R.D.; Reddy, K.V. Identification, cloning, characterization and recombinant expression of an anti-lipopolysaccharide factor from the hemocytes of indian mud crab, *Scylla serrata*. *Fish Shellfish Immunol.* **2009**, *27*, 275–284. [CrossRef] [PubMed]
26. Yang, H.; Li, S.; Li, F.; Lv, X.; Xiang, J. Recombinant expression and functional analysis of an isoform of anti-lipopolysaccharide factors (FcALF5) from Chinese shrimp *Fenneropenaeus chinensis*. *Dev. Comp. Immunol.* **2015**, *53*, 47–54. [CrossRef] [PubMed]
27. Tharntada, S.; Somboonwiwat, K.; Rimphanitchayakit, V.; Tassanakajon, A. Anti-lipopolysaccharide factors from the black tiger shrimp, *Penaeus monodon*, are encoded by two genomic loci. *Fish Shellfish Immunol.* **2008**, *24*, 46–54. [CrossRef]
28. Ponprateep, S.; Tharntada, S.; Somboonwiwat, K.; Tassanakajon, A. Gene silencing reveals a crucial role for anti-lipopolysaccharide factors from *Penaeus monodon* in the protection against microbial infections. *Fish Shellfish Immunol.* **2012**, *32*, 26–34. [CrossRef]
29. Li, S.H.; Zhang, X.J.; Sun, Z.; Li, F.H.; Xiang, J.H. Transcriptome analysis on Chinese shrimp *Fenneropenaeus chinensis* during WSSV acute infection. *PLoS ONE* **2013**, *8*, e58627. [CrossRef]
30. Liu, Y.; Cui, Z.X.; Song, C.W.; Wang, S.Y.; Li, Q. Multiple isoforms of immune-related genes from hemocytes and eyestalk cDNA libraries of swimming crab *Portunus trituberculatus*. *Fish Shellfish Immunol.* **2011**, *31*, 29–42. [CrossRef]
31. Li, S.; Li, F. The anti-lipopolysaccharide factors in crustaceans. *Subcell Biochem.* **2020**, *94*, 63–80. [CrossRef] [PubMed]
32. Li, S.; Guo, S.; Li, F.; Xiang, J. Functional diversity of anti-lipopolysaccharide factor isoforms in shrimp and their characters related to antiviral activity. *Mar. Drugs* **2015**, *13*, 2602–2616. [CrossRef] [PubMed]
33. Matos, G.M.; Schmitt, P.; Barreto, C.; Farias, N.D.; Toledo-Silva, G.; Guzmán, F.; Destoumieux-Garzón, D.; Perazzolo, L.M.; Rosa, R.D. Massive gene expansion and sequence diversification is associated with diverse tissue distribution, regulation and antimicrobial properties of anti-lipopolysaccharide factors in shrimp. *Mar. Drugs* **2018**, *16*, 381. [CrossRef] [PubMed]
34. Zhou, L.; Li, G.; Jiao, Y.; Huang, D.; Li, A.; Chen, H.; Liu, Y.; Li, S.; Li, H.; Wang, C. Molecular and antimicrobial characterization of a group G anti-lipopolysaccharide factor (ALF) from *Penaeus monodon*. *Fish Shellfish Immunol.* **2019**, *94*, 149–156. [CrossRef]
35. Hou, Z.G.; Wang, Y.; Hui, K.; Fang, W.H.; Zhao, S.; Zhang, J.X.; Ma, H.; Li, X.C. A novel anti-lipopolysaccharide factor SpALF6 in mud crab *Scylla paramamosain* exhibiting different antimicrobial activity from its single amino acid mutant. *Dev. Comp. Immunol.* **2017**, *72*, 44–56. [CrossRef]
36. Hoess, A.; Watson, S.; Siber, G.R.; Liddington, R. Crystal structure of an endotoxin-neutralizing protein from the horseshoe crab, *Limulus* antiLPS factor, at 1.5 A resolution. *EMBO J.* **1993**, *12*, 3351–3356. [CrossRef]
37. Tharntada, S.; Ponprateep, S.; Somboonwiwat, K.; Liu, H.; Söderhäll, I.; Söderhäll, K.; Tassanakajon, A. Role of antilipopolysaccharide factor from the black tiger shrimp, *Penaeus monodon*, in protection from white spot syndrome virus infection. *J. Gen. Virol.* **2009**, *90*, 1491–1498. [CrossRef] [PubMed]

38. Pan, C.Y.; Chao, T.T.; Chen, J.C.; Chen, J.Y.; Liu, W.C.; Lin, C.H.; Kuo, C.M. Shrimp (*Penaeus monodon*) antilipopolysaccharide factor reduces the lethality of *Pseudomonas aeruginosa* sepsis in mice. *Int. Immunopharmacol.* **2007**, *7*, 687–700. [CrossRef]
39. Guo, S.Y.; Li, S.H.; Li, F.H.; Zhang, X.J.; Xiang, J.H. Modification of a synthetic LPS-binding domain of anti-lipopolysaccharide factor from shrimp reveals strong structure-activity relationship in their antimicrobial characteristics. *Dev. Comp. Immunol.* **2014**, *45*, 227–232. [CrossRef]
40. Yang, Y.; Boze, H.; Chemardin, P.; Padilla, A.; Moulin, G.; Tassanakajon, A.; Pugniere, M.; Roquet, F.; Destoumieux-Garzon, D.; Gueguen, Y.; et al. NMR structure of rALF-Pm3, an anti-lipopolysaccharide factor from shrimp: Model of the possible lipid A-binding site. *Biopolymers* **2009**, *91*, 207–220. [CrossRef]
41. Andrä, J.; Howe, J.; Garidel, P.; Rössle, M.; Richter, W.; Leiva-León, J.; Moriyon, I.; Bartels, R.; Gutsmann, T.; Brandenburg, K. Mechanism of interaction of optimized Limulus-derived cyclic peptides with endotoxins: Thermodynamic, biophysical and microbiological analysis. *Biochem. J.* **2007**, *406*, 297–307. [CrossRef] [PubMed]
42. Gutsmann, T.; Razquin-Olazarán, I.; Kowalski, I.; Kaconis, Y.; Howe, J.; Bartels, R.; Hornef, M.; Schürholz, T.; Rössle, M.; Sanchez-Gómez, S.; et al. New antiseptic peptides to protect against endotoxin mediated shock. *Antimicrob. Agents Chemother.* **2010**, *54*, 3817–3824. [CrossRef]
43. Ovchinnikova, T.V.; Aleshina, G.M.; Balandin, S.V.; Krasnosdembskaya, A.D.; Markelov, M.L.; Frolova, E.I.; Kokryakov, V.N. Purification and primary structure of two isoforms of arenicin, a novel antimicrobial peptide from marine polychaeta *Arenicola marina*. *FEBS Lett.* **2004**, *577*, 209–214. [CrossRef] [PubMed]
44. Nielsen, A.K.; Sandvang, D.; Neve, S.; Kruse, T.; Kristensen, H.H. Transcriptional profiling indicates a dual mode-of-action of arenicin-3, poster F1-2072. In Proceedings of the 50th Interscience Conference on Antimicrobial Agents and Chemotherapy, Boston, MA, USA, 12–15 September 2010; American Society for Microbiology: Washington, DC, USA, 2010.
45. Ovchinnikova, T.V.; Shenkarev, Z.O.; Nadezhdin, K.D.; Balandin, S.V.; Zhmak, M.N.; Kudelina, I.A.; Finkina, E.I.; Kokryakov, V.N.; Arseniev, A.S. Recombinant expression, synthesis, purification, and solution structure of arenicin. *Biochem. Biophys. Res. Commun.* **2007**, *360*, 156–162. [CrossRef] [PubMed]
46. Elliott, A.; Huang, J.; Neve, S.; Zuegg, J.; Edwards, I.; Cain, A.; Boinett, C.; Barquist, L.; Lundberg, C.; Steen, J.; et al. An amphipathic peptide with antibiotic activity against multidrug-resistant Gram-negative bacteria. *Nat. Commun.* **2020**, *11*, 3184. [CrossRef] [PubMed]
47. Shenkarev, Z.O.; Balandin, S.V.; Trunov, K.I.; Paramonov, A.S.; Sukhanov, S.V.; Barsukov, L.I.; Arseniev, A.S.; Ovchinnikova, T.V. Molecular mechanism of action of β-hairpin antimicrobial peptide arenicin: Oligomeric structure in dodecylphosphocholine micelles and pore formation in planar lipid bilayers. *Biochemistry* **2011**, *50*, 6255–6265. [CrossRef]
48. Ovchinnikova, T.V.; Shenkarev, Z.O.; Balandin, S.V.; Nadezhdin, K.D.; Paramonov, A.S.; Kokryakov, V.N.; Arseniev, A.S. Molecular insight into mechanism of antimicrobial action of the β-hairpin peptide arenicin: Specific oligomerization in detergent micelles. *Biopolym. Orig. Res. Biomol.* **2008**, *89*, 455–464. [CrossRef]
49. Andrä, J.; Jakovkin, I.; Grötzinger, J.; Hecht, O.; Krasnosdembskaya, A.D.; Goldmann, T.; Gutsmann, T.; Leippe, M. Structure and mode of action of the antimicrobial peptide arenicin. *Biochem. J.* **2008**, *410*, 113–122. [CrossRef]
50. Yang, N.; Liu, X.; Teng, D.; Li, Z.; Wang, X.; Mao, R.; Wang, X.; Hao, Y.; Wang, J. Antibacterial and detoxifying activity of NZ17074 analogues with multi-layers of selective antimicrobial actions against *Escherichia coli* and *Salmonella enteritidis*. *Sci. Rep.* **2017**, *7*, 3392. [CrossRef]
51. Wang, X.; Teng, D.; Mao, R.; Yang, N.; Hao, Y.; Wang, J. Combined systems approaches reveal a multistage mode of action of a marine antimicrobial peptide against pathogenic *Escherichia coli* and its protective effect against bacterial peritonitis and endotoxemia. *Antimicrob. Agents Chemother.* **2017**, *61*, e01056-16. [CrossRef]
52. Hoegenhaug, H.H.K.; Mygind, P.H.; Kruse, T.; Segura, D.R.; Sandvang, D.; Neve, S. Antimicrobial Peptide Variants and Polynucleotides Encoding Same. US Patent 20110306750 A1, 10 June 2011.
53. Miyata, T.; Tokunaga, F.; Yoneya, T.; Yoshikawa, K.; Iwanaga, S.; Niwa, M.; Takao, T.; Shimonishi, Y. Antimicrobial peptides, isolated from horseshoe crab hemocytes, tachyplesin II, and polyphemusins I and II. Chemical structures and biological activity. *J. Biochem.* **1989**, *106*, 663–668. [CrossRef] [PubMed]
54. Nakamura, T.; Furunaka, H.; Miyata, T.; Tokunaga, F.; Muta, T.; Iwanaga, S.; Niwa, M.; Takao, T.; Shimonishi, Y. Tachyplesin, a class of antimicrobial peptide from the hemocytes of the horseshoe crab (*Tachypleus tridentatus*): Isolation and chemical structure. *J. Biol. Chem.* **1988**, *263*, 16709–16713. [CrossRef] [PubMed]
55. Muta, T.; Fujimoto, T.; Nakajima, H.; Iwanaga, S. Tachyplesins isolated from hemocytes of southeast Asian horseshoe crabs (*Carcinoscorpius rotundicauda* and *Tachypleus gigas*): Identification of a new tachyplesin, tachyplesin III, and a processing intermediate of its precursor. *J. Biochem.* **1990**, *108*, 261–266. [CrossRef] [PubMed]
56. Shigenaga, T.; Mutat, T.; Toh, Y.; Tokunaga, F.; Iwanaga, S. Antimicrobial tachyplesin peptide precursor cDNA cloning and cellular localization in the horseshoe crab (*Tachypleus tridentatus*). *J. Biol. Chem.* **1990**, *265*, S0021–S9258. [CrossRef]
57. Kawano, K.; Yoneya, T.; Miyata, T.; Yoshikawa, K.; Tokunaga, F.; Terada, Y.; Iwanaga, S. Antimicrobial peptide, tachyplesin I, isolated from hemocytes of the horseshoe crab (*Tachypleus tridentatus*). NMR determination of the beta-sheet structure. *J. Biol. Chem.* **1990**, *265*, 15365–15367. [CrossRef]
58. Laederach, A.; Andreotti, A.H.; Fulton, D.B. Solution and micelle-bound structures of tachyplesin I and its active aromatic linear derivatives. *Biochemistry* **2002**, *41*, 12359–12368. [CrossRef] [PubMed]

59. Powers, J.-P.S.; Rozek, A.; Hancock, R.E.W. Structure-activity relationships for the β-hairpin cationic antimicrobial peptide polyphemusin I. *Biochim. Biophys. Acta* **2004**, *1698*, 239–250. [CrossRef] [PubMed]
60. Vernen, F.; Harvey, P.J.; Dias, S.A.; Veiga, A.S.; Huang, Y.-H.; Craik, D.J.; Lawrence, N.; Troeira Henriques, S. Characterization of tachyplesin peptides and their cyclized analogues to improve antimicrobial and anticancer properties. *Int. J. Mol. Sci.* **2019**, *20*, 4184. [CrossRef] [PubMed]
61. Hirakura, Y.; Kobayashi, S.; Matsuzaki, K. Specific interactions of the antimicrobial peptide cyclic β-sheet tachyplesin I with lipopolysaccharides. *Biochim. Biophys. Acta Biomembr.* **2002**, *1562*, 32–36. [CrossRef]
62. Kushibiki, T.; Kamiya, M.; Aizawa, T.; Kumaki, Y.; Kikukawa, T.; Mizuguchi, M.; Demura, M.; Kawabata, S.; Kawano, K. Interaction between tachyplesin I, an antimicrobial peptide derived from horseshoe crab, and lipopolysaccharide. *Biochim. Biophys. Acta Proteins Proteom.* **2014**, *1844*, 527–534. [CrossRef]
63. Zhang, L.; Scott, M.G.; Yan, H.; Mayer, L.D.; Hancock, R.E. Interaction of polyphemusin I and structural analogs with bacterial membranes, lipopolysaccharide, and lipid monolayers. *Biochemistry* **2000**, *39*, 14504–14514. [CrossRef] [PubMed]
64. Edwards, I.A.; Elliott, A.G.; Kavanagh, A.M.; Zuegg, J.; Blaskovich, M.A.; Cooper, M.A. Contribution of amphipathicity and hydrophobicity to the antimicrobial activity and cytotoxicity of β-hairpin peptides. *ACS Infect. Dis.* **2016**, *2*, 442–450. [CrossRef]
65. Hong, J.; Hu, J.; Ke, F. Experimental induction of bacterial resistance to the antimicrobial peptide tachyplesin I and investigation of the resistance mechanisms. *Antimicrob. Agents Chemother.* **2016**, *60*, 6067–6075. [CrossRef]
66. Otero-Gonzalez, A.J.; Magalhaes, B.S.; Garcia-Villarino, M.; Lopez-Abarrategui, C.; Sousa, D.A.; Dias, S.C.; Franco, O.L. Antimicrobial peptides from marine invertebrates as a new frontier for microbial infection control. *FASEB J.* **2010**, *24*, 1320–1334. [CrossRef]
67. Powers, J.P.S.; Tan, A.; Ramamoorthy, A.; Hancock, R.E.W. Solution structure and interaction of the antimicrobial polyphemusins with lipid membranes. *Biochemistry* **2005**, *44*, 15504–15513. [CrossRef] [PubMed]
68. Saito, T.; Kawabata, S.I.; Shigenaga, T.; Takayenoki, Y.; Cho, J.; Nakajima, H.; Hirata, M.; Iwanaga, S.A. Novel big defensin identified in horseshoe crab hemocytes: Isolation, amino acid sequence, and antibacterial activity. *J. Biochem.* **1995**, *117*, 1131–1137. [CrossRef]
69. Kouno, T.; Fujitani, N.; Mizuguchi, M.; Osaki, T.; Nishimura, S.; Kawabata, S.; Aizawa, T.; Demura, M.; Nitta, K.; Kawano, K. A Novel beta-defensin structure: A potential strategy of big defensin for overcoming resistance by Gram-positive bacteria. *Biochemistry* **2008**, *47*, 10611–10619. [CrossRef] [PubMed]
70. Kouno, T.; Mizuguchi, M.; Aizawa, T.; Shinoda, H.; Demura, M.; Kawabata, S.; Kawano, K.A. novel beta-defensin structure: Big defensin changes its N-terminal structure to associate with the target membrane. *Biochemistry* **2009**, *48*, 7629–7635. [CrossRef]
71. Dhople, V.; Krukemeyer, A.; Ramamoorthy, A. The human beta-defensin-3, an antibacterial peptide with multiple biological functions. *Biochim. Biophys. Acta Biomembr.* **2006**, *1758*, 1499–1512. [CrossRef]
72. Perrigault, M.; Tanguy, A.; Allam, B. Identification and expression of differentially expressed genes in the hard clam, *Mercenaria mercenaria*, in response to quahog parasite unknown (QPX). *BMC Genom.* **2009**, *10*, 377. [CrossRef]
73. Rosa, R.D.; Santini, A.; Fievet, J.; Bulet, P.; Destoumieux-Garzón, D.; Bachère, E. Big defensins, a diverse family of antimicrobial peptides that follows different patterns of expression in hemocytes of the oyster *Crassostrea gigas*. *PLoS ONE* **2011**, *6*, e25594. [CrossRef] [PubMed]
74. Yang, J.; Luo, J.; Zheng, H.; Lu, Y.; Zhang, H. Cloning of a big defensin gene and its response to *Vibrio parahaemolyticus* challenge in the noble scallop *Chlamys nobilis* (Bivalve: Pectinidae). *Fish Shellfish Immunol.* **2016**, *56*, 445–449. [CrossRef] [PubMed]
75. Teng, L.; Gao, B.; Zhang, S. The first chordate big defensin: Identification, expression and bioactivity. *Fish Shellfish Immunol.* **2012**, *32*, 572–577. [CrossRef]
76. Loth, K.; Vergnes, A.; Barreto, C.; Voisin, S.N.; Meudal, H.; Da Silva, J.; Bressan, A.; Belmadi, N.; Bachère, E.; Aucagne, V.; et al. The ancestral N-terminal domain of big defensins drives bacterially triggered assembly into antimicrobial nanonets. *mBio* **2019**, *10*, e01821-19. [CrossRef] [PubMed]
77. Zhao, J.; Song, L.; Li, C.; Ni, D.; Wu, L.; Zhu, L.; Wang, H.; Xu, W. Molecular cloning, expression of a big defensin gene from bay scallop *Argopecten irradians* and the antimicrobial activity of its recombinant protein. *Mol. Immunol.* **2007**, *44*, 360–368. [CrossRef]
78. Zhao, J.; Li, C.; Chen, A.; Li, L.; Su, X.; Li, T. Molecular characterization of a novel big defensin from clam *Venerupis philippinarum*. *PLoS ONE* **2010**, *5*, e13480. [CrossRef]
79. Ding, J.L.; Ho, B. A new era in pyrogen testing. *Trends Biotechnol.* **2001**, *19*, 277–281. [CrossRef]
80. Koshiba, T.; Hashii, T.; Kawabata, S.I. A structural perspective on the interaction between lipopolysaccharide and factor C, a receptor involved in recognition of Gram-negative bacteria. *J. Biol. Chem.* **2007**, *282*, 3962–3967. [CrossRef]
81. Ariki, S.; Koori, K.; Osaki, T.; Motoyama, K.; Inamori, K.I.; Kawabata, S.I. A serine protease zymogen functions as a pattern-recognition receptor for lipopolysaccharides. *Proc. Natl. Acad. Sci. USA* **2004**, *101*, 953–958. [CrossRef]
82. Piehler, M.; Roeder, R.; Blessing, S.; Reich, J. Comparison of LAL and rFC assays-participation in a proficiency test program between 2014 and 2019. *Microorganisms* **2020**, *8*, 418. [CrossRef]
83. Li, P.; Ho, B.; Ding, J.L. Recombinant factor C competes against LBP to bind lipopolysaccharide and neutralizes the endotoxicity. *J. Endotoxin Res.* **2007**, *13*, 150–157. [CrossRef] [PubMed]
84. Nakamura, T.; Morita, T.; Iwanaga, S. Lipopolysaccharide-sensitive serine-protease zymogen (factor C) found in Limulus hemocytes: Isolation and characterization. *Eur. J. Biochem.* **1986**, *154*, 511–521. [CrossRef]

85. Tokunaga, F.; Miyata, T.; Nakamura, T.; Morita, T.; Kuma, K.; Miyata, T.; Iwanaga, S. Lipopolysaccharide-sensitive serine-protease zymogen (factor C) of horseshoe crab hemocytes. Identification and alignment of proteolytic fragments produced during the activation show that it is a novel type of serine protease. *Eur. J. Biochem.* **1987**, *167*, 405–411. [CrossRef] [PubMed]
86. Tokunaga, F.; Nakajima, H.; Iwanaga, S. Further studies on lipopolysaccharide-sensitive serine protease zymogen (factor C): Its isolation from *Limulus polyphemus* hemocytes and identification as an intracellular zymogen activated by α-chymotrypsin, not by trypsin. *J. Biochem.* **1991**, *109*, 150–157. [CrossRef]
87. Ding, J.L.; Navas, M.A.; Ho, B. Two forms of factor C from the amoebocytes of Carcinoscorpius rotundicauda: Purification and characterization. *Biochim. Biophys. Acta Protein Struct. Mol. Enzymol.* **1993**, *1202*, 149–156. [CrossRef]
88. Muta, T.; Miyata, T.; Misumi, Y.; Tokunaga, F.; Nakamura, T.; Toh, Y.; Ikehara, Y.; Iwanaga, S. Limulus factor C. An endotoxin-sensitive serine protease zymogen with a mosaic structure of complement-like, epidermal growth factor-like, and lectin-like domains. *J. Biol. Chem.* **1991**, *266*, 6554–6561. [CrossRef] [PubMed]
89. Trexler, M.; Banyai, L.; Patthy, L. The LCCL module. *Eur. J. Biochem.* **2000**, *267*, 5751–5757. [CrossRef]
90. Norman, D.G.; Barlow, P.N.; Baron, M.; Day, A.J.; Sim, R.B.; Campbell, I.D. Three-dimensional structure of a complement control protein module in solution. *J. Mol. Biol.* **1991**, *219*, 717–725. [CrossRef]
91. Savage, C.R.; Hash, J.H.; Cohen, S. Epidermal growth factor. Location of disulfide bonds. *J. Biol. Chem.* **1973**, *248*, 7669–7672. [CrossRef]
92. Ullner, M.; Selander, M.; Persson, E.; Stenflo, J.; Drakenberg, T.; Teleman, O. Three-dimensional structure of the apo form of the N-terminal EGF-like module of blood coagulation factor X as determined by NMR spectroscopy and simulated folding. *Biochemistry* **1992**, *31*, 5974–5983. [CrossRef]
93. Lu, H.S.; Chai, J.J.; Li, M.; Huang, B.R.; He, C.H.; Bi, R.C. Crystal structure of human epidermal growth factor and its dimerization. *J. Biol. Chem.* **2001**, *276*, 34913–34917. [CrossRef] [PubMed]
94. Nakamura, T.; Tokunaga, F.; Morita, T.; Iwanaga, S.; Kusumoto, S.; Shiba, P.; Kobayashi, T.; Inoue, K. Intracellular serine protease zymogen factor C, from horseshoe crab hemocytes: Its activation by synthetic lipid A analogues and acidic phospholipids. *Eur. J. Biochem.* **1988**, *176*, 89–94. [CrossRef] [PubMed]
95. Nakamura, T.; Tokunaga, F.; Morita, T.; Iwanaga, S. Interaction between lipopolysaccharide and intracellular serine protease zymogen, factor C, from horseshoe crab (*Tachypleus tridentatus*) hemocytes. *J. Biochem.* **1988**, *103*, 370–374. [CrossRef]
96. Tan, N.S.; Ho, B.; Ding, J.L. High-affinity LPS binding domain(s) in recombinant factor C of a horseshoe crab neutralizes LPS-induced lethality. *FASEB J.* **2000**, *14*, 859–870. [CrossRef]
97. Tan, N.S.; Ng, M.L.; Yau, Y.H.; Chong, P.K.; Ho, B.; Ding, J.L. Definition of endotoxin binding sites in horseshoe crab factor C recombinant sushi proteins and neutralization of endotoxin by sushi peptides. *FASEB J.* **2000**, *14*, 1801–1813. [CrossRef] [PubMed]
98. Pristovsek, P.; Feher, K.; Szilagyi, L.; Kidric, J. Structure of a synthetic fragment of the LALF protein when bound to lipopolysaccharide. *J. Med. Chem.* **2005**, *48*, 1666–1670. [CrossRef]
99. Wang, J.; Tan, N.S.; Ho, B.; Ding, J.L. Modular arrangement and secretion of a multidomain serine protease. Evidence for involvement of proline-rich region and N-glycans in the secretion pathway. *J. Biol. Chem.* **2002**, *277*, 36363–36372. [CrossRef]
100. Tobias, P.S.; Mathison, J.C.; Ulevitch, R.J. A family of lipopolysaccharide binding proteins involved in responses to Gram-negative sepsis. *J. Biol. Chem.* **1988**, *263*, 13479–13481. [CrossRef] [PubMed]
101. Levy, O. Antimicrobial proteins and peptides: Anti-infective molecules of mammalian leukocytes. *J. Leukoc. Biol.* **2004**, *76*, 909–925. [CrossRef]
102. Iovine, N.M.; Elsbach, P.; Weiss, J. An opsonic function of the neutrophil bactericidal/permeability-increasing protein depends on both its N- and C-terminal domains. *Proc. Natl. Acad. Sci. USA* **1997**, *94*, 10973–10978. [CrossRef]
103. Weiss, J. Bactericidal/permeability-increasing protein (BPI) and lipopolysaccharide-binding protein (LBP): Structure, function and regulation in host defence against Gram-negative bacteria. *Biochem. Soc. Trans.* **2003**, *31*, 785–790. [CrossRef] [PubMed]
104. Theprungsirikul, J.; Skopelja-Gardner, S.; Rigby, W.F. Killing three birds with one BPI: Bactericidal, opsonic, and anti-inflammatory functions. *J. Transl. Autoimmun.* **2021**, *4*, 100105. [CrossRef] [PubMed]
105. Weiss, J.; Franson, R.C.; Beckerdite, S.; Schmeidler, K.; Elsbach, P. Partial characterization and purification of a rabbit granulocyte factor that increases permeability of *Escherichia coli*. *J. Clin. Investig.* **1975**, *55*, 33–42. [CrossRef]
106. Weiss, J.; Elsbach, P.; Olsson, I.; Odeberg, H. Purification and characterization of a potent bactericidal and membrane active protein from the granules of human polymorphonuclear leukocytes. *J. Biol. Chem.* **1978**, *253*, 2664–2672. [CrossRef]
107. Beamer, L.J.; Carroll, S.F.; Eisenberg, D. Crystal structure of human BPI and two bound phospholipids at 2.4 angstrom resolution. *Science* **1997**, *276*, 1861–1864. [CrossRef]
108. Kleiger, G.; Beamer, L.J.; Grothe, R.; Mallick, P.; Eisenberg, D. The 1.7 A crystal structure of BPI: A study of how two dissimilar amino acid sequences can adopt the same fold. *J. Mol. Biol.* **2000**, *299*, 1019–1034. [CrossRef]
109. Krasity, B.C.; Troll, J.V.; Weiss, J.P.; McFall-Ngai, M.J. LBP/BPI proteins and their relatives: Conservation over evolution and roles in mutualism. *Biochem. Soc. Trans.* **2011**, *39*, 1039–1044. [CrossRef] [PubMed]
110. Gonzalez, M.; Gueguen, Y.; Destoumieux-Garzón, D.; Romestand, B.; Fievet, J.; Pugnière, M.; Roquet, F.; Escoubas, J.-M.; Vandenbulcke, F.; Levy, O.; et al. Evidence of a bactericidal permeability increasing protein in an invertebrate, the *Crassostrea gigas* Cg-BPI. *Proc. Natl. Acad. Sci. USA* **2007**, *104*, 17759–17764. [CrossRef]
111. Little, R.G.; Kelner, D.N.; Lim, E.; Burke, D.J.; Conlon, P.J. Functional domains of recombinant bactericidal/permeability increasing protein (rBPI23). *J. Biol. Chem.* **1994**, *269*, 1865–1872. [CrossRef]

112. Capodici, C.; Weiss, J. Both N- and C-terminal regions of the bioactive N-terminal fragment of the neutrophil granule bactericidal/permeability-increasing protein are required for stability and function. *J. Immunol.* **1996**, *156*, 4789–4796. [CrossRef]
113. Beamer, L.J.; Carroll, S.F.; Eisenberg, D. The BPI/LBP family of proteins: A structural analysis of conserved regions. *Protein Sci.* **1998**, *7*, 906–914. [CrossRef] [PubMed]
114. Horwitz, A.H.; Leigh, S.D.; Abrahamson, S.; Gazzano-Santoro, H.; Liu, P.S.; Williams, R.E.; Carroll, S.F.; Theofan, G. Expression and characterization of cysteine-modified variants of an amino-terminal fragment of bactericidal/permeability-increasing protein. *Protein Expr. Purif.* **1996**, *8*, 28–40. [CrossRef] [PubMed]
115. Weiss, J.; Beckerdite-Quagliata, S.; Elsbach, P. Resistance of Gram-negative bacteria to purified bactericidal leukocyte proteins: Relation to binding and bacterial lipopolysaccharide structure. *J. Clin. Investig.* **1980**, *65*, 619–628. [CrossRef] [PubMed]
116. Capodici, C.; Chen, S.; Sidorczyk, Z.; Elsbach, P.; Weiss, J. Effect of lipopolysaccharide (LPS) chain length on interactions of bactericidal/permeability-increasing protein and its bioactive 23-kilodalton NH2-terminal fragment with isolated LPS and intact *Proteus mirabilis* and *Escherichia coli*. *Infect. Immun.* **1994**, *62*, 259–265. [CrossRef]
117. Zhang, Y.; He, X.; Li, X.; Fu, D.; Chen, J.; Yu, Z. The second bactericidal permeability increasing protein (BPI) and its revelation of the gene duplication in the Pacific oyster, *Crassostrea gigas*. *Fish Shellfish Immunol.* **2011**, *30*, 954–963. [CrossRef]
118. Chen, F.; Krasity, B.C.; Peyer, S.M.; Koehler, S.; Ruby, E.G.; Zhang, X.; McFall-Ngai, M.J. Bactericidal permeability-increasing proteins shape host-microbe interactions. *mBio* **2017**, *8*, e00040-17. [CrossRef]
119. Altincicek, B.; Vilcinskas, A. Analysis of the immune-related transcriptome of a lophotrochozoan model, the marine annelid, *Platynereis dumerilii*. *Front. Zool.* **2007**, *4*, 18. [CrossRef]
120. González-Aravena, M.; Pérez-Troncoso, C.; Urtubia, R.; Branco, P.; Silva, J.R.; Mercado, L.; Lorgeril, J.; Bethke, J.; Paschke, K. Immune response of the Antarctic Sea urchin *Sterechinus neumayeri*: Cellular, molecular and physiological approach. *Rev. Biol. Trop.* **2015**, *63*, 309–320. [CrossRef]
121. Mao, Y.; Zhou, C.; Zhu, L.; Huang, Y.; Yan, T.; Fang, J.; Zhu, W. Identification and expression analysis on bactericidal permeability-increasing protein (BPI)/lipopolysaccharide-binding protein (LBP) of ark shell, *Scapharca broughtonii*. *Fish Shellfish Immunol.* **2013**, *35*, 642–652. [CrossRef]
122. Miki, T.; Holst, O.; Hardt, W. The bactericidal activity of the C-type lectin RegIIIβ against Gram-negative bacteria involves binding to lipid A. *J. Biol. Chem.* **2012**, *287*, 34844–34855. [CrossRef]
123. Kuo, T.H.; Chuang, S.C.; Chang, S.Y.; Liang, P.H. Ligand specificities and structural requirements of two *Tachypleus plasma* lectins for bacterial trapping. *Biochem. J.* **2006**, *393*, 757–766. [CrossRef]
124. Saito, T.; Kawabata, S.I.; Hirata, M.; Iwanaga, S. A novel type of Limulus lectin-L6—Purification, primary structure, and antibacterial activity. *J. Biol. Chem.* **1995**, *270*, 14493–14499. [CrossRef] [PubMed]
125. Okino, N.; Kawabata, S.; Saito, T.; Hirata, M.; Takagi, T.; Iwanaga, S. Purification, characterization, and cDNA cloning of a 27-kDa lectin (L10) from horseshoe crab hemocytes. *J. Biol. Chem.* **1995**, *270*, 31008–31015. [CrossRef]
126. Kawabata, S.-L.; Iwanaga, S. Role of lectins in the innate immunity of horseshoe crab. *Dev. Comp. Immunol.* **1999**, *23*, 391–400. [CrossRef] [PubMed]
127. Inamori, K.; Saito, T.; Iwaki, D.; Nagira, T.; Iwanaga, S.; Arisaka, F.; Kawabata, S. A newly identified horseshoe crab lectin with specificity for blood group A antigen recognizes specific O-antigens of bacterial lipopolysaccharides. *J. Biol. Chem.* **1999**, *274*, 3272–3278. [CrossRef]
128. Saito, T.; Hatada, M.; Iwanaga, S.; Kawabata, S.I. A newly identified horseshoe crab lectin with binding specificity to O-antigen of bacterial lipopolysaccharides. *J. Biol. Chem.* **1997**, *272*, 30703–30708. [CrossRef] [PubMed]
129. Chen, S.C.; Yen, C.H.; Yeh, M.S.; Huang, C.J.; Liu, T.Y. Biochemical properties and cDNa cloning of two new lectins from the plasma of *Tachypleus tridentatus*: Tachypleus plasma lectin 1 and 2. *J. Biol. Chem.* **2001**, *276*, 9631–9639. [CrossRef]
130. Ng, S.K.; Huang, Y.T.; Lee, Y.C.; Low, E.L.; Chiu, C.H.; Chen, S.L.; Mao, L.C.; Chang, M.D.T. A recombinant horseshoe crab plasma lectin recognizes specific pathogen-associated molecular patterns of bacteria through rhamnose. *PLoS ONE* **2014**, *9*, e115296. [CrossRef]
131. Chiou, S.T.; Chen, Y.W.; Chen, S.C.; Chao, C.F.; Liu, T.Y. Isolation and characterization of proteins that bind to galactose, lipopolysaccharide of *Escherichia coli*, and protein A of *Staphylococcus aureus* from the hemolymph of *Tachypleus tridentatus*. *J. Biol. Chem.* **2000**, *275*, 1630–1634. [CrossRef]
132. Beisel, H.G.; Kawabata, S.; Iwanaga, S.; Huber, R.; Bode, W. Tachylectin-2: Crystal structure of a specific GlcNAc/GalNAc-binding lectin involved in the innate immunity host defense of the Japanese horseshoe crab *Tachypleus tridentatus*. *EMBO J.* **1999**, *18*, 2313–2322. [CrossRef]
133. Kawabata, S.; Tsuda, R. Molecular basis of non-self recognition by the horseshoe crab lectins. *J. Endotoxin Res.* **2002**, *8*, 437–439. [CrossRef] [PubMed]
134. Schroder, H.C.; Ushijima, H.; Krasko, A.; Gamulin, V.; Thakur, N.L.; Diehl-Seifert, B.; Muller, I.M.; Muller, W.E. Emergence and disappearance of an immune molecule, an antimicrobial lectin, in basal metazoa: A tachylectin-related protein in the sponge *Suberites domuncula*. *J. Biol. Chem.* **2003**, *278*, 32810–32817. [CrossRef]
135. Lee, S.Y.; Wang, R.; Soderhall, K.; Lee, S.Y.; Wang, R.; Soderhall, K.A. Lipopolysaccharide- and b-1,3-glucan-binding protein from hemocytes of the freshwater crayfish *Pacifastacus leniusculus*: Purification, characterization, and cDNA cloning. *J. Biol. Chem.* **2000**, *275*, 1337–1343. [CrossRef] [PubMed]

136. Chaosomboon, A.; Phupet, B.; Rattanaporn, O.; Runsaeng, P.; Utarabhand, P. Lipopolysaccharide- and β-1,3-glucan-binding protein from *Fenneropenaeus merguiensis* functions as a pattern recognition receptor with a broad specificity for diverse pathogens in the defense against microorganisms. *Dev. Comp. Immunol.* **2017**, *67*, 434–444. [CrossRef]
137. Zhang, X.; Zhu, Y.T.; Li, X.J.; Wang, S.C.; Li, D.; Li, W.-W.; Wang, Q. Lipopolysaccharide and beta-1, 3-glucan binding protein (LGBP) stimulates prophenoloxidase activating system in Chinese mitten crab (*Eriocheir sinensis*). *Dev. Comp. Immunol.* **2016**, *61*, 70–79. [CrossRef] [PubMed]
138. Mu, C.; Song, X.; Zhao, J.; Wang, L.; Qui, L.; Zhang, H.; Zhou, Z.; Wang, M.; Song, L.; Wang, C. A scallop C-type lectin from *Argopecten irradians* (AiCTL5) with activities of lipopolysaccharide binding and Gram-negative bacteria agglutination. *Fish Shellfish Immunol.* **2012**, *32*, 716–723. [CrossRef]
139. Amparyup, P.; Sutthangkul, J.; Charoensapsri, W.; Tassanakajon, A. Pattern recognition protein binds to lipopolysaccharide and β-1,3-glucan and activates shrimp prophenoloxidase system. *J. Biol. Chem.* **2016**, *291*, 10949. [CrossRef]
140. Yang, J.; Qiu, L.; Wang, L.; Wei, X.; Zhang, H.; Zhang, Y.; Liu, L.; Song, L. CfLGBP, a pattern recognition receptor in *Chlamys farreri* involved in the immune response against various bacteria. *Fish Shellfish Immunol.* **2010**, *29*, 825–831. [CrossRef]
141. Bakholdina, S.I.; Naberezhnykh, G.A.; Gorbach, V.I.; Isaeva, M.P.; Solov'eva, T.F. Invertebrates of the Sea of Okhotsk as new sources of lipopolysaccharide binding proteins. *Russ. J. Mar. Biol.* **2014**, *40*, 63–69. [CrossRef]
142. Gopal, R.; Seo, C.H.; Park, Y. The role of biophysical parameters in the antilipopolysaccharide activities of antimicrobial peptides from marine fish. *Mar. Drugs* **2014**, *12*, 1471–1494. [CrossRef]
143. Rosa, R.D.; Vergnes, A.; de Lorgeril, J.; Goncalves, P.; Perazzolo, L.M.; Sauné, L.; Romestand, B.; Fievet, J.; Gueguen, Y.; Bachère, E.; et al. Functional divergence in shrimp anti-lipopolysaccharide factors (ALFs): From recognition of cell wall components to antimicrobial activity. *PLoS ONE* **2013**, *8*, e67937. [CrossRef]
144. Saravanan, R.; Mohanram, H.; Joshi, M.; Domadia, P.N.; Torres, J.; Ruedl, C.; Bhattacharjya, S. Structure, activity and interactions of the cysteine deleted analog of tachyplesin-1 with lipopolysaccharide micelle: Mechanistic insights into outer-membrane permeabilization and endotoxin neutralization. *Biochim. Biophys. Acta.* **2012**, *1818*, 1613–1624. [CrossRef]
145. Bhunia, A.; Domadia, P.N.; Torres, J.; Hallock, K.J.; Ramamoorthy, A.; Bhattacharjya, S. NMR structure of pardaxin, a pore-forming antimicrobial peptide, in lipopolysaccharide micelles: Mechanism of outer membrane permeabilization. *J. Biol. Chem.* **2010**, *285*, 3883–3895. [CrossRef]
146. Rosenfeld, Y.; Lev, N.; Shai, Y. Effect of the hydrophobicity to net positive charge ratio on antibacterial and anti-endotoxin activities of structurally similar antimicrobial peptides. *Biochemistry* **2010**, *49*, 853–861. [CrossRef]
147. Sun, Y.; Shang, D. Inhibitory effects of antimicrobial peptides on lipopolysaccharide-induced inflammation. *Mediat. Inflamm.* **2015**, *2015*, 167572. [CrossRef] [PubMed]
148. Domadia, P.N.; Bhunia, A.; Ramamoorthy, A.; Bhattacharjya, S. Structure, interactions, and antibacterial activities of MSI-594 derived mutant peptide MSI-594F5A in lipopolysaccharide micelles: Role of the helical hairpin conformation in outer-membrane permeabilization. *J. Am. Chem. Soc.* **2010**, *132*, 18417–18428. [CrossRef] [PubMed]
149. Rončević, T.; Puizina, J.; Tossi, A. Antimicrobial peptides as anti-infective agents in pre-post-antibiotic era? *Int. J. Mol. Sci.* **2019**, *20*, 5713. [CrossRef] [PubMed]
150. Maximiano, M.R.; Rios, T.B.; Campos, M.L.; Prado, G.S.; Dias, S.C.; Franco, O.L. Nanoparticles in association with antimicrobial peptides (NanoAMPs) as a promising combination for agriculture development. *Front. Mol. Biosci.* **2022**, *9*, 890654. [CrossRef]
151. Benincasa, M.; Zahariev, S.; Pelillo, C.; Milan, A.; Gennaro, R.; Scocchi, M. PEGylation of the peptide Bac7 (1-35) reduces renal clearance while retaining antibacterial activity and bacterial cell penetration capacity. *Eur. J. Med. Chem.* **2015**, *95*, 210–219. [CrossRef] [PubMed]
152. Korneev, K.V. Mouse Models of Sepsis and Septic Shock. *Mol. Biol.* **2019**, *53*, 799–814. [CrossRef]
153. Mansour, S.C.; Pena, O.M.; Hancock, R.E.W. Host defense peptides: Front-line immunomodulators. *Trends Immunol.* **2014**, *35*, 443–450. [CrossRef] [PubMed]

MDPI
St. Alban-Anlage 66
4052 Basel
Switzerland
www.mdpi.com

Marine Drugs Editorial Office
E-mail: marinedrugs@mdpi.com
www.mdpi.com/journal/marinedrugs

www.ingramcontent.com/pod-product-compliance
Lightning Source LLC
Chambersburg PA
CBHW041947260926
51850CB00018B/1811
* 9 7 8 3 7 2 5 8 0 3 4 7 7 *